T0236592

Lecture Notes in Computer Science 10390

More information about this series at http://www.springer.com/series/7407

Jirong Wen · Jianyun Nie
Tong Ruan · Yiqun Liu
Tieyun Qian (Eds.)

Information Retrieval

23rd China conference, CCIR 2017
Shanghai, China, July 13–14, 2017
Proceedings

Editors
Jirong Wen
Renmin University
Beijing
China

Jianyun Nie
Université de Montréal
Montreal
Canada

Tong Ruan
East China University of Science and Technology
Shanghai
China

Yiqun Liu
Tsinghua University
Beijing
China

Tieyun Qian
Wuhan University
Wuhan, Hubei
China

ISSN 0302-9743 ISSN 1611-3349 (electronic)
Lecture Notes in Computer Science
ISBN 978-3-319-68698-1 ISBN 978-3-319-68699-8 (eBook)
https://doi.org/10.1007/978-3-319-68699-8

Library of Congress Control Number: 2017956723

LNCS Sublibrary: SL1 – Theoretical Computer Science and General Issues

Printed on acid-free paper

This Springer imprint is published by Springer Nature
The registered company is Springer International Publishing AG
The registered company address is: Gewerbestrasse 11, 6330 Cham, Switzerland

Preface

The 2017 China Conference on Information Retrieval (CCIR 2017), co-organized by the Chinese Information Processing Society of China (CIPS) and the Chinese Computer Federation (CCF), was the 23rd instalment of the conference series. The conference was hosted by the East China University of Science and Technology in Shanghai, China on July 13–14, 2017.

The annual CCIR conference constitutes the major forum for researchers and practitioners from both China and other Asian countries/regions to share their ideas, present new research results, and demonstrate new systems and techniques in the broad field of information retrieval (IR). Unlike previous events, CCIR 2017 enjoyed contributions spanning the theory and application of information retrieval, both in English and Chinese.

This year we received 80 submissions from both China and other Asian countries, among which 41 were English papers and 39 were Chinese ones. Submissions were all carefully reviewed by at least three domain experts and the PC chairs made the final decision. The final English program of CCIR 2017 featured 21 papers.

CCIR 2017 included abundant academic activities. Besides keynote speeches delivered by world-renowned scientists from China and abroad, as well as a traditional paper presentation session and a poster session, we also hosted an evaluation workshop on paragraph ranking for question-answering.

CCIR 2017 featured four keynote speeches: "Neural Models for Modeling and Predicting Information Interaction Behavior" by Maarten De Rijke (University of Amsterdam), "Fighting Misinformation in Social Media through New IR Techniques" by Qiaozhu Mei (University of Michigan), "The Future of Information Streaming and the Opportunities for AI" by Weiying Ma (Toutiao), and "Scalable Learning with Generative Models and Applications" by Jun Zhu (Tsinghua University).

The conference and program chairs of CCIR 2017 extend their sincere gratitude to all authors and contributors to this year's conference. We are also grateful to the Program Committee for their great reviewing effort, which guaranteed that CCIR 2017 could feature a quality program of original and innovative research in information retrieval. Special thanks go to our sponsors for their generosity: TRS, Sogou, Toutiao, Huawei, Baidu, and Alibaba Group. We also thank Springer for supporting the best paper award of CCIR 2017.

Jirong Wen
Jian-Yun Nie
Yiqun Liu
Tong Ruan
Tieyun Qian

Organization

Steering Committee

Shuo Bai	Shanghai Stock Exchange, China
Xueqi Cheng	Institute of Computing Technology, Chinese Academy of Sciences, China
Shoubin Dong	East China University of Science and Technology, China
Xiaoming Li	Beijing University, China
Hongfei Lin	Dalian University of Technology, China
Ting Liu	Harbin Institute of Technology, China
Jun Ma	Shandong University, China
Shaoping Ma	Tsinghua University, China
Shuicai Shi	Beijing TRS Information Technology Co., Ltd., China
Mingwen Wang	Jiangxi Normal University, China

Conference General Chairs

Jirong Wen	Renmin University of China, China
Jianyun Nie	Université de Montréal, Canada

Program Committee Chairs

Yiqun Liu	Tsinghua University, China
Tong Ruan	East China University of Science and Technology, China

Tutorial and Youth Forum Chairs

Yanyan Lan	Institute of Computing Technology, Chinese Academy of Sciences, China
Qi Zhang	Fudan University, China

Publication Chairs

Tieyun Qian	Wuhan University, China
Jiafeng Guo	Institute of Computing Technology, Chinese Academy of Sciences, China

Publicity Chairs

Kang Liu	Institute of Automation, Chinese Academy of Sciences, China
Xianpei Han	Institute of Software, Chinese Academy of Sciences, China

Sponsorship Chair

Jianhua Li East China University of Science and Technology, China

Organizing Chairs

Qi Ye East China University of Science and Technology, China
Dongdong Li East China University of Science and Technology, China
Xiang Feng East China University of Science and Technology, China

Organizing Committee

Gaoqi He East China University of Science and Technology, China
Jianhua Huang East China University of Science and Technology, China
Xuli Liu East China University of Science and Technology, China
Jiahui Qiu East China University of Science and Technology, China
Chenglin Sun East China University of Science and Technology, China
Yubo Yuan East China University of Science and Technology, China
Jie Zhai East China University of Science and Technology, China
Haipeng Zhang East China University of Science and Technology, China
Xiaoqin Zhang East China University of Science and Technology, China

Program Committee

Shuo Bai Shanghai Stock Exchange, China
Deng Cai Zhejiang University, China
Chong Chen Beijing Normal University, China
Yueguo Chen Renmin University of China, China
Xueqi Chen Institute of Computing Technology, Chinese Academy of Sciences, China
Shoubin Dong South China University of Technology, China
Yajun Du Xihua University, China
Shicong Fen Beijing Siming Software System Co., Ltd., China
Jiafeng Guo Institute of Computing Technology, Chinese Academy of Sciences, China
Xiaofei He Zhejiang University, China
Xuanjing Huang Fudan University, China
Donghong Ji Wuhan University, China
Yanyan Lan Institute of Computing Technology, Chinese Academy of Sciences, China
Chenliang Li Wuhan University, China
Guoliang Li Tsinghua University, China
Jihong Li Shanxi University, China
Ru Li Shanxi University, China
Xun Liang Renmin University of China, China
Hongfei Lin Dalian University of Technology, China

Yuan Lin	Dalian University of Technology, China
Kang Liu	Institute of Automation, Chinese Academy of Sciences, China
Ting Liu	Harbin Institute of Technology, China
Yiqun Liu	Tsinghua University, China
Xueqiang Lv	Beijing Information Science and Technology University, China
Jun Ma	Shandong University, China
Shaoping Ma	Tsinghua University, China
Dunlu Peng	University of Shanghai for Science and Technology, China
Bin Qin	Harbin Institute of Technology, China
Tong Ruan	East China University of Science and Technology, China
Shuicai Shi	Beijing Tuoer Si Information Technology Co., Ltd., China
Bin Sun	Peking University, China
Jie Tang	Tsinghua University, China
Xiaojun Wan	Peking University, China
Bin Wang	Institute of Information Engineering, Chinese Academy of Sciences, China
Lihong Wang	National Computer Network and Information Security Management Center, China
Hongjun Wang	Beijing Tuoer Si Information Technology Co., Ltd., China
Mingwen Wang	Jiangxi Normal University, China
Xiao Chuan	Sohu, China
Tong Xiao	Northeastern University, China
Hongbo Xu	Institute of Computing Technology, Chinese Academy of Sciences, China
Weiran Xu	Peking University of Posts and Telecommunications, China
Hongfei Yan	Peking University, China
Hua Yuan	South China University of Technology, China
Jianmin Yao	Suzhou University, China
Mianzhu Yi	Luoyang Foreign Language Institute, China
Xiaohui Yu	Shandong University, China
Keliang Zhang	Luoyang Foreign Language Institute, China
Ming Zhang	Peking University, China
Shichao Zhang	Guangxi Normal University, China
Yan Zhang	Peking University, China
Chengzhi Zhang	Nanjing University of Science and Technology, China
Jun Zhao	Institute of Automation, Chinese Academy of Sciences, China
Guodong Zhou	Suzhou University, China
Jingbo Zhu	Northeastern University, China
Yude Bi	Luoyang Foreign Language Institute, China
Dongfeng Cai	Shenyang Aerospace University, China
Lijiang Chen	HP China Research Institute, China
Zhumin Chen	Shandong University, China
Bin Cui	Peking University, China
Zhicheng Dou	Renmin University of China, China
Lei Duan	Sichuan University, China

Yi Guan	Harbin Institute of Technology, China
Qing He	Institute of Computing Technology, Chinese Academy of Sciences, China
Yu Hong	Suzhou University, China
Jinlong Hu	South China University of Technology, China
Xiaolong Jin	Institute of Computing Technology, Chinese Academy of Sciences, China
Jingsheng Lei	Shanghai Electric Power Institute, China
Fang Li	Shanghai Jiaotong University, China
Hang Li	Microsoft Asia Research Institute, China
Juanzi Li	Tsinghua University, China
Wenjie Li	Hong Kong Polytechnic University, China
Xiangwen Liao	Fuzhou University, China
Kunhui Li	Xiamen University, China
Yue Liu	Institute of Computing Technology, Chinese Academy of Sciences, China
Peiyu Liu	Shandong Normal University, China
Yang Liu	Shandong University, China
Jiaheng Lu	Renmin University of China, China
Jianming Lv	South China University of Technology, China
Bo Peng	Peking University, China
Weining Qian	East China Normal University, China
Liyun Ru	Beijing Sogou Technology Development Co., Ltd., China
Huawei Shen	Institute of Computing Technology, Chinese Academy of Sciences, China
Dawei Song	Tianjing University, China
Le Sun	Institute of Software, Chinese Academy of Sciences, China
Songbo Tan	Tongbao Fortune Beijing Science and Technology Co., Ltd., China
Wenyong Wang	University of Electronic Science and Technology of China, China
Ting Wang	National University of Defense Technology, China
Haifeng Wang	Baidu, China
Jianyong Wang	Tsinghua University, China
Suge Wang	Shanxi University, China
Jirong Wen	Renmin University of China, China
Deyi Xiong	Suzhou University, China
Yong Xu	South China University of Technology, China
Guirong Xue	Alibaba, China
Muyun Yang	Harbin Institute of Technology, China
Zhihao Yang	Dalian University of Technology, China
Tianfang Yao	Shanghai Jiaotong University, China
Zhengtao Yu	Kunming University of Science and Technology, China
Jiali Zuo	Jiangxi Normal University, China
Min Zhang	Tsinghua University, China
Qi Zhang	Fudan University, China

Weinan Zhang	Harbin Institute of Technology, China
Yu Zhang	Harbin Institute of Technology, China
Jiakui Zhao	China Electric Power Research Institute, China
Shiqi Zhao	Baidu, China
Jianke Zhu	Zhejiang University, China
Lei Zou	Peking University, China

Contents

Recommendation

Understanding Users

NLP for IR

Recommendation

Neural or Statistical: An Empirical Study on Language Models for Chinese Input Recommendation on Mobile

Hainan Zhang(✉), Yanyan Lan, Jiafeng Guo, Jun Xu, and Xueqi Cheng

CAS Key Lab of Network Data Science and Technology,
Institute of Computing Technology, Chinese Academy of Sciences,
Beijing 100190, China
zhanghainan1990@163.com

Abstract. Chinese input recommendation plays an important role in alleviating human cost in typing Chinese words, especially in the scenario of mobile applications. The fundamental problem is to predict the conditional probability of the next word given the sequence of previous words. Therefore, statistical language models, i.e. n-grams based models, have been extensively used on this task in real application. However, the characteristics of extremely different typing behaviors usually lead to serious sparsity problem, even n-gram with smoothing will fail. A reasonable approach to tackle this problem is to use the recently proposed neural models, such as probabilistic neural language model, recurrent neural network and word2vec. They can leverage more semantically similar words for estimating the probability. However, there is no conclusion on which approach of the two will work better in real application. In this paper, we conduct an extensive empirical study to show the differences between statistical and neural language models. The experimental results show that the two different approach have individual advantages, and a hybrid approach will bring a significant improvement.

Keywords: Neural network · Deep learning · Language model · Machine learning · Sequential prediction

1 Introduction

Chinese input recommendation is a useful technology in reducing users' human cost in typing Chinese words, especially when users are using some mobile applications, since the typing costs are heavier than those on the computer. Typically, the recommender system will recommend some possible next words based on a user' typing history, and the user can directly select the word which he or she would like to input rather than directly typing it. This task can usually be formulated as follows: given a user's typing history, i.e. a word sequence $w_1, \cdots, w_{t-1}$, the fundamental problem is to estimate the conditional probability of the next word w_t given the word sequence $w_1, \cdots, w_{t-1}$, i.e. $P(w_t|w_1, \cdots, w_{t-1})$.

J. Wen et al. (Eds.): CCIR 2017, LNCS 10390, pp. 3–16, 2017.
https://doi.org/10.1007/978-3-319-68699-8_1

Traditional n-gram based statistical language models can be a natural choice to this task. Benefiting from the simplicity and stability, n-gram based statistical language models have been widely used in the real application of Chinese input recommendation. However, it cannot fully solve this problem since the task of Chinese input recommendation has some intrinsic characteristics. (1) The representation of Chinese input can be quite diverse, even though users refer to the same meaning. (2) Users' typing behaviors are quite different. Some users input Chinese sentences character by character, while others prefer the way of word by word, or even directly typing the Pinyin of the whole sentence. These characteristics pose great challenges of sparsity to traditional n-gram based models, even smoothed n-gram [1,2] will fail.

Recently in academic community, a new approach named neural language models has been proposed to further taking into account the 'similarity' between words. Examples include neural language model (NLM) [3], word2vec [4,5], recurrent neural network [6] (RNN) and long short term memory [7] (LSTM). Specifically, these models represent each word as a dense vector. Therefore, different words can be connected by their semantic representations to alleviate the sparsity problem. So far however, there has been no conclusion on which one of the two different approach will perform better and should be used in real application, to the best of our knowledge.

In this paper, we conduct an extensive empirical study on real data of Chinese input recommendation to compare the two different approaches. Specifically, we collect a large scale data set, named CIR from a commercial company focusing on Chinese input recommendation to facilitate our study. The experimental results show that when using a single language model, the statistical language model can give the best results, the neural probabilistic language model is a little worse, while the other neural language models perform the worst. This result tells us that the exact matching approach (n-gram based models) can find accurate results, while semantic matching approach (neural language models) can give further candidates but also introduce noises.

We also find that the overlap between the results given by two different approaches are relatively small. Therefore, it motivates us to use a hybrid model to combine both, and thus better recommendation results can be obtained than using the single model. Specifically, the combination of NLM with n-gram model is better than that of other neural models and n-grams. We also find that word2vec is different with NLM and RNN due to the negative sampling, and the combination of n-gram, word2vec and NLM can obtains further improvement.

The rest of the paper is organized as follows. Section 2 discusses the background of Chinese input recommendation task, the statistical language model and the evaluation measures. Section 3 describes the neural language models, including NLM, word2vec, RNN, and LSTM. Section 4 shows the experimental results and discussions. The conclusion is made in Sect. 5.

2 Backgrounds on Chinese Input Recommendation

2.1 Task Description

Inputing Chinese words is difficult on the mobile device. Taking the most popular Pinyin [8] typing methods for example, people first need to type the right Pinyin (composed of different English characters), and then choose which words he/she wants to input since different words can share a same pinyin representation. This is more difficult than English words input task on mobile. Based on users' typing history, the recommender system can give a ranking list of possible words user want to type in the next step. With this ranking list, users can directly input the words by clicking rather than typing if the words is recommended on top of the ranking list. By this way, the typing efforts can be largely reduced and the typing error can be avoided at the same time.

Fig. 1. An example of Chinese input recommendation on the mobile phone.

Figure 1 gives an illustration of the real system of Chinese input recommendation. In the example, we are given a user's typing history, i.e. a sequence of words '我 (I)- 明天(tomorrow)-要 (will) 要 (go to)'. The recommender system gives a recommendation list containing five words: '超市 (supermarket) 青海 (Qinghai) 医院 (hospital) 上班 (work) 学校 (school)'. If the user's next word exactly lies in the recommendation list, he/she can directly click the word (e.g. '学校'(school)), rather than typing the exact words. Therefore, we can see that users' cost is largely reduced. This is extremely useful in the scenario of mobile.

Therefore, the task of Chinese input recommendation can be formulated as a language modeling problem. Given a user's typing history, i.e. a word sequence $w_1, \cdots, w_{t-1}$, a recommendation list can be given out by estimating the conditional probability of each next word $P(w_t|w_1, \cdots, w_{t-1})$. The goal is to rank the required words on top of ranking lists, i.e. $P(w_t^*|w_1, \cdots, w_{t-1}) = \max_{w_t} P(w_t|w_1, \cdots, w_{t-1})$, where w_t* is the next word the user want to input.

2.2 Statistical Language Models

N-gram based models [9] are the most successful statistical language models, which has also been widely used in the real application of Chinese input recommendation due to its simplicity in implementation and explanation. Therefore, the probability generated for a specific word sequence

$w_1, \cdots, w_t$ is calculated as: $P(w_1, \cdots, w_t) \approx \prod_{i=1}^{t} P(w_i|w_{i-n+1}, \cdots, w_{i-1})$. where $P(w_i|w_{i-n+1}, \cdots, w_{i-1}) = \frac{count(w_{i-n+1}, \cdots, w_i)}{count(w_{i-n+1}, \cdots, w_{i-1})}$. However, the n-gram has severe sparsity problems. In this paper, we use a popular smoothing method, namely the interpolated Kneser-Ney smoothing [2,3,10,11], for comparison.

2.3 Evaluation Measures

Typical measures for recommendation task is Precision, Recall, and F1 score [12]. We give their formal definitions as follows. Given context $w_1, \cdots, w_{t-1}$, the recommendation results are denoted as $r_1, \cdots, r_{K_t}$, and the ground-truth words are denoted as $w_t^{(1)}, \cdots, w_t^{(U)}$, which are words that aggregated from different users typing behavior. The precision score is defined as:

$$P@K = \frac{\sum_{u=1}^{U} \sum_{k=1}^{K} \delta(r_k = w_t^{(u)})}{\sum_{t=1}^{U} \delta(K^t > 0)},$$

where $\delta(\cdot)$ is the indicator function and K^t is the number of the recommendation words. That is, if A is true, $\delta(A) = 1$, otherwise $\delta(A) = 0$.

Recall reflects the fraction of relevant instances that are retrieved:

$$R@K = \frac{1}{U} \sum_{u=1}^{U} \sum_{k=1}^{K} \delta(r_k = w_t^{(u)}).$$

Since the cost of recommendation is much lower than the cost of input, this task value recall more important than precision. That is, the user prefer to glancing at the recommendation list rather than inputting the whole word on the mobile phone. F1 both considers the precision and recall to compute the testing score, defined as $F1@K = \frac{P@K \times R@K}{\beta P@K + (1-\beta) R@K}$.

Regarding the Chinese input recommendation task as a ranking problem rather than a classification problem, we take the mean average precision (MAP) [12] as ranking measure for evaluation.

$$MAP = \frac{1}{U} \sum_{u=1}^{U} \frac{1}{rank(w_t^{(u)})},$$

where $rank(w_t^{(u)})$ is the ranking position of $w_t^{(u)}$ in recommendation list.

Specially for the task of Chinese input recommendation, we would like to evaluate from the perspective of how much the recommender system reduce the users' typing costs. We introduce two new evaluation measures in this paper, namely saved words(SW) and saved characters(SC). The measure of SW and SC is defined as percentage of words and characters that users can directly select from the recommendation lists rather than typing on the mobile. We give the precise mathematical formulas as follows:

$$SW = \frac{\sum_{u=1}^{U} \sum_{k=1}^{K} \delta(r_k = w_t^{(u)})}{U},$$

$$SC = \frac{\sum_{u=1}^{U}\sum_{k=1}^{K}\delta(r_k = w_t^{(u)})lenc(w_t^{(u)})}{\sum_{u=1}^{U}lenc(w_t^{(u)})},$$

where $lenc(\cdot)$ stands for the length of the word $(\cdot)$.

As we can see, the measure of SW is the same as $R@K$. Therefore, the measure of Recall is more important than Precision in this task. Considering the above issue, we set the β in $F1$ measure as 2/3 in Recall.

3 Neural Language Models

Though statistical language models such as n-gram have been widely used in the real systems, it usually encounters serious sparsity problem because Chinese input recommendation task has some intrinsic characteristics. (1) The representation of Chinese input can be quite diverse, even though users refer to the same meaning. Since the n-gram have no generalization to other sequence of n words and no cross-generalization between different n-tuples, the n-gram has poor generalization and heavy sparse problem [13]. (2) Users' typing behaviors are quite different. Some users input Chinese sentences character by character, while others prefer the way of word by word, or even directly typing the Pinyin of the whole sentence.

Though smoothed n-gram language models [1,14] can alleviate the sparsity problem [15], there are at least two characteristics which beg to be improved upon. It is not taking into account (1) contexts farther than 1 or 2 words[1] and (2) the 'similarity' between words. Therefore, the effect of sparsity can not be well tackled with the smoothed statistical language models. Recently, the approach of neural language models has been proposed to tackle these problems [16,17].

3.1 Probabilistic Neural Language Model

Neural probabilistic language model (NLM for short) is proposed by Bengio et al. [3]. Specifically, v is a mapping from each element $i \in V$ to a real vector $v(i)$, where i is a word and V is the vocabulary. The input vectors are first concatenate to form a word features layer activation vector $x = (v(w_{t-1}), v(w_{t-2}), \cdots, v(w_{t-n+1}))$. The hidden layer after an activation function will output $tanh(b_h + Hx)$, where H is the hidden layer weights with h hidden units, b_h is the hidden bias, and $tanh$ stands for the hyperbolic tangent activation function. The hidden layer output is further combined with the linear compositions of word features to form the input of softmax layer, i.e. $y = b + Wx + Utanh(b_h + Hx)$, where U is the hidden-to-output weight, W is the weight of word features to output, and b is the output bias. Finally, a softmax output layer is defined to output the probability as follows:

$$P(w_t|w_{t-1}, \cdots, w_{t-n+1}) = \frac{e^{y_{w_t}}}{\sum_i e^{y_i}},$$

where y_i stands for the score of word i.

[1] N-grams with up to 5 (i.e. 4 words of context) have been reported, though, but due to data scarcity, most predictions are made with a much shorter context.

3.2 Word2vec

NLM is computationally expensive for training. Among these neural methods, word2vec is the most successful one in terms of both efficiency and effectiveness. When applying word2vec for language modeling, we can modify the above context $C(w_t)$ to indicate the previous $n-1$ words, i.e. $C'(w_t) = (w_{t-n+1}, \cdots, w_{t-1})$. The others are kept the same as the original word2vec. In this paper we only use CBOW for our study, since it is more accordant with the language modeling task than skip-gram.

Instead of using softmax for optimization, word2vec propose to use hierarchical softmax or negative sampling [18,19] to save computation complexity. In this paper, we use the negative sampling approach. The basic idea is to randomly select some words as the negative instances instead of all other words in language.

3.3 Recurrent Neural Network

Both NLM and word2vec have two shortcomings. (1) They directly compress the history to a single vector, without distinguishing the orders of different words in the history. However, the word order is usually crucial for this task. For example, the representations of '上天 (up to the skies)' and '天上 (heaven)' are quite different. (2) They can not capture the long term dependencies of contexts, which is usually important for the next word prediction. For example, given a context '在 (on) 海边 (seaside) 旅游 (travel), 我(I) 去 (go to) 抓 (catch)', the next word is '螃蟹 (crab)' in the real data. We can see that the context '海边 (seaside)' is crucial information for predicting the correct next word '螃蟹 (crab)'. However, the NLM and word2vec are usually modeling the dependencies of contexts with a fixed-length window size, which may lose the long term dependencies.

To tackle this problem, Mikolov [6] used recurrent neural networks (RNN) as a language model. It can not only learn to compress the whole history to a low dimensional space, but also capture the sequence information and long term dependencies of the whole sentence.

3.4 Long Short Term Memory

Though RNN is capable to capture the long term dependencies of contexts, it usually faces the problem of gradient vanishing and gradient explosion for long sentences [7,20]. The long short term memory (LSTM) is an advanced type of Recurrent Neural Network by using memory cells and gates to learn long term dependencies [7,21].

4 Empirical Settings

4.1 Data Set

In this paper, we collected a large scale data set from a commercial Chinese input recommendation(CIR) engine to facilitate our experiments. The data set, named CIR, contains 20,000,000 word sequences. As we do not want to be biased by any artificial segmentation in the real application, we directly use the data segmented by the users themselves. The CIR corpus has been preprocessed by replacing all numbers with one notation 'NUM' and ignoring the English words. For our study, we randomly select 80% sequences from the whole CIR data as training set, 10% sequences as validation set and the rest 10% as test set.

4.2 Parameter Settings

For the two statistical language models, pure n-grams (n-gram for short), and n-grams with interpolated Kneser-Ney smoothing (n-gram-KN for short) [3], we use unigram, bigram and trigram for counting. For neural models, we use sigmoid function in experiments.

The performance results of P@1,10,R@10,F1@10,MAP with different window size, different embedding dimension and hidden nodes size are reported in Figs. 2, 3 and 4. We set embedding dimension, window size and hidden units as 100, 6, and 200 in NLM and we set the dimension as 200 in RNN and 300 in LSTM. For word2vec, the dimension of word embedding, window size and the number of hidden units is set to be 200, 5, and 200.

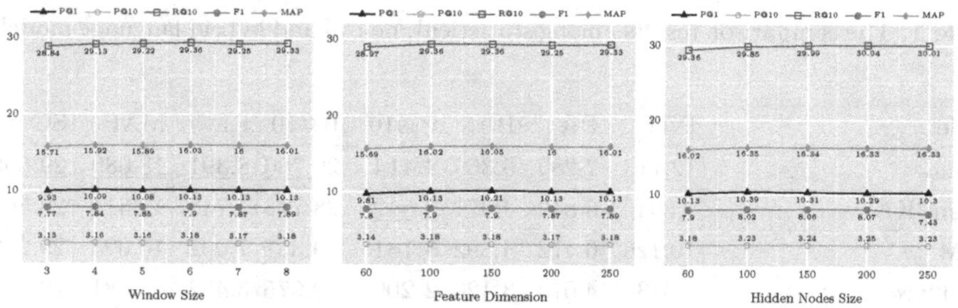

Fig. 2. Influences of window size, feature dimension and hidden nodes size for NLM.

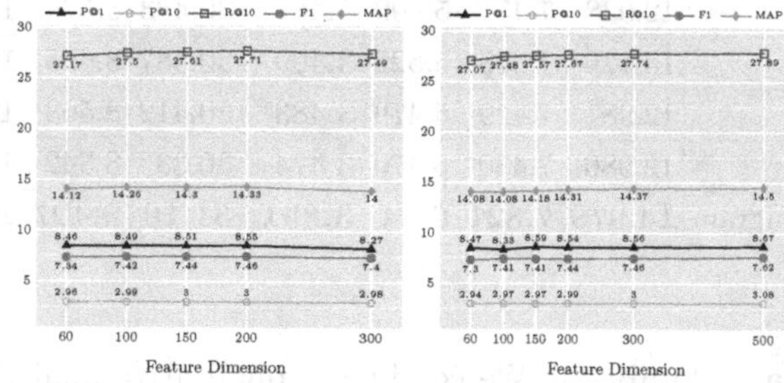

Fig. 3. Influences of feature dimension and hidden nodes for RNN (left), LSTM (right).

5 Experimental Results

In this section, we conducted experiments to study the comparison of different approaches at first. Secondly, we combined the statistical and neural language models, and found that better results can be obtained than using a single one. Thirdly, we conducted a further discussion on different neural methods.

5.1 Comparison Between Statistical and Neural Language Models

We compare all the models on CIR for fair comparison, and the experimental results are shown in Table 1. From the results, we can see that two statistical methods (i.e. n-gram and n-gram-KN) obtain comparable results. Similar results can be obtained for the four neural methods. As for the comparisons between statistical and neural approaches, we can see that the statistical language models performs much better than that of the neural ones. Taking MAP as an example, the best statistical language model (i.e. n-gram/18.09) can improve the best neural language model (i.e. NLM/16.03) by 12.85%. An exception is R@10, we can see that NLM performs slightly better than n-gram. We make a further analysis to show the differences between statistical and neural language models.

Table 1. The comparison results among statistical, neural and hybrid language models.

model	P@1	P@3	P@5	P@10	R@10	F1	MAP	SC
ngram	12.713	7.289	5.369	3.414	29.766	8.391	18.089	29.596
ngram-KN	9.804	6.842	4.971	3.542	28.775	7.743	15.314	28.564
NLM	10.126	6.472	4.892	3.181	29.357	7.902	16.025	29.226
word2vec	6.313	4.073	3.120	2.209	19.675	5.452	10.181	19.454
RNN	8.547	5.796	4.509	3.002	27.710	7.417	14.334	27.586
LSTM	8.672	5.766	4.517	3.078	27.887	7.381	14.499	27.363
NLM+ngram	13.154	7.536	5.554	3.467	30.683	8.545	18.777	30.508
word2vec+ngram	12.938	7.451	5.499	3.487	30.483	8.576	18.473	30.312
weighted+ngram	13.126	7.618	5.522	3.499	30.587	8.605	18.545	30.471
RNN+ngram	12.981	7.429	5.479	3.483	30.412	8.563	18.468	30.234
LSTM+ngram	12.986	7.430	5.470	3.474	30.333	8.542	18.457	30.156
NLM+word2vec+ngram	**14.979**	**7.821**	**6.34**	**3.891**	**31.18**	**9.407**	**19.587**	**30.938**

Overlapping Rate Analysis. We conduct a qualitative analysis between n-gram and neural language models.

Specifically, we make a statistics on the overlapping rates of the recommendation results between each two models, among n-gram, n-gram-KN, NLM, word2vec, RNN and LSTM. The experimental results are shown in Table 2.

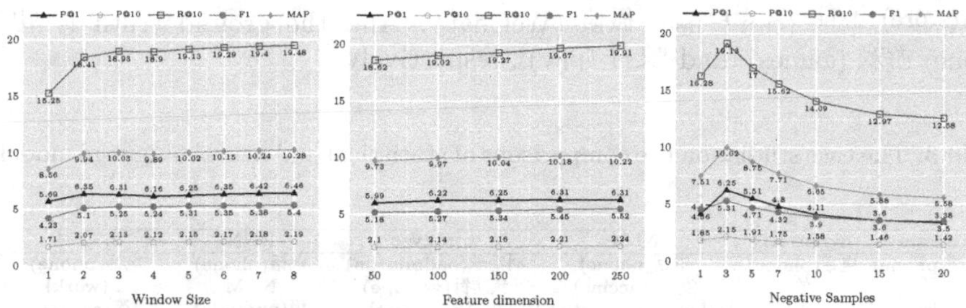

Fig. 4. Influences of feature dimension and negative samples for word2vec.

Table 2. The overlapping rates of the recommendation results between each two models.

overlap	ngram	ngram-KN	NLM	word2vec	RNN	LSTM
ngram	1	0.590726	0.1855	0.143099	0.198572	0.200481
ngram-KN		1	0.201453	0.145458	0.224591	0.228102
NLM			1	0.0858913	0.532115	0.525754
word2vec				1	0.0794111	0.0792779
RNN					1	0.727284
LSTM						1

From the results, there is large overlap between n-gram and n-gram-KN (i.e. 59.07%). This is understandable since n-gram-KN is just a smoothed version of n-gram. Secondly, there is large overlap rate among some neural language models, such as NLM, RNN and LSTM. For example RNN and NLM has 53.21% overlap, LSTM and NLM (i.e. 52.58%), RNN and LSTM (i.e. 72.73%). All these neural language models may predict the similar results.

However, the word2vec is different from other neural language models, since the overlapping rate between word2vec and others is quite small, i.e. 8.5%, 7.9% and 7.9%, respectively. This may be caused by the negative sampling strategy used in word2vec. For both NLM and RNN, the likelihood function is utilized as the loss function. That is to say, the words which is popular will be encouraged. However, they will be penalized in word2vec when using negative sampling. Thirdly, there is small overlap between statistical and neural language models. For example, the overlap between n-gram and NLM is 18.55%, while that between n-gram and word2vec is 14.31%. Therefore, statistical and neural language models will provide different recommendations. These results indicate that we can combine different approaches to obtain a better results. We will give further investigations on this issue later.

Case Studies. Table 3 gives some cases of different prediction results for a given word sequence. The proceeding context is '我(I) 打(I) (play)' and

'商店(store) 不能 (can't use) 微信 (WeChat)'. And the ground truth is '电话 (phone) 游戏 (games)' and '支付 (pay)', respectively.

Table 3. The case studies on the comparisons of statistical and neural language models.

ngram	ngram-KN	NLM	word2vec	RNN	LSTM
电话(phone)	电话(phone)	电话(phone)	羽毛球(badminton)	电话(phone)	电话(phone)
了	的	疫苗(vaccine)	预防针(vaccine)	NUM	工(work)
的	了	针(have an injection)	小报告(report)	级(upgrade)	牌(cards)
你(your)	NUM	交道(contact)	麻将(mahjong)	没人接(no answer)	麻将(mahjong)
我(my)	你(your)	点滴(transfuse)	游戏(games)	短信(message)	你(you)
号(ID)	是(is)	好友(friend)	密码(password)	密码(password)	号(ID)
上(sign in)	的	群(group)	账号(ID)	NUM	了
红包(money)	好友(friend)	支付(pay)	下载(download)	给你(give you)	发(give)
了	群(group)	红包(money)	支付(pay)	付款(pay)	NUM
群(group)	红包(money)	你(you)	转账(virement)	下单(order)	红包(money)

From the results, we can see that statistical language models can only give some frequent result, since they are estimating the probability based on the counting. For example, the n-gram of '打游戏 (play games)' and '微信支付 (WeChat Pay)' are rare in the training data, therefore statistical language models will miss these positive words, and can only output some popular positive words such as '电话 (phone)' and '群 (group)'. While for neural language model, we can see that they can output many more reasonable words, such as '游戏 (games)', '疫苗 (vaccine)' for '我打 (I play(do))', and '密码 (password)', '账号 (ID)' for '微信 (WeChat)'. This is mainly because they can leverage the distributed representations to construct connections between similar words, thus help to expand the candidates.

Therefore, statistical and neural language models both have their own advantages: statistical ones can accurately predict the popular words, while neural ones have the ability to provide more chances to target the rare words to tackle the sparse problem. Furthermore, the influence of smoothed strategy on the sparse problem is limited, since n-gram-KN provide exactly the same results with n-gram model, only with different order.

5.2 Combination of Statistical and Neural Language Models

According to the above analysis that statistical and neural language models usually give different results, we give a hybrid model as follows. We first study the influences of different λ ranging from 0 to 1 with step 0.1, on the validation set of CIR to see whether the combination will help for the Chinese input recommendation task. Figure 5 show the results of different λ when combing n-gram with NLM, word2vec, RNN and LSTM respectively. From the results, we can see that the performances are changing in a similar trend, i.e. first increase and then drop. The best λ for combing n-gram with NLM is 0.5, with word2vec is 0.8, with RNN is 0.9, and with LSTM is 0.9. Therefore, we can see that the hybrid model can indeed obtain better results. These parameters are used for further comparison.

$$P(w_t|C) = \lambda P_n(w_t|C) + (1-\lambda) P_d(w_t|C),$$

where P_n and P_d stands for the conditional probability produced by statistical and neural language models, respectively. C is the context of w_t, i.e. $C = (w_1, \cdots, w_{t-1})$. λ is a tradeoff factor in the range of [0,1]. When $\lambda = 1$, it reduces to statistical language model. While if $\lambda = 0$, it reduces to neural language model.

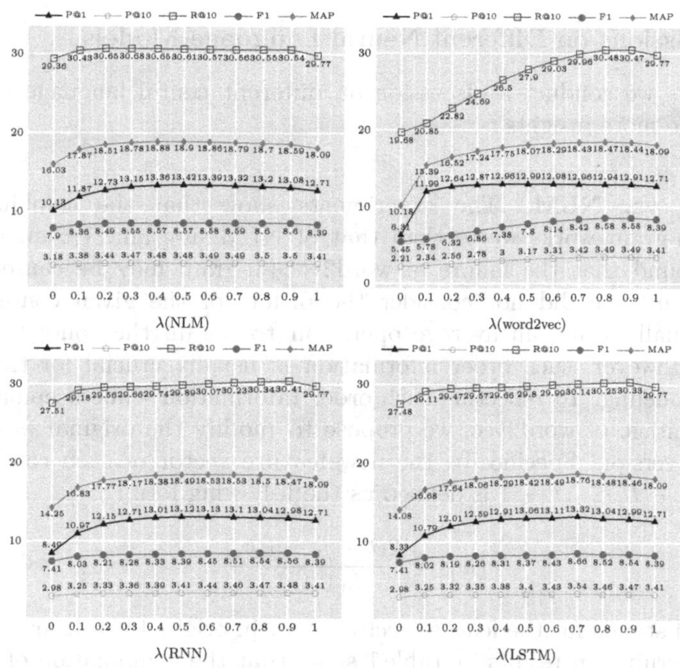

Fig. 5. Influences of different λ in combining n-gram and neural model.

Table 1 gives the comparison results between the combination approach with the single ones. We can see that the combination approach significantly improve the results. Taking MAP as an example, the improvement of the best combination method (i.e. NLM+n-gram/18.78) over the best statistical method (i.e. n-gram/18.09) is 3.8%. While the improvement over the best neural method (i.e. NLM/16.03) is 17.2%. This is accordant with the experimental findings that neural language models can provide suitable similar words for candidate to tackle the sparse problem, as shown in the above case studies.

Considering the above results that word2vec is different from other neural language models, and the overlapping is very small, we propose to combine n-gram, word2vec, and other neural language models. Therefore, we can obtain the following hybrid model:

$$P(w_t|C)=\lambda_1 P_n(w_t|C)+\lambda_2 P_w(w_t|C) + (1-\lambda_1-\lambda_2)P_d(w_t|C),$$

where P_n, P_w and P_d stands for the conditional probability produced by n-gram ,word2vec and other neural language models, respectively. C is the context of w_t, i.e. $C = (w_1, \cdots, w_{t-1})$. λ_1, λ_2 are tradeoff parameters in the range of [0,1]. In our experiments, we also tune these parameters in the validation set, and only report the best results with $\lambda_1 = 0.3$ and $\lambda_2 = 0.2$. The experimental results are shown in Table 1. We can see that the performance is improved significantly. Taking MAP as an example, the improvement (i.e. NLM+n-gram+word2vec/19.59) over the best statistical method (i.e. n-gram/18.09) is 8.3%.

5.3 Discussions on Different Neural Language Models

Furthermore, we conduct a discussion on different neural language models to provide some more insights.[2]

Word2Vec vs. NLM. The experiments show that the combination of word2vec+n-gram works worse than that of NLM+n-gram. We analyzed the data and found that the failure of word2vec+n-gram may be caused by the fact that word2vec did not consider the orders of the given context, since they are usually using an average operation to obtain the context vector in word2vec. However, such order information is usually crucial for the task of language modeling. To reflecting such order information while keeping the efficiency advantage of word2vec, we propose to modify the original word2vec to a weighted version. Specifically, the new feature vector of each context word $v(w_j), j = t - L, \cdots, t - 1$ is defined as the following form:

$$v(w_j) = \frac{2(t-j) \cdot v(w_j)}{L(1+L)},$$

where $v(w_j)$ stands for the feature vector of w_j in original word2vec.

The experimental results in Table 1 show that the combination of weighted word2vec and n-gram can perform better than that of word2vec and n-gram. Specifically, we give a specific case for further explanation. Given a word sequence '你(you)没有 (not have it) 就(is)-行 (okay)', if we adopt weighted word2vec, the predicted words will be '行了 (okay) 可以了 (alright) 行(okay)可以 (alright) 好了 (good)'. Compared with the results produced by word2vec '这样 (doing this)那么 (that)必要 (need)发言权 (right to speak) 这么 (so)', we can see that the words related to the latter word '就 (is)' has been ranked on top positions. Therefore, the order information of contexts have been taken into account in the model.

[2] Since word2vec is proposed as a simplified version of NLM, and RNN can be viewed as more complicated than NLM, we conduct the discussions on word2vec vs. NLM, and RNN vs. NLM, respectively.

NLM vs. RNN. The experiments show that RNN+ngram is worse than NLM+ngram. We give some explanations as follows. RNN compresses the context information into a hidden layer, while NLM has a fully connected with the context. Therefore the advantage of RNN lies in the modeling for long context, since it can capture the long term dependencies. But in our data, each sentence has 4.9 words on average and NLM has a window size 6 in our experiments. That is to say, NLM has ability to control most situations.

6 Conclusions and Future Work

In this paper, we conduct an empirical study on a large scale real data of Chinese input recommendation task. The experimental findings show that: (1) Statistical language models can provide more accurate results, while neural ones can alleviate the sparsity problem by providing more similar results; (2) The combination of these approachs will improve the results; For the future work, we will also study the efficiency issue of different neural language models, which is more important for the application on mobile.

Acknowledgments. The work was funded by 973 Program of China under Grant No. 2014CB340401, the National Key R&D Program of China under Grant No. 2016QY02D0405, the National Natural Science Foundation of China (NSFC) under Grants No. 61232010, 61472401, 61433014, 61425016, and 61203298, the Key Research Program of the CAS under Grant No. KGZD-EW-T03-2, and the Youth Innovation Promotion Association CAS under Grants No. 20144310 and 2016102.

References

1. Chen, S.F., Goodman, J.: An empirical study of smoothing techniques for language modeling. In: ACL, 310–318 (1996)
2. Reinhard, K., Hermann, N.: Improved backing-off for m-gram language modeling. In: Acoustics, Speech, and Signal Processing, pp. 181–184 (1995)
3. Bengio, Y., Ducharme, R., Vincent, P., Jauvin, C.: A neural probabilistic language model. J. Mach. Learn. Res. **3**, 1137–1151 (2003)
4. Mikolov, T., Chen, K., Greg, C., Jeffrey, D.: Efficient estimation of word representations in vector space. arXiv preprint arXiv:1301.3781 (2013)
5. Mikolov, T., Sutskever, I., Chen, K., Corrado, G.S., Dean, J.: Distributed Representations of Words and Phrases and their Compositionality. In: NIPS, pp. 3111–3119 (2013)
6. Mikolov, T., Martin, K., Lukas, B., Jan, C., Sanjeev, K.: Recurrent neural network based language model. In: INTERSPEECH 2010, pp. 1045–1048 (2010)
7. Hochreiter, S., Schmidhuber, J.: Long short-term memory. Neural Comput. **9**(8), 1735–1780 (1997)
8. Chen, S.-Y., Wang, R., Zhao, H.: Neural Network Language Model for Chinese Pinyin Input Method Engine (2015)
9. Katz, S.: Estimation of probabilities from sparse data for the language model component of a speech recognizer. ASSP IEEE Trans. **35**(3), 400–401 (1987)

10. Moore, R.C., Quirk, C.: Improved smoothing for N-gram language models based on ordinary counts. In: ACL, pp. 349–352 (2009)
11. Stolcke, A.: Srilm-an extensible language modeling toolkit. In: INTERSPEECH 2002, pp. 257–286 (2002)
12. Chen, H.: Machine learning for information retrieval: Neural networks, symbolic learning, and genetic algorithms. ASIS **46**(46), 194–216 (1995)
13. Bengio, Y.: Deep learning of semantics for natural language. In: Twitter Boston (2016)
14. Zhai, C., John, L.: A study of smoothing methods for language models applied to ad hoc information retrieval. In: ACM SIGIR, pp. 334–342 (2001)
15. Trnka, K.: Adaptive language modeling for word prediction. In: ACL, pp. 61–66 (2008)
16. Zheng, X., Chen, H., Tianyu, X.: Deep learning for chinese word segmentation and POS tagging. In: EMNL, pp. 647–657 (2013)
17. Zou, W.Y., Socher, R., Cer, D.M., Manning, C.D.: Bilingual word embeddings for phrase-based machine translation. In: EMNL, pp. 1393–1398 (2013)
18. Goldberg, Y., Levy, O.: word2vec Explained: deriving Mikolov et al'.s negative sampling word embedding method. arXiv preprint arXiv:1402.3722 (2014)
19. Levy, O., Goldberg, Y.: Neural word embedding as implicit matrix factorization. In: NIPS, pp. 2177–2185 (2014)
20. Bengio, Y., Simard, P., Frasconi, P.: Learning long-term dependencies with gradient descent is difficult. IEEE Trans. Neural Netw. **5**(2), 157–166 (1994)
21. Gers, F.A., Schmidhuber, J., Cummins, F.: Learning to forget: continual prediction with LSTM. Neural Comput. **12**, 2451–2471 (2000)

Dynamic-K Recommendation with Personalized Decision Boundary

Yan Gao(✉), Jiafeng Guo, Yanyan Lan, and Huaming Liao

CAS Key Lab of Network Data Science and Technology Institute of Computing Technology, Chinese Academy of Sciences, Beijing, China
gaoyan@software.ict.ac.cn, {guojiafeng,lanyanyan,lhm}@ict.ac.cn

Abstract. In this paper, we investigate the recommendation task in the most common scenario with implicit feedback (e.g., clicks, purchases). State-of-the-art methods in this direction usually cast the problem as to learn a personalized ranking on a set of items (e.g., webpages, products). The top-N results are then provided to users as recommendations, where the N is usually a fixed number pre-defined by the system according to some heuristic criteria (e.g., page size, screen size). There is one major assumption underlying this fixed-number recommendation scheme, i.e., there are always sufficient relevant items to users' preferences. Unfortunately, this assumption may not always hold in real-world scenarios. In some applications, there might be very limited candidate items to recommend, and some users may have very high relevance requirement in recommendation. In this way, even the top-1 ranked item may not be relevant to a user's preference. Therefore, we argue that it is critical to provide a dynamic-K recommendation, where the K should be different with respect to the candidate item set and the target user. We formulate this dynamic-K recommendation task as a joint learning problem with both ranking and classification objectives. The ranking objective is the same as existing methods, i.e., to create a ranking list of items according to users' interests. The classification objective is unique in this work, which aims to learn a personalized decision boundary to differentiate the relevant items from irrelevant items. Based on these ideas, we extend two state-of-the-art ranking-based recommendation methods, i.e., BPRMF and HRM, to the corresponding dynamic-K versions, namely DK-BPRMF and DK-HRM. Our experimental results on two datasets show that the dynamic-K models are more effective than the original fixed-N recommendation methods.

Keywords: Implicit feedback · Dynamic-K recommendation

1 Introduction

Recommender systems have been widely used in many applications, such as Amazon, YouTube and so on. In this paper, we address the most common recommendation scenario with implicit feedbacks, e.g., clicks or purchases from

J. Wen et al. (Eds.): CCIR 2017, LNCS 10390, pp. 17–29, 2017.
https://doi.org/10.1007/978-3-319-68699-8_2

users. Most methods in this direction cast the problem as to learn a personalized ranking on a set of items [6], e.g., webpages or products. The top-N ranked items are then provided to users as recommendations [1,8,11], where N is usually a fixed number pre-defined by the system according to the recommendation space (e.g., page size, screen size) or some heuristic criteria.

There is an underlying assumption for such fixed-N recommendation scheme, i.e., there are always sufficient relevant items to users' preferences. Unfortunately, this assumption may not always hold in real-world scenarios. For example, if one aims to recommend newly uploaded papers on arxiv[1] for academic readers everyday, he/she may face the following two problems. Firstly, there might be very limited new papers updated on arxiv everyday. Secondly, some academic readers might have very high relevance requirement in recommendation since paper reading is time-consuming. In this way, even the top-1 ranked paper may not be appropriate to be recommended to such readers on some day. However, by using a fixed-N recommendation scheme, each reader will constantly receive N recommended papers, and it is very likely to annoy the readers in this situation.

To avoid the above problem, we argue that it is critical to provide a dynamic-K recommendation, where the K should be different with respect to the candidate item set and the target user. If there are sufficient relevant items or if the user likes receiving diverse recommendations (i.e., relatively low relevance requirement), the K could be large. In contrary, if there are limited relevant items or if the user likes receiving recommendations unless they are highly relevant, the K should be small or even no recommendation should be provided sometimes. Ideally, a good recommender system should be able to learn from users' implicit feedbacks to present such dynamic-K recommendations for different candidate item set and target users.

In this work, we formulate this dynamic-K recommendation task as a joint learning problem with both ranking and classification objectives. Specifically, the ranking objective is the same as many existing ranking-based methods, i.e., to create a ranking list of items according to users' interests. The classification objective, which is unique in this work, aims to learn a personalized rather than a global decision boundary to differentiate the relevant items from irrelevant items. We apply the above joint learning idea over two state-of-the-art ranking-based recommendation methods, i.e., BPRMF and HRM, and extend them to the corresponding dynamic-K versions, namely DK-BPRMF and DK-HRM, respectively.

We conduct empirical experiments over two publicly available datasets, i.e., a transaction dataset named Ta-Feng and a movie recommendation dataset named MovieLens. Our experimental results show that the dynamic-K models are more effective than the corresponding fixed-N recommendation methods as well as those existing hybrid recommendation methods.

[1] https://arxiv.org/.

2 Related Work

Many methods have been developed in literature to build implicit feedback recommendation systems. With repsect to the objective function, the recommendation methods can be further divided into three folds, namely ranking-based methods, classification-based methods and hybrid methods.

Ranking-based recommendation methods, which aim to correctly rank items rather than to correctly predict their ratings have demonstrated good performance for top-N recommendation system. Ranking-based recommendation methods can be categoried into two folds, i.e., pair-wise approach and list-wise approach. Pair-wise approach aims to optimize the pair-wise loss. For example, Rendle et al. [10] and Aiolli et al. [2] optimized AUC score; Yun et al. [18] explored the connection between the Discounted Cumulative Gain (DCG) and the binary classification to change the ranking problem into binary classification problems; Park et al. [9] proposed a large-scale collaborative ranking method to minimize the ranking risk in the reconstructed recommendation matrix. List-wise approach optimizes the preference of each user to a list of items. An important branch in this category is designed to directly optimize evaluation merics, such as Mean Average Precision(MAP), Mean Reciprocal Rank(MRR) and Normalized Discounted Cumulative Gain (NDCG), which are usually list-wise ranking metrics. Typical methods include TFMAP [14], CLiMF [15] and CofiRank [17]. Different types of ranking approaches have different strengths in producing the ranking list, but all these ranking-based methods are proposed under the fixed-N recommendation scenario.

Classification-based recommendation methods attempt to predict whether a user would like to interact with an item. Mnih et al. [7] provide a probabilistic framework for the implicit case where they model the probability of a user choosing an item according to a normalized exponential function. They avoid linear time computation and approximate the normalization to the distribution by traversing a tree structure. Goplan et al. [4] introduced a factorization model that factorizes users and items by Poisson distribution. More recently, Johnson [5] proposed a new probabilistic framework for the implicit case in which they model the probability of a user choosing an item by a logistic function. Nevertheless, these methods do not learn a personalzied classifier for each user, thus can't be used for dynamic-K recommendation scenario.

Hybrid recommendation methods try to optimize several objectives simultaneously which can integrate the complementary strengths of different types of losses. Zhao et al. [3] propose a personalized top-N recommendation approach that minimizes a combined heterogeneous loss based on linear self-recovery models. The heterogeneous loss integrates the strengths of both pair-wise ranking loss and point-wise recovery loss to provide more informative recommendation predictions. Compared with their heterogeneous loss based on linear self-recovery models, we formulate our dynamic-K recommendation task as a joint learning problem with both ranking and classification objectives. Specifically, our classification objective aims to learn a personalized decision boundary for each user to differentiate the relevant items from irrelevant items.

3 Our Approach

In this section, we first introduce the problem formalization of recommendation with implicit feedback. We then describe the joint learning approach in detail. After that, we show how it can be applied to extend two state-of-the-art ranking-based recommendation methods, i.e., BPRMF and HRM, to the corresponding dynamic-K versinons, namely DK-BPRMF and DK-HRM.

3.1 Problem Definition

Let U be a set of users and I be a set of items. In our scenario, implicit feedback $X \subseteq U \times V$ is available. Each instance $(u, i) \in X$ is a pair which means an interaction between user u and item i. Formally, we define the set of items with which user u has interactions is $B_u^+ := \{i \in I | (u, i) \in X\}$. With the learned model, we generate a list of all candidate items for each user. The top K ranked items are then provided to users as recommendations, where K is different with repsect to the candidate item set and the target user. That is called dynamic-K recommendation system.

3.2 Joint Learning Approach

In this work, we formulate the dynamic-K recommendation task as a joint learning problem with both ranking and classification objectives. Specifically, the ranking objective could use any existing ranking-based methods, i.e., to create a ranking list of items according to user's interests. Since we are dealing with implicit feedback data, we follow the rationale from Rendle et al. [10] assuming that user u prefers item i over item j if $i \in B_u^+ \wedge j \notin B_u^+$. Formally, we define the set P as the set of tuples $\{(u, i, j)\}$ selected from dataset X as follows: $P = \{(u, i, j) | i \in B_u^+ \wedge j \notin B_u^+\}$. Therefore, the ranking objective can be formazlied as:

$$L(P; \theta)_{rank} = \sum_{(u,i,j) \epsilon P} \ell(s_{u,i} \succ s_{u,j}; \theta) \tag{1}$$

where θ represents the parameter vector of an arbitrary model (e.g. matrix factorization). $s_{u,i}$, $s_{u,j}$ means the predicted score of item i to user u and item j to user u respectively. $\ell(\cdot)$ can be any arbitrary ranking loss function.

The classification objective, which is unique in this work, aims to learn a personalized decision boundary to differentiate the relevant items from irrelevant items. Intuitively, if the predicted score of the item is above user's decision boundary, the item is relevant. If the predicted score is under user's decision boundary, the item is irrelevant. We use a variable called $margin_{ui}$ to measure the confidence of the predicted class of item i to user u (i.e., if item i is relevant or irrevelant to user u):

$$margin_{ui} = y_{ui}(s_{u,i} - t_u) \tag{2}$$

where $y_{ui} \in \{-1, 1\}$ is the target class, t_u is the personazlied decision boundary for each user. $s_{u,i}$ is the predicted score of item i to user u with an arbitary model (e.g. matrix factorification). $Margin_{ui} < 0$ indicates item i is misclassified to user u, while $margin_{ui} > 0$ indicates item i is correctly classified to user u. $Margin_{ui}$ represents the "margin of safety" by which the prediction for item i to user u is correct. Furthermore, we assume the personalized decision boundary is regularized by a global constraint to avoid over-fitting. For implicit feedback data, we define the dataset D as a set of tuples $\{(u, i, y_{ui})\}$ selected from $U \times I$, and y_{ui} is defined as follows:

$$y_{ui} = \begin{cases} 1 & (u, i) \in X \\ -1 & (u, i) \notin X \end{cases} \tag{3}$$

Therefore, the learning objective in classification is to find the best θ and t_u to minimize classification loss over the training data D as following:

$$L(D; \theta, t_u)_{cf} = \sum_{(u,i,y_{ui}) \in D} \ell(margin_{ui}; \theta) + \lambda_t R(t_u - t) \tag{4}$$

where $\ell(\cdot)$ is defined as a loss function of the margin for each data, t is the global constrain we set for t_u and $R(\cdot)$ is a regularizer (typically L2 or L1). $\lambda_t > 0$ is a co-efficient controling the regularization strength. As noted above, any classification loss function could be applied in this framework. In particular, we explored the use of logisitc loss and hinge loss in preliminary experiments, but found that logistic loss performed better on the data sets in this paper. Logistic loss is $\ell_{log} = ln(1 + e^{-margin_{ui}})$.

Finally, by combining (1) and (4), we obtain our joint learning approach as follows:

$$L_{hybrid} = \alpha L_{cf} + (1 - \alpha) L_{rk} \tag{5}$$

where the parameter $\alpha \in [1, 0]$ denotes the trade-off between optimizing ranking loss and classification loss. Note that by setting $\alpha = 1$, the model reduces to a classification-based model; By setting $\alpha = 0$, we obtain a ranking-based model.

A direct algorithm for optimizing the joint objective function would enumerate the full set P of candidate pairs. Because $|P|$ is quadratic in $|D|$, this would be intractable for large-scale data sets. Instead, following similar idea with Sculley [12], we take the sampling approach from P rather than constructing P explicitly. Algorithm 1 gives a method for efficiently solving the joint learning optimization problem using stochastic gradient descent (SGD) [13].

In prediction stage, with the learned model, the dynamic-K recommendation with our joint learning approach is as following. Given a user u and an itemset I, for each candidate item $i \in I$, we calculate the predicted score s_{ui}. We then rank the items according to their predicted score, and select the top K results above the threshold t_u as the final recommendation results.

Algorithm 1. Joint Learning Approach. Given: tradeoff parameter α, regularization parameter λ_t, iterations N

$\theta, t_u \leftarrow \emptyset, t$
for $n = 1$ to N **do**
  pick z uniformly at random from $[0, 1]$
  if $z < \alpha$ **then**
    $(u, i, y_{ui}) \leftarrow RandomExample(D)$
    $L_{cf} = \ell(margin_{ui}) + \lambda_t R(t_u - t)$
    $\theta^n \leftarrow StochasticGradientStep(\theta^{n-1}, \frac{\partial L_{cf}}{\partial \theta}, \eta)$
    $t_u^n \leftarrow StochasticGradientStep(t_u^{n-1}, \frac{\partial L_{cf}}{\partial t_u}, \lambda_t, \eta)$
  else
    $((u, i, j)) \leftarrow RandomCandidatePair(P)$
    $L_{rk} = \ell(s_{u,i} \succ s_{u,j}; \theta)$
    $\theta^n \leftarrow StochasticGradientStep(\theta^{n-1}, \frac{\partial L_{rk}}{\partial \theta}, \eta)$
  end if
end for

3.3 Implementation of the Joint Approach

We can see that the proposed joint approach is a general framework which can be appied to any existing learning to rank methods. In the following, we adopt two state-of-the-art ranking-based recommendation methods, i.e., BPRMF and HRM, and extend them to the corresponding dynamic-K versions, namely DK-BPRMF and DK-HRM, respectively.

DK-BPRMF. Bayesian personalized ranking maxtrix factoriaztion(BPRMF) is a matrix factorization model with BPR [10] as optimize criteria. With matrix factorization, the target matrix X is approximated by the matrix product of two low-rank matrices $P : |U| \times f$ and $Q : |I| \times f$:

$$\hat{X} := PQ^t \tag{6}$$

where f is the dimensionality/rank of the approximation. Thus the prediction formula can also be written as:

$$\hat{x}_{ui} =< p_u, q_i >= \sum_{f=1}^{F} p_{uf} \cdot q_{if} \tag{7}$$

BPR is a state-of-the-art pairwise ranking framework for the implicit feedback data, specifically, it use the maximum posterior estimator that is derived from a Bayesian analysis of the problem. Therefore, the ranking objective can be written as:

$$\ell_{rk} = \sum_{(u,i,j) \in P} \sigma(x_{ui} - x_{uj}) + \lambda_\theta \|\theta\|^2 \tag{8}$$

where σ is the logistic sigmoid $\sigma(x) := \frac{1}{1+e^{-x}}$, and λ_θ is a regularization parameter that controls the complexity of the model. According to our joint approach, the classification objective with logistic loss can be written as:

$$\ell_{cf} = \sum_{(u,i,y_{ui}) \in D} ln(1 + e^{-y_{ui}(x_{ui} - t_u)}) + \lambda_t \|t_u - t\|^2 \tag{9}$$

Therefore, the dynamic-K version of BPRMF is:

$$\ell_{DK\text{-}BPRMF} = \alpha \cdot \ell_{cf} + (1-\alpha) \cdot \ell_{rk} \tag{10}$$

DK-HRM. HRM is a state-of-the-art model for next basket recommendation which can capture both sequential behavior and users' general taste, and meanwhile model complicated interactions among these factors in prediction.

For each user u, a purchase history T^u is given by $T^u := (T_1^u, T_2^u, ..., T_{t_u-1}^u)$, where $T_t^u \subseteq I$, $t \in [1, t_u - 1]$. Given this history, the task of next basket recommendation is to recommend items that user u would probably buy at the next visit. We define the set H as the set of quaternion $\{(u,t,i,j)\}$ selected from dataset as follows: $H = \{(u,t,i,j) | i \in T_t^u \wedge j \notin T_t^u\}$. The set D is a set of quaternion $\{(u,t,i,y_{u,t,i})\}$ selected from $U \times T \times I$, and $y_{u,t,i}$ is defined as follows:

$$y_{u,t,i} = \begin{cases} 1 & i \subset T_t^u \\ -1 & i \notin T_t^u \end{cases} \tag{11}$$

Given a user u and his/her two consecutive transactions T_{t-1}^u and T_t^u, HRM defines the score of buying next item i as following:

$$x_{u,t,i} = v_i \cdot v_{u,t-1}^{Hybrid} \tag{12}$$

where v_i denotes the representation of item i and $v_{u,t-1}^{Hybrid}$ denotes the hybrid representation obtained from the hierarchical aggregation according to T_{t-1}^u.

In the learning process, the ranking objective with logistic loss is written as follows:

$$\ell_{rk} = \sum_{(u,t,i,j) \in H} \frac{1}{1 + e^{-(x_{u,t,i} - x_{u,t,j})}} \tag{13}$$

The classification objective is as follows:

$$\ell_{cf} = \sum_{(u,t,i,y_{u,t,i}) \in D} ln(1 + e^{-y_{u,t,i}(x_{u,t,i} - t_u)}) + \lambda_t \|t_u - t\|^2 \tag{14}$$

Finally, the dynamic-K version of HRM is:

$$\ell_{DK\text{-}HRM} = \alpha \cdot \ell_{cf} + (1-\alpha) \cdot \ell_{rk} \tag{15}$$

4 Evaluation

In this section, we conduct empirical experiments to demonstrate the effectiveness of our proposed joint learning models DK-BPRMF and DK-HRM for dynamic-K recommendation. We first introduce the dataset, baseline methods, and the evaluation metrics employed in our experiments. Then we compare the performance of hyper parameters for dynamic-K recommendation. After that, we compare our DK-BPRMF and DK-HRM to the corresponding fixed-N versions to demonstrate effectiveness.

4.1 Datasets

We use two datasets. The Movielens100K dataset and a retail datasets Ta-Feng.

- The Movielens100K dataset, it contains 100,000 ratings (1–5) from 943 users on 1682 movies. Each user has rated at least 20 movies.
- The Ta-Feng dataset is a public dataset released by RecSys conference, which covers products from food, office supplies to furniture. It contains 817, 741 trans- actions belonging to 32, 266 users and 23, 812 items.

We first conduct some pre-process on these datasets. For Movielens100K, as we want to solve an implicit feedback task, we removed the rating scores from the dataset. Now the task is to predict if a user is likely to rate a movie. For Ta-Feng dataset, we remove all the items bought by less than 10 users and users that has bought in total less than 10 items. In order to simulate the real world recommendation scenatio, finally, we seprate the training data and testing data by time. For Ta-Feng, the testing set contains only the last transaction of each user, and for Movielens100K, the testing set contains only the last rated movies of each user, while all the remaining interactions are put into the training set. The models are then learned on S_{train} and their predicted personalized ranking is evaluated on the test set S_{test}.

4.2 Evaluation Methodology

The performance is evaluated for each user u in the testing set. For each recommendation method, we generate a list of K items for each user u, denoted by $R(u)$, where $R_i(u)$ stands for the item recommended in the i-th position. We use the following measures to evaluate the recommendation lists.

- F1-score: F1-score is the harmonic mean of precision and recall, which is a widely used measure in recommendation:

$$Precision(R(u), S^u_{test}) := \frac{|R(u) \cap S^u_{test}|}{|R(u)|} \tag{16}$$

$$Recall(R(u), S^u_{test}) := \frac{|R(u) \cap S^u_{test}|}{|S^u_{test}|} \tag{17}$$

$$F1\text{-}score := \frac{2 \times Precision \times Recall}{Precision + Recall} \tag{18}$$

- NDCG@k: Normalized Discounted Cumulative Gain (NDCG) is a ranking based measure which takes into account the order of recommended items in the list [11], and is formally given by:

$$NDCG@k = \frac{1}{N_k} \sum_{j=1}^{k} \frac{2^{I(R_j(u) \in T^u_{t_u})} - 1}{log_2(j+1)} \tag{19}$$

where $I(\cdot)$ is an indicator function and N_k is a constant which denotes the maximum value of NDCG@k given $R(u)$.

- Cover-Ratio: In dynamic-K recommendation, Cover-Ratio measures how many users can get recommendation results.

$$Cover\text{-}Ratio = \frac{\sum I(|R(u)| > 0)}{|U|} \tag{20}$$

4.3 Baseline and Parameters Setting

The baselines include

- LogMF [5]: LogMF is a probabilistic model for matrix factorization with implicit feedback.
- BPRMF [10]: BPRMF is a matrix factorization model with BPR as optimize criteria. BPR is a state-of-the-art pairwise ranking framework for the implicit feedback data.
- CRRMF [12]: CRRMF is a matrix factorization model with CRR as optimize criteria. CRR is a hybrid framework combining ranking and regression together.
- HRM [16]: HRM is a state-of-the-art hybrid model on next basket recommendation. Both sequential behavior and users' general taste are taken into account for prediction and meanwhile modeling complicated interactions among these factors in prediction.

For all the models, we fixed the dimension of latent factors in U and I to be 50. In LogMF, similar to Johnson [5], for each triple $(u, i, j) \in P$, i represents positive observations, j represents negative observations and we tune the parameter c to balance the positive and negative observations. In CRRMF, for each triple $(u, i, j) \in P$, we regress i and j to 1 and 0 respectively. In HRM and DK-HRM, we use the average pooling. Parameter tuning was performed using cross validation on the training data for each method. We update parameters utilizing Stochastic Gradient Ddescent(SGD) until converge.

4.4 Comparison Among Different Hyper Parameters

For the proposed method, we have three hyper parameters, t, λ_t and α. Because the limitation of space, we only report the result of DK-HRM over two datasets Ta-Feng and Movielens100K. Similar results can also be obtained from DK-BPRMF. The results over two datasets are shown in Fig. 1.

Firstly, we explored how the parameter t affects the performance of DK-HRM. By setting λ_t and α to be 1.0 and 0.5 respectively, we selected t from {0.5, 1.0, 1.5, 2.0, 2.5}. As we can see, as t increases, F1-score increases but Cover-ratio descreses, indicating that the higher global constraint of personalized decision boundary we set, the higher F1-score we get, but the lower Cover-ratio we get. Take Ta-Feng as an example, when compared with $t = 0.5$, the relative improvement of F1-score by $t = 2.0$ is around 30%, and the decrease of Cover-ratio is 35%.

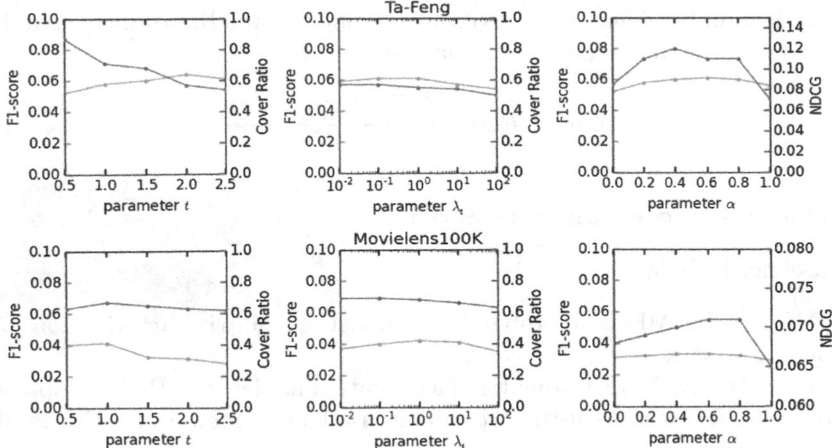

Fig. 1. Performance comparison among hyper parameters of DK-HRM over two datasets.

Then, we fixed t as 2.0 and compared different value of λ_t from {0.01, 0.1, 1, 10, 100}. As we can see, with the increase of λ_t, which means the increase of regularization strength, F1-score gradually increases, reaches the peak and then starts to decrease. It indicates that appropriate regularization strength can improve metric performance. Take Ta-Feng as example, F1-score reaches peak when $\lambda_t = 1.0$, then starts to decrease.

At last, we fixed t as 2.0 and λ_t as 1.0 and selected α from {0, 0.2, 0.4, 0.6, 0.8, 1.0}. α denotes the trade-off between optimizing ranking loss and classification loss. As we can see, as α increases, F1 score and NDCG increase first, and then decrease, indicating that the two types of losses can complement each other and achieve the best performance at the same time.

4.5 Comparison Against Baselines

We further compare our DK-BPRMF and DK-HRM methods to the state-of-the-art baseline methods. Since LogMF, BPRMF, CRRMF and HRM get top-N recommendation results, in order to make a fair comparison between top-N and dynamic-K, we vary the number N and get the F1-score of top-N method from top-1 to top-20, shown in Fig. 2. We find that as the number of recommendation items increases, F1-score increases and then keeps steady. Take Ta-Feng as example, for HRM, F1-score increases to 5.2% and then keeps steady.

Then, we choose the best performed DK-BPRMF (t is set 1.0, α is set 0.5 and λ_t is set 1.0) and DK-HRM (t is set 2.0, α is set 0.5 and λ_t is set 1.0) and the best performed baselines from Fig. 2 as the representative for fair comparison. Results over Ta-Feng and Movielens100K are shown in Table 1. Three main observations can be drawn from the results. (1) BPRMF performs better in NDCG and LogMF performs better in F1-score respectively, due to their dif-

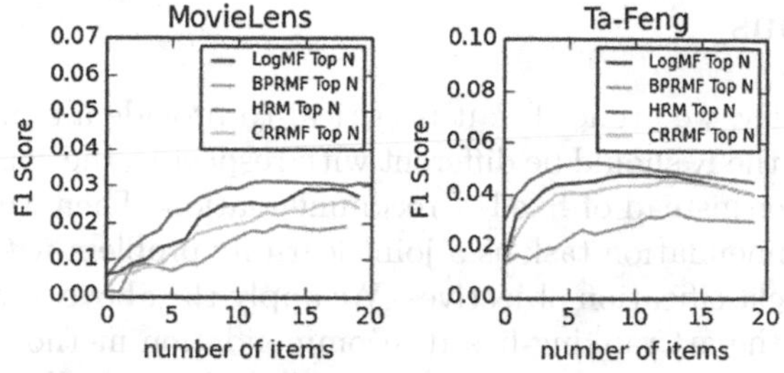

Fig. 2. The top-N F1-score of the Ta-Feng dataset and MovieLens dataset. The number of recommended items is set 1 to 20 on Ta-Feng and MovieLens.

Table 1. Performance comparison of DK-HRM and DK-BPRMF among LogMF, BPRMF, CRRMF and HRM over two datasets. Significant improvement of our model (DK-BPRMF and DK-HRM) with respect to their original baseline methods (BPR and HRM respectively) is indicated as '+' (p-value ≤ 0.05).

Models	Ta-Feng			MovieLens		
	F1-score	Cover-ratio	NDCG	F1-score	Cover-ratio	NDCG
BPRMF	0.033	1.0	0.075	0.018	1.0	0.067
LogMF	0.042	1.0	0.071	0.028	1.0	0.058
CRRMF	0.041	1.0	0.074	0.026	1.0	0.068
DK-BPRMF	**0.048$^+$**	**0.69**	**0.11$^+$**	**0.021$^+$**	**0.9**	**0.135$^+$**
HRM	0.052	1.0	0.08	0.03	1.0	0.117
DK-HRM	**0.06$^+$**	**0.7**	**0.12$^+$**	**0.035$^+$**	**0.68**	**0.12$^+$**

ferent optimization goal for ranking and classification respectively. As a hybrid model, CRRMF can achieve both good ranking and classification performance compared with BPRMF and LogMF. (2) HRM outperforms BPRMF, LogMF and CRRMF in all metrics since it can capture both sequential behavior and users' general taste. Take Ta-Feng as example, F1-score and NDCG of HRM reach 5.2% and 0.08 respectively. (3) Compared with top-N recommenders, DK-BPRMF and DK-HRM for dynamic-K recommendation can consistently outperform their original baseline methods respectively (i.e. BPR and HRM) in terms of F1-score and NDCG over the two datasets. Take the Ta-Feng dataset as an example, when compared with the second best performed baseline method (i.e. HRM), the relative improvement by DK-HRM (t set as 2.0, $alpha$ set as 0.5 and λ_t set as 1.0) is around 33.1%, 50%, in terms of F1-score and NDCG, respectively.

It indicates that our joint learning approach for dynamic-K recommendation is more effective than top-N recommendation methods.

5 Conclusions

In this paper, firstly, we argue that it is critical to provide a dynamic-K recommendation where the K should be different with respect to the candidate item set and the target user instead of fixed-N recommendation. Then we formulate this dynamic-K recommendation task as a joint learning problem with both ranking and personazlied classification objectives. We apply the above joint learning idea over two state-of-the-art ranking-based recommendation methods, i.e., BPRMF and HRM, and extend them to the corresponding dynamic-K versions, namely DK-BPRMF and DK-HRM, respectively. Our experimental results show that the dynamic-K models are more effective than the corresponding original fixed-N recommendation methods.

Acknowledgements. The work was funded by 973 Program of China under Grant No. 2014CB340401, the National Key RD Program of China under Grant No. 2016QY02D0405, the National Natural Science Foundation of China (NSFC) under Grants No. 61232010, 61472401, 61433014, 61425016, and 61203298, the Key Research Program of the CAS under Grant No. KGZD-EW-T03-2, and the Youth Innovation Promotion Association CAS under Grants No. 20144310 and 2016102.

References

1. Adomavicius, G., Tuzhilin, A.: Toward the next generation of recommender systems: a survey of the state-of-the-art and possible extensions. IEEE Trans. Knowl. Data Eng. **17**(6), 734–749 (2005)
2. Aiolli, F.: Convex auc optimization for top-n recommendation with implicit feedback. In: Proceedings of the 8th ACM Conference on Recommender Systems, pp. 293–296. ACM (2014)
3. Feipeng, Z., Yuhong, G.: Improving top-n recommendation with heterogeneous loss. J. Artif. Intell. Res. (2016)
4. Gopalan, P., Hofman, J.M., Blei, D.M.: Scalable recommendation with poisson factorization. arXiv preprint arXiv:1311.1704 (2013)
5. Johnson, C.C.: Logistic matrix factorization for implicit feedback data. In: Advances in Neural Information Processing Systems 27 (2014)
6. Karatzoglou, A., Baltrunas, L., Shi, Y.: Learning to rank for recommender systems. In: Proceedings of the 7th ACM Conference on Recommender Systems, pp. 493–494. ACM (2013)
7. Mnih, A., Teh, Y.W.: Learning item trees for probabilistic modelling of implicit feedback. arXiv preprint arXiv:1109.5894 (2011)
8. Oard, D.W., Kim, J., et al.: Implicit feedback for recommender systems. In: Proceedings of the AAAI Workshop on Recommender Systems, pp. 81–83 (1998)
9. Park, D., Neeman, J., Zhang, J., Sanghavi, S., Dhillon, I.S.: Preference completion: Large-scale collaborative ranking from pairwise comparisons. Statistics **1050**, 16 (2015)
10. Rendle, S., Freudenthaler, C., Gantner, Z., Schmidt-Thieme, L.: Bpr: Bayesian personalized ranking from implicit feedback. In: Proceedings of the Twenty-Fifth Conference on Uncertainty in Artificial Intelligence, pp. 452–461. AUAI Press (2009)

11. Ricci, F., Rokach, L., Shapira, B.: Introduction to Recommender Systems Handbook. Springer, Heidelberg (2011)
12. Sculley, D.: Combined regression and ranking. In: Proceedings of the 16th ACM SIGKDD International Conference on Knowledge Discovery and Data Mining, pp. 979–988. ACM (2010)
13. Shalev-Shwartz, S., Singer, Y., Srebro, N.: Pegasos: Primal estimated sub-gradient solver for svm. In: Proceedings of the 24th International Conference on Machine Learning, pp. 807–814. ACM (2007)
14. Shi, Y., Karatzoglou, A., Baltrunas, L., Larson, M., Hanjalic, A., Oliver, N.: Tfmap: optimizing map for top-n context-aware recommendation. In: Proceedings of the 35th International ACM SIGIR Conference on Research and Development in Information Retrieval, pp. 155–164. ACM (2012)
15. Shi, Y., Karatzoglou, A., Baltrunas, L., Larson, M., Oliver, N., Hanjalic, A.: Climf: learning to maximize reciprocal rank with collaborative less-is-more filtering. In: Proceedings of the Sixth ACM Conference on Recommender Systems, pp. 139–146. ACM (2012)
16. Wang, P., Guo, J., Lan, Y., Xu, J., Wan, S., Cheng, X.: Learning hierarchical representation model for next basket recommendation. In: Proceedings of the 38th International ACM SIGIR Conference on Research and Development in Information Retrieval, pp. 403–412. ACM (2015)
17. Weimer, M., Karatzoglou, A., Le, Q.V., Smola, A.: Maximum margin matrix factorization for collaborative ranking. In: Advances in Neural Information Processing Systems, pp. 1–8 (2007)
18. Yun, H., Raman, P., Vishwanathan, S.: Ranking via robust binary classification. In: Advances in Neural Information Processing Systems, pp. 2582–2590 (2014)

The Impact of Profile Coherence on Recommendation Performance for Shared Accounts on Smart TVs

Tao Lian(✉), Zhengxian Li, Zhumin Chen, and Jun Ma

School of Computer Science and Technology, Shandong University, Jinan, China
liantao1988@gmail.com, im.renascence@gmail.com,
{chenzhumin,majun}@sdu.edu.cn

Abstract. Most recommendation algorithms assume that an account represents a single user, and capture a user's interest by what he/she has preferred. However, in some applications, e.g., video recommendation on smart TVs, an account is often shared by multiple users who tend to have disparate interests. It poses great challenges for delivering personalized recommendations. In this paper, we propose the concept of profile coherence to measure the coherence of an account's interests, which is computed as the average similarity between items in the account profile in our implementation. Furthermore, we evaluate the impact of profile coherence on the quality of recommendation lists for coherent and incoherent accounts generated by different variants of item-based collaborative filtering. Experiments conducted on a large-scale watch log on smart TVs conform that the profile coherence indeed impact the quality of recommendation lists in various aspects—accuracy, diversity and popularity.

Keywords: Profile coherence · Shared account · Recommendation performance · Collaborative filtering · Smart TV

1 Introduction

Recommender systems [15] have become an essential tool to help us overcome the information overload problem by automatically identifying items that suit our interests, such as products, videos, music and social media accounts. Most recommendation algorithms assume that an account represents a single user, and capture a user's interest by the items that are previously preferred by him/her. For example, item-based collaborative filtering (CF) provides users with recommendations that are similar to what they have already preferred.

However, in some applications, an account is often shared by multiple users. For instance, different people in a household usually watch videos on the same smart TV. Thus the observations are the mixed behavior of multiple users. What's worse, their interests may be disparate or even conflict with each other. Take for example a household where three generations live together: the children

J. Wen et al. (Eds.): CCIR 2017, LNCS 10390, pp. 30–41, 2017.
https://doi.org/10.1007/978-3-319-68699-8_3

loves animations; the father likes sports while the mother likes variety shows; the grandparents prefers TV dramas. Therefore, it is challenging to provide personalized recommendations for shared accounts.

There are two major problems in the presence of shared accounts [20]. (i) The *dominance* problem arises when almost all recommendations are relevant to only some users in a shared account but at least one user does not get any relevant recommendation. (ii) The *generality* problem arises when the recommendations are only a little bit relevant to all users in a shared account but not appealing to any of them. When the diverse interests of multiple users are mixed together, the recommendations tend to be comprised of overly general or popular items that are not very bad for most people. Therefore, it seems reasonable to conjecture that the composition of the account profile should influence the recommendation performance in various aspects such as accuracy, diversity and popularity.

In this paper, we propose the concept of profile coherence to measure the coherence of an account's interests, which is computed as the average similarity between items in the account profile. We conjecture that the profile coherence should have an impact on the recommendation performance, especially in applications where an account is shared by multiple users. Let us make an analogy between recommender systems and search engines. The profile coherence of an account is like the clarity of an query [2]. The less clear the query, the worse the retrieval results. In a similar vein, the less coherent the account profile, the worse the recommendation results. Though it is possible that a user has a broad interest in traditional settings where an account represents a single user, the problem is more severe in the presence of shared accounts. Being able to know when a recommender system performs worse can shed light on the possible avenues to improve it. Therefore, we are interested in the question: how profile coherence impact the quality of recommendation lists for coherent and incoherent accounts generated by collaborative filtering algorithms in various aspects—accuracy, diversity and popularity?

The contributions of this paper are summarized as follows:

- We notice an important peculiarity of video viewing behavior on smart TVs—an account is shared by multiple users in a household.
- We propose the concept of profile coherence to measure the coherence of an account's interests in order to discriminate between coherent and incoherent accounts.
- We formulate four different variants of item-based CF that differ in the neighbor selection policy and the similarity aggregation function.
- We evaluate the impact of profile coherence on the quality of recommendation lists for coherent and incoherent accounts generated by different variants of item-based CF in various aspects—accuracy, diversity and popularity.

2 Related Work

2.1 Collaborative Filtering

A well-known class of recommendation algorithms is collaborative filtering [4], which can be further classified into neighborhood-based methods (e.g., user-based CF [9] and item-based CF [3,12,14,17]) and model-based methods (e.g., matrix factorization [10,13]), among others. Some of them [9,13,17] perform rating predictions on explicit feedback datasets, the others [3,10,12,14] perform top-N recommendations (on implicit feedback datasets). They all capture a user's preferences based on the items that are already preferred by him/her.

2.2 Performance Variation

Some scholars attempt to explain the performance variation by the characteristics of user profiles and further predict how well or bad the recommender system would perform for a given user. For explicit feedback datasets, the following characteristics of the rating profile of a user are found to have an influence on the performance of collaborative filtering algorithms to different extent depending on the datasets [8]: the number/popularity/quality of rated items, the standard deviation of provided ratings, the quality of neighborhood, etc.

It is well acknowledged that collaborative filtering suffers from the cold-start problem. Users with only a few consumed items could not get accurate recommendations. But some users with quite a few consumed items still get inaccurate recommendations. They are referred to as gray sheep users [1,18] whose preferences do not consistently agree or disagree with any group of users. Several methods have been developed to identify gray sheep users prior to the recommendation process [6,7], then they are handled by other techniques such as content-based methods [5].

Recently, Saia et al. [16] proposed a semantic approach to remove incoherent items from a user's profile in order to improve the recommendation accuracy. It is reasonable in conventional settings where an account represents a single user, though some users may have a broad interest. However, it is useless in settings where an account is shared by multiple users. When the behavior of multiple users are mixed together, the observations are likely to be widely scattered in the item space. If we remove some items in the account's profile, some users in the shared account might not receive customized recommendations.

2.3 Shared Account Recommendation vs. Group Recommendation

Verstrepen and Goethals [20] were the first to tackle the challenge of recommendation for shared accounts in the absence of contextual information. Shared account recommendation is different from group recommendation. (i) The individual profiles of the users in the group are typically known in group recommendation, whereas they are unknown in shared account recommendation. In the case of recommending videos on smart TVs, we do not know how many people are living in a household, let alone their individual preferences. (ii) In group recommendation, the recommendations will be consumed by all users in the group. But in shared account recommendation, the recommendations are supposed to be consumed individually, and every user in the shared account should be able to identify the recommendations meant for him/her. For example, when recommending videos on smart TVs, a video is qualified if it matches the interest of any single user in the shared account rather than their common interests, but there should be at least one video recommendation for each of them.

3 Our Work

In this paper, we aim to investigate the impact of profile coherence on the recommendation performance of collaborative filtering algorithms. We choose item-based CF as the experimental algorithm, which is currently employed in the Hisense Cloud Platform where our data comes from. It generates recommendations for an account by finding other items similar to those previously consumed by the account. Thus, we conjecture that the profile coherence of an account will influence the quality of recommendation list generated by item-based CF.[1]

Table 1 lists the notations frequently used in this paper. We consider the top-N recommendation task based on positive-only implicit feedback. Throughout the paper, an item corresponds to a video, and an account denotes a smart TV which are shared by multiple users in a household. If a video has been played on a smart TV, we say that the account has consumed the corresponding item.

3.1 Profile Coherence

If we treat an account in a recommender system as a query in a search engine, the more similar the items in its profile, the more clear the information need. We measure the profile coherence of an account by the average similarity of all pairs of items in its profile, which is defined as

$$coh\left(a\right) = \frac{\sum_{i \in \mathcal{I}_a} \sum_{j \in \mathcal{I}_a \setminus \{i\}} sim\left(i, j\right)}{|\mathcal{I}_a| * \left(|\mathcal{I}_a| - 1\right)} . \tag{1}$$

In applications with only implicit feedback, the similarity between items i and j can be measured by the binary cosine similarity[2], given by

$$sim\left(i, j\right) = \frac{|\mathcal{A}_i \cap \mathcal{A}_j|}{\sqrt{|\mathcal{A}_i|}\sqrt{|\mathcal{A}_j|}} , \tag{2}$$

where $\mathcal{A}_i$ denotes the subset of accounts who have consumed the item i.

[1] The profile coherence are expected to influence the recommendation performance of other collaborative filtering algorithms too, even content-based filtering methods.

[2] We have also experimented with the Jaccard similarity, and the qualitative conclusions are the same.

Table 1. Notations

Symbol	Description
$\mathcal{A}$	the set of accounts
$\mathcal{I}$	the set of items
$\boldsymbol{P} \in \{0,1\}^{\lvert\mathcal{A}\rvert \times \lvert\mathcal{I}\rvert}$	the preference matrix
a	an account
i, j	two items
$p_{a,i}$	$p_{a,i} = 1$ if the account a has consumed the item i, otherwise $p_{a,i} = 0$
$\mathcal{I}_a = \{i \in \mathcal{I} \mid p_{a,i} = 1\}$	the set of items consumed by the account a
$\mathcal{A}_i = \{a \in \mathcal{A} \mid p_{a,i} = 1\}$	the set of accounts who have consumed the item i
$sim\,(i, j)$	the similarity between the items i and j
$\mathrm{KNN}\,(i)$	the K most similar items of the item i
$r\,(a, i)$	the predicted ranking score of the item i for the account a
$\mathcal{L}_a$	the top-N recommendation list for the account a

There are other alternatives to measuring the profile coherence of an account. For example, we can compute the entropy of the distribution of consumed items among predefined categories/genres if the meta data of an item is available. But we adopt the Jaccard similarity based only on consumption history for the following reasons: (i) it is domain independent without the need to collect other information, and (ii) collaborative filtering also relies only on the ratings or consumption history [4,12,17].

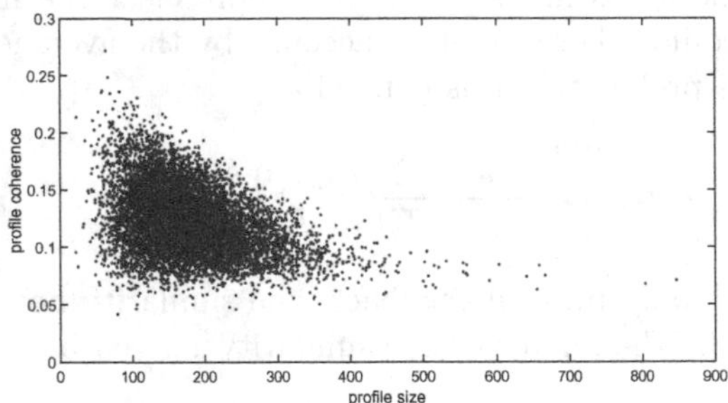

Fig. 1. Profile coherence vs. profile size

The scatter plot in Fig. 1 shows the relationship between the profile coherence (i.e., $coh\,(a)$) and the profile size (i.e., $|\mathcal{I}_a|$). We can make several observations. (i) For accounts with a large profile size, the coherence scores are very low. It means that accounts that have consumed a large number of items are generally incoherent, probably because items in their profiles are consumed by different users in the shared account. (ii) But for accounts with a relatively small profile size, the coherence scores vary a lot. That is to say, their profiles can be coherent, incoherent, or in between. (iii) There exist incoherent accounts everywhere in the spectrum of profile size, since most households consist of more than one people who tend to have different preferences.

Then we obtain two subsets of accounts in the two ends of the spectrum of profile coherence in a manner similar to box plot[3]:

Coherent Accounts: the 25% accounts with the highest coherence scores;
Incoherent Accounts: the 25% accounts with the lowest coherence scores.

Another alternative is the threshold-based method, but it is hard to determine the thresholds. What matters is the relative magnitude of the coherence scores rather than the absolute values. In a similar vein, Gras et al. [7] considered the $x\%$ users with the highest abnormality scores as atypical users whose ratings tend to be different from the community.

3.2 Variants of Item-Based CF

Item-based CF analyzes the user-item preference matrix to identify relations between items, and recommend items that are similar to what the account has already consumed. The general process of item-based top-N recommendation is listed in Algorithm 1.

Algorithm 1. The general process of item-based top-N recommendation

Input: the preference matrix $\boldsymbol{P}$, the set of accounts $\mathcal{A}$, the set of items $\mathcal{I}$, the neighborhood size K, the length of the recommendation list N
Output: the top-N recommendation list for each account

1 Compute the similarity $sim(i, j)$ between each pair of items i and j // (2)
2 **foreach** $i \in \mathcal{I}$ **do**
3   store the K most similar neighbors $\mathrm{KNN}(i)$ and their similarities
4 **foreach** $a \in \mathcal{A}$ **do**
5   **foreach** $i \in \mathcal{I} \setminus \mathcal{I}_a$ **do**
6   predict its ranking score $s(a, i)$ // ⋆
7   sort the items in $\mathcal{I} \setminus \mathcal{I}_a$ in descending order of the predicted ranking scores
8   only retain the top-N items $\mathcal{L}_a$

It is of great importance how the ranking scores are predicted (line 6). Generally speaking, the ranking score of an item i for an account a is computed based on the similarities between the item i and items in the profile $\mathcal{I}_a$:

$$s(a, i) = sim(\mathcal{I}_a, i) . \tag{3}$$

However, there are many variants with regard to how to aggregate the similarities between the item i and items in $\mathcal{I}_a$ to obtain the final ranking score $s(a, i)$.

[3] https://en.wikipedia.org/wiki/Box_plot.

When the item-based CF algorithm was first proposed for predicting ratings on explicit feedback datasets [17], the predicted rating on the item i for the account a is given by

$$\hat{r}_{a,i} = \frac{\sum_{j \in \mathcal{I}_a} \mathbb{1}_{j \in \text{KNN}(i)} * sim\,(i,j) * r_{a,j}}{\sum_{j \in \mathcal{I}_a} \mathbb{1}_{j \in \text{KNN}(i)} * |sim\,(i,j)|}\,, \tag{4}$$

where $\mathbb{1}.$ is the indicator function. The rating given by the account a on an item $j \in \mathcal{I}_a$ contributes to the weighted *average* only if $j \in \text{KNN}\,(i)$. If there are no items in $\mathcal{I}_a$ satisfying $j \in \text{KNN}\,(i)$, $\hat{r}_{a,i}$ is predicted to be the global or account-specific mean. Later on, the item-based CF was adapted to the top-N recommendation task on implicit feedback datasets [12]. The predicted ranking score of the item i for the account a is given by

$$s\,(a,i) = \sum_{j \in \mathcal{I}_a} \mathbb{1}_{i \in \text{KNN}(j)} * sim\,(i,j)\,, \tag{5}$$

where the similarity between an item $j \in \mathcal{I}_a$ and the item i contributes to the *sum* only if $i \in \text{KNN}\,(j)$. Thus, an item i can receive a non-zero score only if it is among the K most similar neighbors of at least one of the items in $\mathcal{I}_a$, otherwise it is not a qualified candidate. This speeds up the prediction by greatly reducing the set of candidate items, which is also adopted by other works [3,14,20].

There are two points worthy of attention here: the *neighbor selection policy* and the *similarity aggregation function*. The *neighbor selection policy* determines which items in the profile contribute to the predicted ranking score of a candidate item and also indirectly determines the subset of qualified candidate items. We refer to $\mathbb{1}_{i \in \text{KNN}(j)}$ as `KNN` which is widely adopted in the top-N recommendation task on implicit feedback datasets, and the policy $\mathbb{1}_{j \in \text{KNN}(i)}$ as `iKNN`, which means the inverted neighborhood [19] and is rarely adopted in the top-N recommendation task on implicit feedback datasets. The *similarity aggregation function* determines how to aggregate the similarities between a candidate item and items in the profile selected by either policy. The two different similarity aggregation functions are abbreviated to `SUM` and `AVG`. Thus, we can formulate four different variants of item-based CF for the top-N recommendation task on implicit feedback datasets.

KNN-SUM: $s\,(a,i) = \sum_{j \in \mathcal{I}_a} \mathbb{1}_{i \in \text{KNN}(j)} * sim\,(i,j)$
KNN-AVG: $s\,(a,i) = \frac{1}{\sum_{j \in \mathcal{I}_a} \mathbb{1}_{i \in \text{KNN}(j)}} \sum_{j \in \mathcal{I}_a} \mathbb{1}_{i \in \text{KNN}(j)} * sim\,(i,j)$
iKNN-SUM: $s\,(a,i) = \sum_{j \in \mathcal{I}_a} \mathbb{1}_{j \in \text{KNN}(i)} * sim\,(i,j)$
iKNN-AVG: $s\,(a,i) = \frac{1}{\sum_{j \in \mathcal{I}_a} \mathbb{1}_{j \in \text{KNN}(i)}} \sum_{j \in \mathcal{I}_a} \mathbb{1}_{j \in \text{KNN}(i)} * sim\,(i,j)$

4 Experiments

Dataset. We conduct our experiments on a large-scale watch log provided by a well-known smart TV manufacturer, Hisense. On Hisense smart TVs, users can stream a variety of videos via an app named JuHaoKan[4]. Each video is classified

[4] http://www.juhaokan.org/.

as one of the following categories: animation, movie, TV drama, sports, children's program, variety show, music, news, lifestyle, education, documentary, entertainment, autos, info, short film, and others. The watch log spans over a four-month period from 2015-12-07 to 2016-04-24. In our experiments, we only include 10 000 relatively active accounts and videos in the six categories—animation, movie, TV drama, sports, children's program, variety show—which receive 91.4% of total views. In addition, we remove videos that have been watched by less than 20 accounts. Note that different episodes of the same program are denoted by the same video ID. Table 2 shows the statistics of the final dataset.

Table 2. Dataset statistics

Num of accounts	10000
Num of videos	6747
Max/avg/min num of videos per account	847/178.7/21
Max/avg/min num of accounts per video	6535/264.9/20
Data sparsity	0.0265

Methodology. We perform 5-fold cross validation by randomly partitioning the observed entries in the preference matrix $\boldsymbol{P}$ into five folds. The top-N recommendation list ($N = 10$ in our experiments) for each account is generated by different variants of item-based CF based on the consumed items in any four folds, and is evaluated against the consumed items in the remaining one fold. We evaluate the quality of a recommendation list in three aspects—accuracy, diversity and popularity. Then the metrics are averaged over the accounts. Finally, the results averaged over the five folds are reported.

4.1 Experiment Results

Accuracy. We evaluate the accuracy of a top-N recommendation list $\mathcal{L}_a$ by the fraction of relevant items, given by

$$precision\,(\mathcal{L}_a) = \frac{1}{|\mathcal{L}_a|} \sum_{i \in \mathcal{L}_a} \mathbb{1}_{p_{a,i}=1}. \tag{6}$$

Figure 2 shows the precision values for coherent and incoherent accounts achieved by different variants of item-based CF with various neighborhood size k. We can obtain a number of interesting insights.[5] (i) Coherent accounts get more accurate recommendations than incoherent ones. (ii) Among the two neighbor selection policies, `iKNN` outperforms `KNN` in accuracy, except when $k = 1, 5$ using `SUM`. (iii) Among the two similarity aggregation functions, `SUM` achieves more accurate recommendations than `AVG`. (iv) As the neighborhood size k increases, the four variants of item-based CF exhibit different variations. (v) They all level off after k exceeds 50. Thereafter, `iKNN-SUM` performs the best while `KNN-AVG` performs the worst.

[5] The qualitative conclusions are the same when we evluate the accuracy of a recommendation list by the other measures such as recall, MRR and MAP.

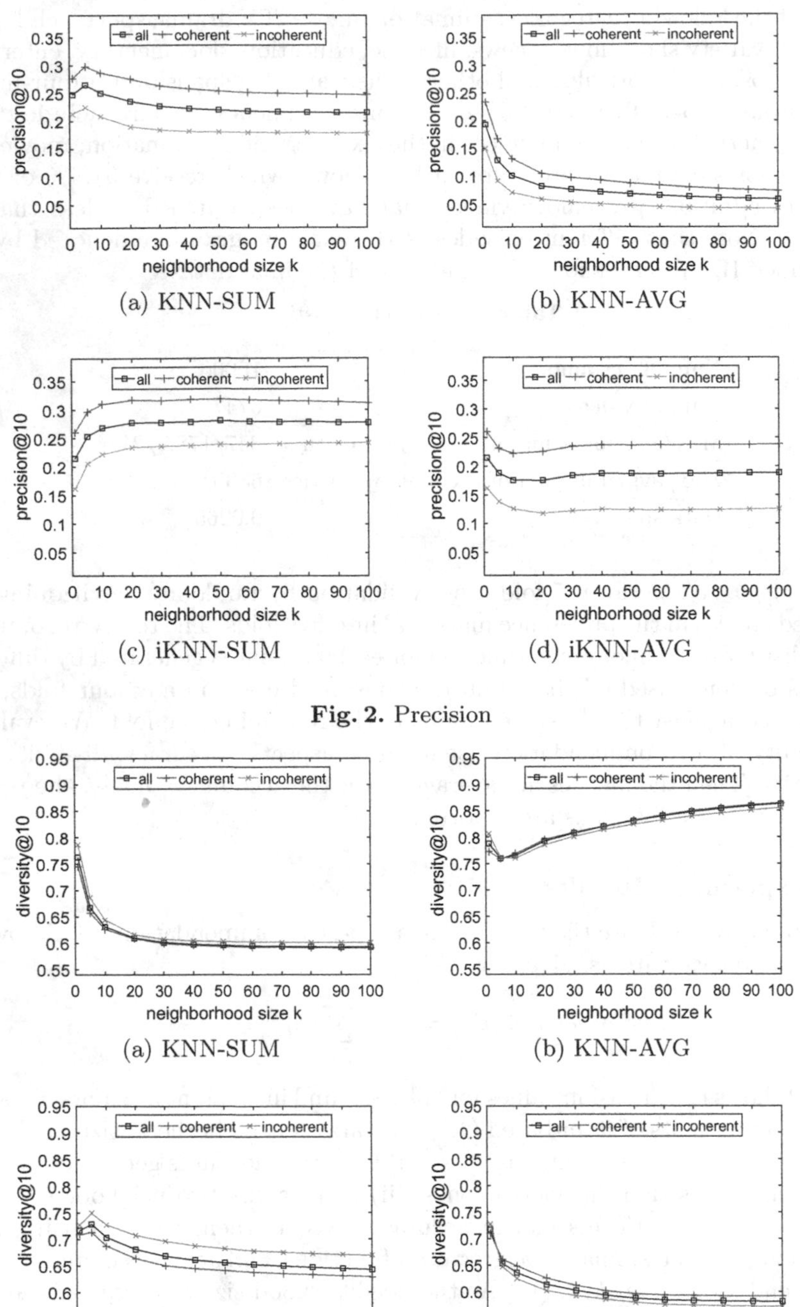

Fig. 2. Precision

Fig. 3. Diversity

Diversity. We evaluate the diversity of a top-N recommendation list $\mathcal{L}_a$ by the average dissimilarity of all pairs of recommended items [21], given by:

$$diversity(\mathcal{L}_a) = \frac{1}{|\mathcal{L}_a| * (|\mathcal{L}_a| - 1)} \sum_{i \in \mathcal{L}_a} \sum_{j \in \mathcal{L}_a \setminus \{i\}} 1 - sim(i,j). \quad (7)$$

The diversity values of the recommendation lists for coherent and incoherent accounts generated by different variants of item-based CF are shown in Fig. 3. We can make the following observations. (i) If we adopt `SUM` as the similarity aggregation function (especially in combination with `iKNN`), incoherent accounts get more diverse recommendations than coherent ones; if we adopt `AVG`, incoherent accounts get slightly less diverse recommendations than coherent ones (except when $k = 1$). (ii) As k increases (except when $k = 1$), the diversity achieved by `KNN-AVG` increases whereas the diversity achieved by the other three variants decreases. (iii) Though `KNN-AVG` performs the worst in accuracy, `KNN-AVG` generates the most diverse recommendations.

Popularity. We evaluate the popularity of a top-N recommendation list $\mathcal{L}_a$ by the average popularity of recommended items, given by:

$$APop(\mathcal{L}_a) = \frac{1}{|\mathcal{L}_a|} \sum_{i \in \mathcal{L}_a} \frac{|\mathcal{A}_i|}{|\mathcal{A}|}, \quad (8)$$

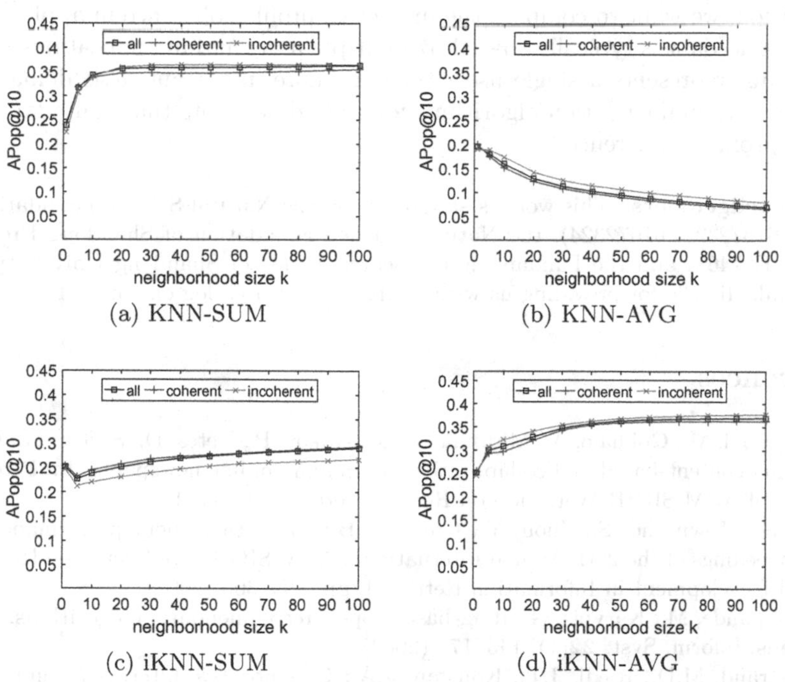

(a) KNN-SUM (b) KNN-AVG

(c) iKNN-SUM (d) iKNN-AVG

Fig. 4. Popularity

which is opposite to the novelty [11] of a recommendation list. Figure 4 illustrates how the average popularity of the recommendation list generated by different variants of item-based CF varies with respect to the neighborhood size. (i) Generally speaking, as k increases, the recommended items are more biased towards popular items, except in the case of `KNN-AVG`. (ii) On average, incoherent accounts get more popular recommendations than coherent ones, except in the case of `iKNN-SUM`.

5 Conclusion and Future Work

In this paper, we identify a novel profile characteristic—profile coherence—that impacts the quality of recommendations on smart TVs, where an account is shared by multiple users. Experiments conducted on a large-scale watch log on smart TVs conform that incoherent accounts indeed get less accurate recommendations. But the recommendation lists for coherent and incoherent accounts generated by different variants of item-based CF exhibit different characteristics in diversity and popularity. We believe our findings are especially valuable for practical applications, since many commercial recommender systems are item-based. The findings may provide guidance for tweaking the recommender systems according to the business goals.

In the future, we plan to conduct more extensive experiments to evaluate the impact of profile coherence on more advanced collaborative filtering algorithms. In addition, we want to compare the impact of profile coherence in applications where an account is typically shared by multiple users and in applications where an account represents a single user. What is more important, it is demanding to develop recommendation algorithms for shared accounts that can adaptively handle profile incoherence.

Acknowledgements. This work is supported by the Natural Science Foundation of China (61672322, 61672324), the Natural Science Foundation of Shandong Province (2016ZRE27468) and the Fundamental Research Funds of Shandong University. We also thank Hisense for providing us with a large-scale watch log on smart TVs.

References

1. Claypool, M., Gokhale, A., Miranda, T., Murnikov, P., Netes, D., Sartin, M.: Combining content-based and collaborative filters in an online newspaper. In: Proceedings of ACM SIGIR Workshop on Recommender Systems (1999)
2. Cronen-Townsend, S., Zhou, Y., Croft, W.B.: Predicting query performance. In: Proceedings of the 25th Annual International ACM SIGIR Conference on Research and Development in Information Retrieval, pp. 299–306 (2002)
3. Deshpande, M., Karypis, G.: Item-based top-N recommendation algorithms. ACM Trans. Inform. Syst. **22**(1), 143–177 (2004)
4. Ekstrand, M.D., Riedl, J.T., Konstan, J.A.: Collaborative filtering recommender systems. Found. Trends Hum. Comput. Interact. **4**(2), 81–173 (2011)

5. Ghazanfar, M.A., Prügel-Bennett, A.: Leveraging clustering approaches to solve the gray-sheep users problem in recommender systems. Expert Syst. Appl. **41**(7), 3261–3275 (2014)
6. Gras, B., Brun, A., Boyer, A.: Identifying grey sheep users in collaborative filtering: a distribution-based technique. In: Proceedings of the 2016 Conference on User Modeling Adaptation and Personalization, pp. 17–26 (2016)
7. Gras, B., Brun, A., Boyer, A.: When Users with preferences different from others get inaccurate recommendations. In: Monfort, V., Krempels, K.-H., Majchrzak, T.A., Turk, Ž. (eds.) WEBIST 2015. LNBIP, vol. 246, pp. 191–210. Springer, Cham (2016). doi:10.1007/978-3-319-30996-5_10
8. Griffith, J., O'Riordan, C., Sorensen, H.: Investigations into user rating information and predictive accuracy in a collaborative filtering domain. In: Proceedings of the 27th Annual ACM Symposium on Applied Computing, pp. 937–942 (2012)
9. Herlocker, J.L., Konstan, J.A., Borchers, A., Riedl, J.: An algorithmic framework for performing collaborative filtering. In: Proceedings of the 22nd Annual International ACM SIGIR Conference on Research and Development in Information Retrieval, pp. 230–237 (1999)
10. Hu, Y., Koren, Y., Volinsky, C.: Collaborative filtering for implicit feedback datasets. In: Proceedings of the 2008 Eighth IEEE International Conference on Data Mining, pp. 263–272 (2008)
11. Kaminskas, M., Bridge, D.: Diversity, serendipity, novelty, and coverage: a survey and empirical analysis of beyond-accuracy objectives in recommender systems. ACM Trans. Interact. Intell. Syst. **7**(1), 2:1–2:42 (2016)
12. Karypis, G.: Evaluation of item-based top-N recommendation algorithms. In: Proceedings of the Tenth International Conference on Information and Knowledge Management, pp. 247–254 (2001)
13. Koren, Y., Bell, R., Volinsky, C.: Matrix factorization techniques for recommender systems. Computer **42**(8), 30–37 (2009)
14. Linden, G., Smith, B., York, J.: Amazon.com recommendations: Item-to-item collaborative filtering. IEEE Internet Comput. **7**(1), 76–80 (2003)
15. Ricci, F., Rokach, L., Shapira, B.: Recommender Systems Handbook, 2nd edn. Springer, US (2015)
16. Saia, R., Boratto, L., Carta, S.: A semantic approach to remove incoherent items from a user profile and improve the accuracy of a recommender system. J. Intell. Inform. Syst. **47**(1), 111–134 (2016)
17. Sarwar, B., Karypis, G., Konstan, J., Riedl, J.: Item-based collaborative filtering recommendation algorithms. In: Proceedings of the 10th International Conference on World Wide Web, pp. 285–295 (2001)
18. Su, X., Khoshgoftaar, T.M.: A survey of collaborative filtering techniques. Adv. Artif. Intell. **2009** (2009)
19. Vargas, S., Castells, P.: Improving sales diversity by recommending users to items. In: Proceedings of the 8th ACM Conference on Recommender Systems, pp. 145–152 (2014)
20. Verstrepen, K., Goethals, B.: Top-N recommendation for shared accounts. In: Proceedings of the 9th ACM Conference on Recommender Systems, pp. 59–66 (2015)
21. Zhang, M., Hurley, N.: Avoiding monotony: improving the diversity of recommendation lists. In: Proceedings of the 2008 ACM Conference on Recommender Systems, pp. 123–130 (2008)

Academic Access Data Analysis for Literature Recommendation

Yixing Fan(✉), Jiafeng Guo, Yanyan Lan, Jun Xu, and Xueqi Cheng

CAS Key Lab of Network Data Science and Technology, Institute of Computing Technology, Chinese Academy of Sciences, Beijing 100190, China
fanyixing@software.ict.ac.cn,
{guojiafeng,lanyanyan,junxu,cxq}@ict.ac.cn

Abstract. Academic reading plays an important role in researchers' daily life. To alleviate the burden of seeking relevant literature from rapidly growing academic repository, different kinds of recommender systems have been introduced in recent years. However, most existing work focused on adopting traditional recommendation techniques, like content-based filtering or collaborative filtering, in the literature recommendation scenario. Little work has yet been done on analyzing the academic reading behaviors to understand the reading patterns and information needs of real-world academic users, which would be a foundation for improving existing recommender systems or designing new ones. In this paper, we aim to tackle this problem by carrying out empirical analysis over large scale academic access data, which can be viewed as a proxy of academic reading behaviors. We conduct global, group-based and sequence-based analysis to address the following questions: (1) Are there any regularities in users' academic reading behaviors? (2) Will users with different levels of activeness exhibit different information needs? (3) How to correlate one's future demands with his/her historical behaviors? By answering these questions, we not only unveil useful patterns and strategies for literature recommendation, but also identify some challenging problems for future development.

Keywords: Academic access data · Literature recommendation · User study

1 Introduction

A major part of researchers' daily life is academic reading, which can help them acquire new knowledge, find related work in their domains and keep them up-to-date with the research frontier [17]. Traditionally, academic users rely on keyword-based search or browsing through proceedings of conferences and journals to find interested literature. However, such information seeking process becomes more and more difficult since a huge number of academic papers are coming out from a lot of conferences and journals [14].

J. Wen et al. (Eds.): CCIR 2017, LNCS 10390, pp. 42–54, 2017.
https://doi.org/10.1007/978-3-319-68699-8_4

To ease this difficulty, various literature recommender systems have been introduced for academic users, such as TechLens [18], CiteULike [3] SciRecSys [1], Refseer [11] and so on. Although different types of recommendation techniques have been adopted in literature recommender systems, little work has yet been done on analyzing the academic reading behaviors to understand the reading patterns and information needs of real-world academic users. There are some early work on analysis of information seeking behaviors of academic users [2,7,10,14,16,20]. For example, Hemminger et al. [10] conducted a census survey to quantify the transition to electronic communications and how this affects different aspects of information seeking. Luis et al. [20] tried to understand the most frequent type of academic search conducted by different users through transaction log analysis. However, these work either relied on questionnaires to survey very limited faculty members [10,14,16], or only focused on analyzing users' query patterns based on some search logs [2,7,9,20].

In this work, we propose to conduct some in-depth analysis over academic reading behaviors using real-world scholarly usage data, to unveil the underlying reading patterns and information needs of academic users. We argue that this type of research would be useful for improving existing literature recommender systems or designing new ones. Specifically, we take the large scale academic access data from OpenURL[1] as a good proxy of users' academic reading behaviors, with the assumption that papers accessed by a user are those he/she read or would like to read[2]. We then match all the access records to a large academic repository to identify the corresponding papers, and extract different types of meta-data (e.g., author, venue, and publication time) which would be useful in detailed analysis.

We then conduct different types of analysis over this user behavioral dataset, including global analysis, group-based analysis and sequence-based analysis. By these analysis, we aim to address the following research questions:

Q1: Are there any regularities in users' academic reading behaviors? (global analysis)
Q2: Will users with different levels of activeness exhibit different reading patterns? (group-based analysis)
Q3: How to correlate one's future reading with his/her reading history? (sequence-based analysis)

In brief, here are some take-away conclusions based on our analysis:

1. The regularities in both frequency and time show that literature recommendation is in general a difficult problem due to the long-tail phenomena and diverse reading patterns in user behaviors. A good point is that systems can obtain sufficient recommendation resources by only focusing on a small set of important authors and venues.

[1] http://www.openurl.ac.uk/doc/index.html.
[2] Note that the assumption here is reasonable since accessing is a strong signal of users' reading interests. Many existing studies have leveraged the academic access data as signals of reading interests to improve literature recommender systems [13,15,19] and show its effectiveness.

2. Users with different levels of activeness exhibit different reading patterns in terms of recency and popularity of the papers. Therefore, a smart recommender system needs to emphasize different factors for different groups of users rather than to employ a unified model.
3. Both content-based filtering and collaborative filtering can cover partial future readings based on historical behaviors with different efficiencies. However, there are a good number of future readings difficult in finding by these means, which will be a critical challenge for designing a better literature recommender system.

The rest of the paper is organized as follows: In Sect. 2, we describe the dataset and some pre-processing steps. We then present the global analysis, group-based analysis, and sequence-based analysis in Sects. 3, 4 and 5, respectively. Section 6 discusses the related work while conclusions are made in Sect. 7.

2 Dataset Description

Although user behavior analysis should be the foundation of designing a good recommender system, this is often difficult to be conducted due to the lack of user data or some privacy issues. In this work, we make use of publicly available paper access logs as well as a large scale academic repository to conduct the analysis on academic reading behaviors.

The public paper access logs are from OpenURL router data. Typically, each entry in the log records who (encryptedUserIP) accessed which paper (a title) at what time (logData, logTime). Note that the encryptedUserIP (anonymised IP address or session identifier) can be taken as the unique ID to identify a user. we collected 17,040,154 records between April 1, 2011 and December 31, 2014 from OpenURL. By removing duplicated records which a same user generates in a short time interval and some portal IPs[3], we obtained 8,675,280 distinct records over 1,089,916 users containing 6,667,942 papers.

We then map these papers to a large scale academic repository which contains over 10 million papers. In this way, 285,541 distinct papers were successfully matched, covering 441,840 records and 118,436 users in total. We extract all the meta-data over these papers, including author, venue, publication time, keywords, domain and citations for later analysis. This matched dataset are then taken as a proxy of academic reading behaviors in the following analysis.

3 Global Analysis

In this section, we conduct some global analysis over this dataset, with the purpose to identify the regularities in users' academic reading behaviors to guide system design. Here we call the whole sequence of reading records of a distinct user as the user's reading profile. By this analysis, we try to obtain some idea of the overall interests and behavior of users in reading papers.

[3] Note here we treat those IPs with extremely high volume of accessed papers (i.e., more than 3000 a year) as portals rather than real users.

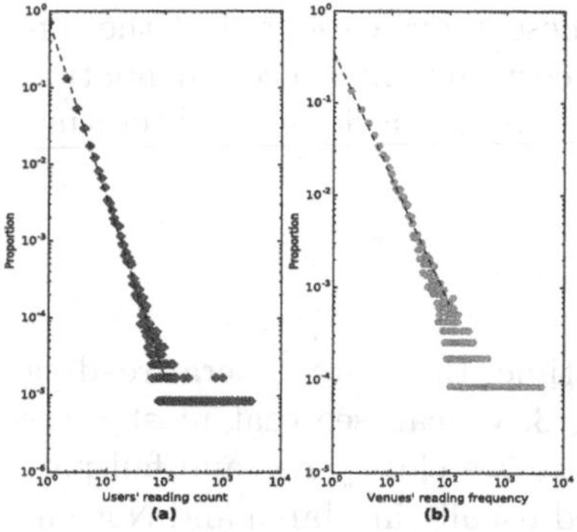

Fig. 1. (a) Distribution of users' reading number; (b) Distribution of venue' reading frequency.

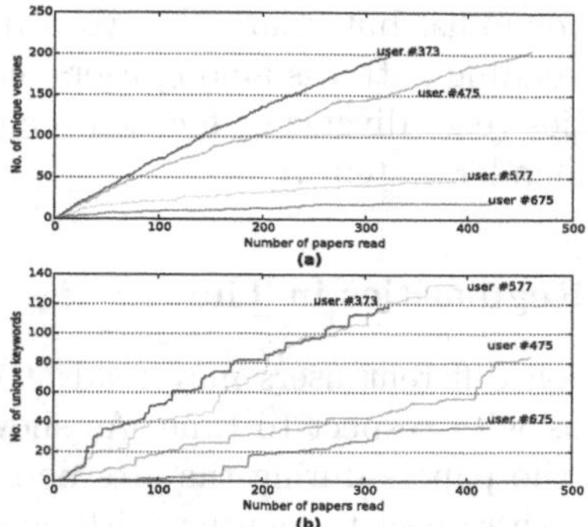

Fig. 2. (a) Venue growth of users(#373, #475, #577, #675) as they read more papers. (b) Keyword growth of users as they read more papers.

3.1 Regularities in Frequency

As might be expected, users vary greatly in the frequency of their academic reading. As shown in Fig. 1(a), user's reading count follows the power law distribution, indicating most users read very few papers while a few users read very frequently. Specifically, about 92.14% users read less than 4 papers and average reading count is about 3.93. It shows that the reading behavioral data is extremely sparse due to the inactivity of large proportion of users, which is a typical phenomenon in recommender systems. This makes the literature recommendation problem very challenging, since it is usually difficult to achieve good recommendation performance on long-tail users. We further analyze the reading frequencies with respect to venue. As shown in Fig. 1(b), which also follows the Zipf's law. The results show that most readings focus on a small set of venues' publications, e.g., 84.39% readings are within 20% venues. This Zipf distribution is in contrary good news for recommender systems. It indicates that the recommender system can obtain sufficient paper resources for recommendtion by only collecting publications from a small set of important venues.

As users read more and more papers, the venue and keyword lists corresponding to the papers grow over time. These lists may exhibit very different growth rates, however, reflecting how users' interests develop and change over time. As shown in Fig. 2, some users' (#675,#577) venue lists grow steadily, reflecting concentrated reading behaviors. While some users' (#475,#373) venue lists grow rapidly, reflecting very broad reading behaviors. The different growth rates can also be viewed with respect to keywords, reflecting very focused interests (#675) and diverse interests (#373) in reading respectively. It is more interesting to find that users who read broadly may have very focused interests (e.g., user #475 has a large venue list but a relatively small keyword list), while users who read in a narrow range may exhibit diverse interests (e.g., user #577 has a relatively small venue list but a large keyword list). All these above cases reflect the different reading patterns among users, calling for corresponding recommendation strategies (e.g., diverse or focused recommendations) when designing literature recommender systems.

3.2 Regularities in Time

Although different users may read at different time, there are general reading patterns with respect to time. As shown in Fig. 3, we can see that most users access the papers during the afternoon, and there are clearly rise-and-fall patterns with respect to months, with peaks reached usually in March and November. This phenomenon might be related to users' work time. Since most academic users are faculty members and students, it is not surprising to see the low reading intensity between June to September when summer vacation takes place.

We further examine the time intervals of academic readings of different users. We find that there are two interesting temporal patterns as shown in Fig. 4. The user #2583 in the top read 302 papers from 2012/05/16 to 2014/12/04 (932 days), and show a *burst* phenomenon in temporal patterns, i.e., short time

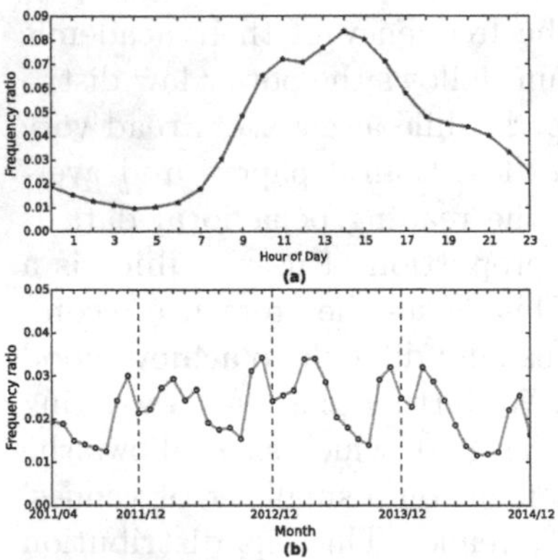

Fig. 3. (a) Distribution of reading intensity in a day; (b) Distribution of reading intensity over years.

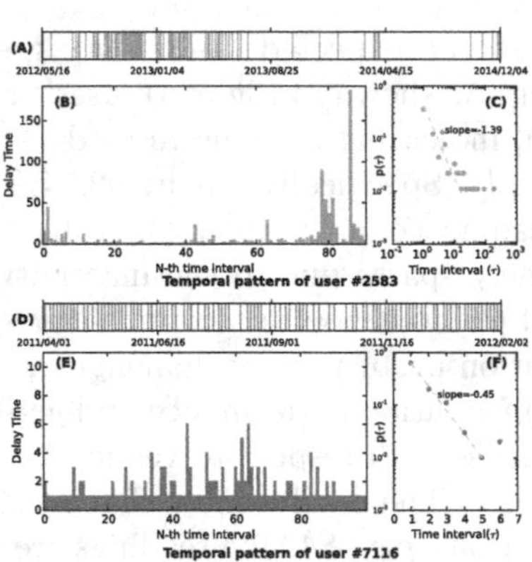

Fig. 4. Different reading patterns from two specific users (#2583, #7116). (A, D) Time stamps of reading events. Each vertical line represents a reading event occurring at the time stamp. (B, E) The time intervals between two adjacent events, where the height corresponds to the interval. (C) Log-log plot of the distribution of time intervals, which follows a power law distribution $P(\tau) \simeq \tau^{-1.39}$; (F) Log-linear plot of the distribution of time intervals, which follows an exponential distribution $P(\tau) \simeq 10^{-0.45\tau}$.

frames of intense readings are separated by long idle periods. The existence of bursty reading behaviors can also be supported by the evidence that the reading time interval follows the power law distribution as depicted in Fig. 4(C). While the user #7116 in the bottom who have read 509 papers from 2011/04/01 to 2012/02/02 (306 days) exhibits quite different reading patterns. Readings happen regularly in a Poisson process, resulting in an exponential distribution of time interval as shown in Fig. 4(F).

To further check the proportions of users that exhibit power-law and exponential temporal distributions, we randomly sampled 1,000 users who have read more than 100 papers. We then apply the *goodness-of-fit test* based on bootstrapping and the Kolmogorov-Smirnov (K-S) statistic to check which hypothesis is plausible [6].

As a result, we find that 51.21% users follow a power law temporal distribution, while 25.65% users follow a exponential temporal distribution. Besides, there are 23.14% users that cannot be modeled by either distribution confidently, which might be mixed behaviors and require further investigation in the future. From the results we can see that there are quite different temporal patterns in academic reading, and literature recommender systems should take these patterns into account to design better recommendation strategies. For example, most recommender systems would like to push recommendations regularly to users via email or notification. This might be acceptable for users with regular reading behaviors (i.e., exponential temporal distribution), but annoying for users with bursty reading behaviors (i.e., power law distribution).

4 Group-Based Analysis

In this section, we aim to analyze this problem by conducting finer group-based analysis. Specifically, we divide users into three groups. Users who have read very few (i.e., less than 4 papers) are categorized as *inactive* users. Users who have read a lot (i.e., more than 100 papers) are categorized as *active* users. The rest are categorized as *normal* users. We then analyze what kind of papers these three groups of users prefer to read in terms of recency and popularity.

4.1 Recency

We first investigate the question whether users prefer newly published papers or some older ones. To answer this question, we define the recency of a reading as the time interval between users' reading time and the paper's publication time (i.e., the year a user read the paper minus the year it published), and plot the average distribution of recency over different groups of users in Fig. 5.

As we can see, in general users prefer reading newly published papers. In average, more than 53% readings are within five years of the paper's publication over all the users. Therefore, a good literature recommender system should assign priority to recent publications. However, when we compare different groups, we can see that inactive users read much less newly published papers but more older papers as compared with normal and active users. For example, the percentage of reading within five years is about 50.13% for inactive users, while 56.5% for active users. We conduct the K-S test among the distributions of the three groups and find the differences are significant ($p \prec 0.05$).

The possible reason of the above observation is that many inactive users might be fresh men in research or occasional users, who would be more likely to survey or read some existing old work. While the active users are more likely to be serious academic users (e.g., Ph.D. students, researchers or faculties), who read frequently and would take more attention to the up-to-date work since they have already been familiar with their domains.

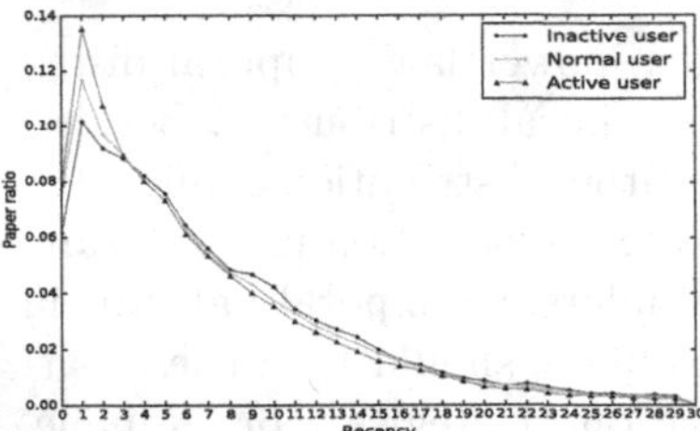

Fig. 5. Distributions of recency in terms of year over different groups of users.

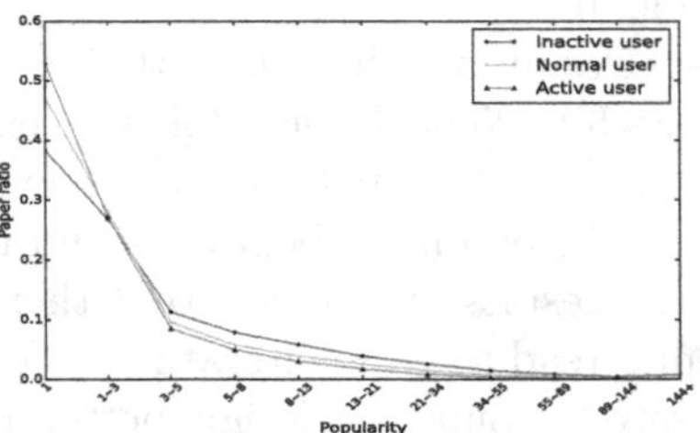

Fig. 6. Distributions of papers' popularity over different groups of users.

4.2 Popularity

Finally, we check the popularity of the papers read by different groups of users. Here we define the popularity of a paper as the number of distinct users who have read it. We then depict the average distributions of paper popularity read by users in these three groups in Fig. 6. As we can see, users' most readings are non-popular papers (i.e. papers with popularity less than 5). However, inactive users seem to be more likely to read popular papers than normal and active users. Specifically, in average more than 23.58% of readings are papers with popularity larger than 5 for inactive users, while the corresponding percentage is 15.19% for normal users and 11.25% for active users, respectively. The results indicate that recommending popular papers would be more effective for inactive users than normal and active users.

5 Sequence-Based Analysis

In literature recommendation, one aims to recommend a set of papers for each user based on his/her historical behaviors. In real recommender systems, A common practice is to score over a proper candidate set rather than the entire academic repository for recommendation. Therefore, it would be of great importance to generate the candidate set effectively. In this section, we conduct sequence based analysis over users' reading behaviors to investigate this correlation.

Specifically, we partition each user's readings into two parts according to the time stamp, and take the latest paper read by the user as the future reading and the rest as the historical readings. We then examine the connections between these two parts from both the content-based filtering and collaborative filtering views. We aim to investigate whether these filtering methods can generate proper candidate sets covering the future readings.

5.1 Coverage of Content-Based Filtering

Firstly, we investigate whether the content-based filtering can generate the candidate set with good coverage over users' future readings. Given users' reading history, we analyze the following five ways for candidate generation based on the meta-data of the historical readings.

* **Co-venue**: users would like to follow the same venues of the historical readings to find the next reading.
* **Co-author**: users would like to follow the same authors in the historical readings to find the next reading.
* **Co-keyword**: users would like to use the same keywords in the historical readings to find the next reading.
* **Ref & Co-ref**: users would like to follow the references of the historical readings to find the next reading. Note here we take the direct references of the historical readings, as well as papers that share at least one same reference of the historical readings (co-reference) as the candidates.
* **Cite & Co-cite**: users would like to follow the citations of the historical readings to find the next reading. Note here we take the direct citations of the historical readings, as well as papers that share at least one same citation of the historical readings (co-citation) as the candidates.

For each way of candidate generation, we calculate the coverage ratio as follows

$$R_{cover} = \frac{\#(Hit)}{\#(User)} \tag{1}$$

where $\#(Hit)$ denotes the number of users whose future reading is within the generated candidate set, and $\#(User)$ denotes the total number of users in calculation. The analysis is conducted over different groups of users and also the overall users, and the results are depicted in Fig. 7. From the results we observe that: (1) The coverage ratio increases with the richness of users' historical readings. For example, the coverage ratio with all the five ways on inactive users is only 43.7% while that on active users is about 96.1%. (2) Among the five ways, Co-venue and Co-keyword can cover more users' future readings than others. This is not surprising since the candidate sets generated in these two ways are much larger than the others, and we will further discussed this in Sect. 5.3. (3) With all the five ways of candidate generation methods, the coverage ratio over all the users reaches 52.5%, which represents the upper-bound of user coverage of the content-based filtering methods. In other words, there are about 47.5% users whose future readings cannot even be included in the candidate sets from content-based filtering methods.

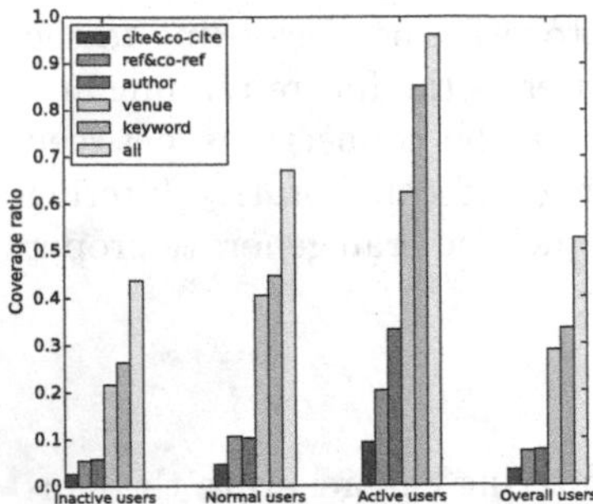

Fig. 7. Coverage ratios via content-based filtering over different groups of users as well as the overall users.

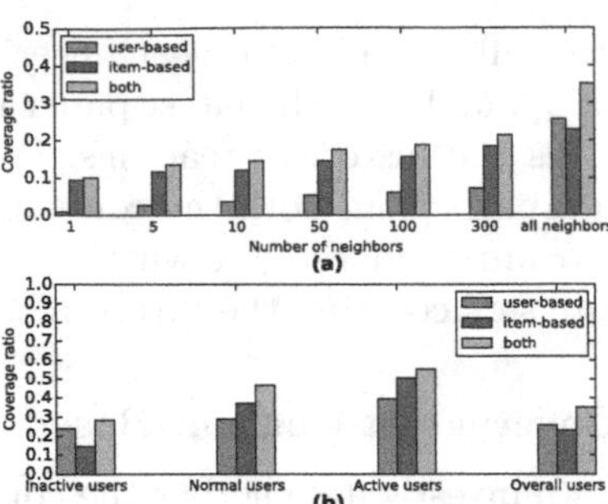

Fig. 8. (a) Coverage ratio variation against neighbor size in CF; (b) Coverage ratio via CF over different groups of users as well as overall users.

5.2 Coverage of Collaborative Filtering

Another widely used recommendation method is collaborative filtering (CF), which can be further categorized as user-based CF and item-based CF. Here we analyze the coverage ratio of users over these two types of CF methods.

Specifically, for each CF method, we vary the size of nearest neighbor to see whether a user's future reading is within the corresponding candidate set generated by the CF method. Note here Jaccard similarity is used to compute the similarity between users/items since the user-item relation matrix is typically binary. We depict the coverage ratio over different groups of users and overall users in Fig. 8.

We have the following observations over the results: (1) When neighbor size is small, the coverage ratio of item-based CF is much better than user-based CF. However, when taking all the neighbors into account, the user-based CF get better coverage since its neighbor size is much larger than that of item-based CF. (2) The coverage ratio of user-based CF is higher than item-based CF for inactive users but lower for normal and active users. Overall, user-based CF wins a little bit since there are more inactive users. (3) The overall coverage ratio with both two types of CF methods reaches 32.25%, leaving a large proportion of users uncovered at all.

5.3 Integrated Analysis and Discussion

Based on the above analysis, we can see that both content-based filtering and collaborative filtering can cover some proportions of users. We further compared the set of users covered by these two types of methods by computing the Jaccard similarity between them. We find that the Jaccard similarity is 35.45%, indicating that there is some overlap between the user set covered by content-based filtering and collaborative filtering. If we integrate both filtering methods, the overall coverage ratio can reach 64.68%.

As mentioned previously, it is somehow unfair to only compare the coverage of different methods since the candidate sets generated by different methods vary largely. A method that can generate a large candidate set, like Co-venue in content-based filtering, would naturally be able to cover more users. However, by introducing a larger and inevitably noisier candidate set, it will increase the computational complexity and also the prediction difficulty. To take these factors into account, here we define the efficiency of a method as follows

$$E = \frac{R_{Cover}}{\bar{R}_{Shrink}} \tag{2}$$

where R_{cover} is the coverage ratio defined in Eq. 1, and $\bar{R}_{Shrink}$ denotes the average shrinkage ratio of the method with respect to the whole academic repository

$$\bar{R}_{Shrink} = \frac{1}{\#(User)} \sum_{i} \frac{\#(Candidate_i)}{\#(C)} \tag{3}$$

where $\#(Candidate_i)$ denotes the candidate size for the i-th user generated by the method, $\#(C)$ denotes the repository size, and the summation is taken over all the users. From the above definition we can see, a method is efficient if it can generate a small candidate set but cover a large proportion of users' future reading.

We compare the efficiency of different methods as show in Table 1. As a result, we can see that although Co-venue and Co-keyword can achieve better coverage ratios as shown in Fig. 7, they are not the most efficient ways since they will generate very large candidate sets. Meanwhile, the efficiencies of the two CF methods are not so high since their coverage ratios are low. The most efficient ways for candidate generation are Cite & Co-cite and Ref & Co-ref, followed by Co-author. For literature recommender systems, the efficiency would then be a good reference metric when choosing which methods for candidate generation given limited computational resources.

Table 1. Efficiency of different filtering methods.

	Co-keyword	Co-venue	Co-author	Ref & Co-ref	Cite & Co-cite	User-based CF	Item-based CF
$\bar{R}_{Shrink}$	9.35%	1.03%	0.09%	0.053%	0.017%	1.8%	1.09%
R_{cover}	33.26%	28.73%	7.5%	7.32%	3.32%	25.67%	22.9%
E	3.56	27.89	83.33	138.11	195.29	14.26	21.01

6 Related Work

In this section, we briefly review the related work of academic information seeking behavior analysis and usage of academic access data.

6.1 Academic Information Seeking Behavior Analysis

There have been many previous studies [2,7,10,14,16,17,20] on analyzing the academic information seeking behaviors to help design library systems or academic search engines. For example, Niu et al. [14] examined the relationships between scientists' information-seeking behaviors and their personal and environmental factors. Tenopir et al. [16,17] analyzed article seeking and reading patterns in academic faculty readers through surveys. They found that subject discipline of the reader influences many patterns, including amount of reading, format of reading, and average time spent per reading. However, these work either relied on questionnaires to survey very limited faculty members [10,14,16,17], or only focused on analyzing users' query patterns based on some search logs [2,7,20].

6.2 Usage of Academic Access Data

The public paper access logs from OpenURL have been utilized in different ways for literature recommendation [4,13,15,19]. For example, Pohl et al. [15] used the access data to identify related papers for literature recommendation. The BibTip [13] recommended related literatures based on the observation of user patterns and the statistical evaluation of the usage data. Although academic access data have been leveraged for literature recommendation, there is little work on analyzing the data for better understanding the reading patterns and information needs of academic users.

Usage log data, defined as a collection of individual usage events recorded for a given period time [12], has drawn a lot of attention in recent years [4,5,8,12,19]. Bollen et al. [4] presents a technical, standard-based architecture for sharing +usage information. They found the generated relationship networks encode which journals are related in their usage to can be used to recommend documents. Gorraiz et al. [8] focused on the disciplinary differences observed for the behavior of citations and downloads. They pointed the fact that citations can only measure the impact in the 'publish or perish' community.

7 Conclusion

In this paper, we present some in-depth analysis of academic reading behaviors based on a large scale scholarly usage data. We perform the global analysis, group-based analysis as well as sequence-based analysis to unveil the underlying reading patterns and information needs of real-world academic users.

The analysis of user behaviors is a foundation of designing any interactive systems. All the findings in our work may provide some guidelines to improve existing literature recommender systems as well as designing new ones. One of our future work is to build an literature recommender system to integrate the factors revealed in this paper and verify these conclusions from users' feedbacks.

Acknowledgements. The work was funded by 973 Program of China under Grant No. 2014CB340401, the National Key R&D Program of China under Grant No. 2016QY02D0405, the National Natural Science Foundation of China (NSFC) under Grants No. 61232010, 61472401, 61433014, 61425016, and 61203298, the Key Research Program of the CAS under Grant No. KGZD-EW-T03-2, and the Youth Innovation Promotion Association CAS under Grants No. 20144310 and 2016102.

References

1. Le Anh, V., Hoang Hai, V., Tran, H.N., Jung, J.J.: SciRecSys: a recommendation system for scientific publication by discovering keyword relationships. In: Hwang, D., Jung, J.J., Nguyen, N.-T. (eds.) ICCCI 2014. LNCS (LNAI), vol. 8733, pp. 72–82. Springer, Cham (2014). doi:10.1007/978-3-319-11289-3_8
2. Asunka, S., Chae, H.S., Hughes, B., Natriello, G.: Understanding academic information seeking habits through analysis of web server log files. J. Acad. Librarianship **35**(1), 33–45 (2009)
3. Bogers, T., van den Bosch, A.: Recommending scientific articles using citeulike
4. Bollen, J., de Sompel, H.V.: An architecture for the aggregation and analysis of scholarly usage data. In: Proceedings of ACM/IEEE Joint Conference on Digital Libraries. Chapel Hill, pp. 298–307, 11–15 June 2006
5. Bollen, J., Van de Sompel, H., Rodriguez, M.A.: Towards usage-based impact metrics: first results from the mesur project. JCDL 2008, pp. 231–240. ACM, New York (2008)
6. Clauset, A., Shalizi, C.R., Newman, M.E.J.: Power-law distributions in empirical data. SIAM Review **51**(4), 661–703 (2009)
7. Dogan, R.I., Murray, G.C., Névéol, A., Lu, Z.: Understanding pubmed® user search behavior through log analysis. Database, 2009 (2009)
8. Gorraiz, J., Gumpenberger, C., Schlögl, C.: Usage versus citation behaviours in four subject areas. Scientometrics **101**(2), 1077–1095 (2014)
9. Han, H., Jeong, W., Wolfram, D.: Log analysis of academic digital library: user query patterns. iConference 2014 Proceedings (2014)
10. Hemminger, B.M., Lu, D., Vaughan, K.T.L., Adams, S.J.: Information seeking behavior of academic scientists. JASIST **58**(14), 2205–2225 (2007)
11. Huang, W., Wu, Z., Mitra, P., Giles, C.L.: Refseer: a citation recommendation system. In: JCDL. U.K, pp. 371–374, 8–12 Sep 2014
12. Kurtz, M.J., Bollen, J.: Usage bibliometrics. CoRR, abs/1102.2891 (2011)
13. Mönnich, M., Spiering, M.: Adding value to the library catalog by implementing a recommendation system. D-Lib Magazine **14**(5), 4 (2008)
14. Niu, X., Hemminger, B.M.: A study of factors that affect the information-seeking behavior of academic scientists. JASIST **63**(2), 336–353 (2012)
15. Pohl, S., Radlinski, F., Joachims, T.: Recommending related papers based on digital library access records. In: JCDL 2007, Proceedings. Vancouver, Canada, pp. 417–418, 18–23 June 2007
16. Tenopir, C., King, D.W., Bush, A.: Medical faculty's use of print and electronic journals: changes over time and in comparison with scientists. J. Med. Libr. Assoc. **92**(2), 233 (2004)
17. Tenopir, C., Volentine, R., King, D.W.: Article and book reading patterns of scholars: findings for publishers. Learn. Publish. **25**(4), 279–291 (2012)

18. Torres, R., McNee, S.M., Abel, M., Konstan, J.A., Riedl, J.: Enhancing digital libraries with techlens+. In: ACM/IEEE Joint Conference on Digital Libraries, JCDL 2004, Proceedings. Tucson, pp. 228–236, 7–11 June 2004
19. Vellino, A.: A comparison between usage-based and citation-based methods for recommending scholarly research articles. ASIST **47**(1), 1–2 (2010)
20. Villn-Rueda, L., Senso, J.A., de Moya-Anegn, F.: The use of OPAC in a large academic library: a transactional log analysis study of subject searching. J. Acad. Librarianship **33**(3), 327–337 (2007)

Understanding Users

Incorporating Position Bias into Click-Through Bipartite Graph

Rongjie Cai, Cheng Luo, Yiqun Liu(✉), Shaoping Ma, and Min Zhang

State Key Laboratory of Intelligent Technology and Systems,
Tsinghua National Laboratory for Information Science and Technology,
Department of Computer Science and Technology,
Tsinghua University, Beijing 100084, China
jayjay7@163.com, c-luo12@mails.tsinghua.edu.cn,
{yiqunliu,msp,z-m}@tsinghua.edu.cn
http://www.thuir.cn

Abstract. Click-through bipartite graph has been regarded as an effective method in search user behavior analysis researches. In most existing bipartite graph construction studies, user clicks are treated as equally important. However, considering the existence of position bias factor in user click-through behavior, clicks on results in different ranking positions should be treated separately. In this work, we choose a classical click-through bipartite graph model, which named label propagation model, and evaluate the improvement of performance by considering the effect of position bias. We propose three hypotheses to explain the influence of position bias, and modify the formulas of label propagation algorithm. We use AUC as the evaluation metric, which express the effectiveness of spam URLs identification by label propagation algorithm and its improved methods. The experimental results demonstrate that the proposed methods work better than the baseline method.

Keywords: Position bias · Click-through bipartite graph · Label propagation model

1 Introduction

Click-through bipartite graph propagation is an effective method in many Information Retrieval (IR) tasks such as relevance ranking and similarity calculation between queries and documents [2]. Given a query with its initial score in bipartite graph, the method propagate the score to user clicked documents in this query session, and then propagate back to other queries in whose sessions those documents are clicked. After many rounds of iteration, queries and documents gain different scores. A larger score usually reflects that the similarity between the initial query and the query/document is higher. Click-through

R. Cai—This work is supported by Natural Science Foundation of China (Grant No. 61622208, 61532011, 61672311) and National Key Basic Research Program (2015CB358700).

J. Wen et al. (Eds.): CCIR 2017, LNCS 10390, pp. 57–68, 2017.
https://doi.org/10.1007/978-3-319-68699-8_5

bipartite graph is also used in Web spam identification [3,4], which shows performance improvement compared to existing supervised learning methods.

Actually, click-through bipartite graph isn't a simple two-dimensional connected graph. The nodes of URLs in graph have position information, which corresponds to the rank on Search Engine Results Page (SERP). The position factor should not be ignored, since for results on different positions, the possibility of clicking and viewing are quite different, which is referred to Position Bias. However, most existing studies could not solve this problem effectively. Thus, we choose a classical click-through bipartite graph model, which named label propagation model, and analyze the influence of taking position factor into consideration.

Position bias plays an important role in search user behavior studies. Joachims [5] were among the first to describe the influence of this factor in search examination and click behaviors. Craswell [6] proposed the examination hypothesis to reduce its influence and generated a more justified estimation of result relevance based on user clicks. Most studies in click-through bipartite graph model ignored position bias problem [2,7–10] or used simple methods to decrease the influence of this problem [11,12]. To the best of our knowledge, our study is among the first to consider the influence of position bias in label propagation with bipartite graph.

In this paper, we try to investigate the influence of position bias, and improve the performance of click-through bipartite graph model. We propose three hypotheses to describe position bias and improve the label propagation algorithm. We use the query logs of a commercial search engine to evaluate the effectiveness of spam URLs identification by label propagation algorithm and its improved methods. The experimental results demonstrate that the proposed methods perform better than baseline algorithm.

The rest of the paper is organized as follows: in Sect. 2, we conduct a survey on related work. Then, we describe the construction of click-through bipartite graph and the algorithm design of label propagation model in Sect. 3. After that, we explain our methods utilize position bias in Sect. 4. Finally, we report experimental results in Sect. 5, and conclude this study in Sect. 6.

2 Related Work

Click-through bipartite graph is widely used in IR tasks such as query recommendation [2,7,9,10,15] and search result optimizing [2,7,8]. Li [9] presented the use of regularized click graphs in improving query intent classifiers, and Yi [10] tried to discover query clusters from a click-through graph of web search logs, which were both helpful to query recommendation. Antonellis [13] focused on the problem of query rewriting for sponsored search and presented two enhanced versions of Simrank [14], which exploited weights on click graph edges. Cao [15] proposed a novel context-aware query suggestion approach by mining click-through and session data, which contained offline model-learning step online query suggestion step. Xue [8] proposed an iterative reinforced algorithm to utilize user click-through data to improve search performance. Wu [2] and Jiang [7] learned query

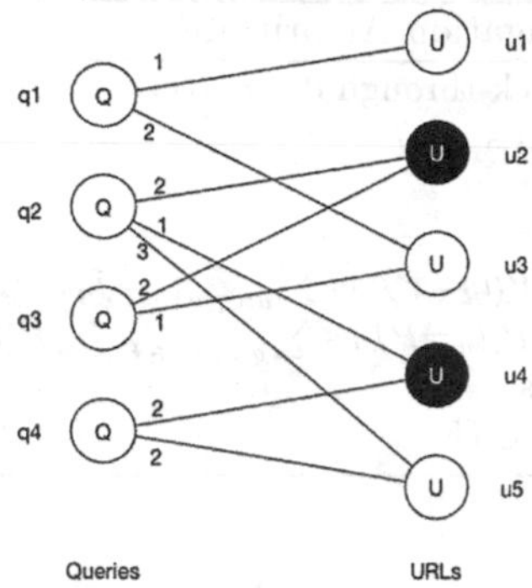

Fig. 1. A sample portion of query-URL bipartite graph.

and document similarities from click graph which can be used in both query recommendation and relevance ranking. However, those studies didn't consider the influence of position bias in click graph. Empirically, decreasing position bias with reasonable method would improve the performance in most of IR tasks.

Position-bias in click model was discovered since 2005 [5]. Richardson [11] proposed a model to explain position bias in the ranked list of ads, whose main point is that each rank of search result has a certain probability of being examined. Agichtein [12] corrected for position bias by subtracting the background click distribution from this query's click distribution. Craswell [6] focused on modelling how probability of click depends on position by examination hypothesis. Those studies either used simple methods to decrease the effect of position bias, or tried to explain the phenomenon of position bias. But they had not been explored the improvement of performance by considering position bias in click graph model. It is necessary to research the effectiveness when eliminating position bias.

3 Label Propagation Algorithm

3.1 Problem Formulation

Click-through data C and bipartite graph G. Click-through data consists of queries and URLs with their co-click information. We use a triple $< q, u, n_{qu} >$ to present click log, where q is a query, u is a URL, and n_{qu} is the times that URL u is clicked when users search for query q. We define $Q = \{q|q\, appears\, in\, C\}$ and $U = \{u|u\, appears\, in\, C\}$. Click-through data C can be presented as a click-through bipartite graph form *G=(Q, U, E)*. Queries and URLs are two type of nodes in this graph. For an edge $(q, u) \in E$, q/u is assigned a value p_q/p_u, which denotes the probability that query/URL is a spam query/page.

Labelled Seed URL set L. URLs in L are randomly selected from G and are manually labelled as spam. More formally, $L = \{u|u\, is\, labelled\, as\, a\, spam\, page\}$.

Algorithm 1. Label Propagation Algorithm

Input: labeled seed set L, click-through data $C(G)$
Output: $P(l_u = \boldsymbol{P})$ for all URLs in G
repeat
  for $u \in L$, set $P(l_u = \boldsymbol{P}) = 1$
  for all $q \in Q$, calculate $P(l_q = \boldsymbol{P})$ as $\sum_{u:(q,u)\in E} f_{qu} P(l_u = \mathbf{P})$
  for all $u \in U$, calculate $P(l_u = \boldsymbol{P})$ as $\sum_{q:(q,u)\in E} f_{uq} P(l_q = \mathbf{P})$
until Algorithm converges
Output $P(l_u = \boldsymbol{P})$ for all $u \in U$

URL result set RU **and query result set** QU**.** RU and QU contain the $< u, p_u >$ and $< q, p_q >$ pairs. After algorithms ends, each URL u or query q in G will obtain a probability p_u/p_q, which denotes how likely this URL/query is to be a spam page/query. Let $G=(Q, U, E)$ and $L \subset U$, the goal of spam pages mining is to obtain the results set RU, which contains all of the possible spam pages in G.

3.2 Algorithm Design

We propose a label propagation algorithm to mining the spam pages. For each URL u, we can calculate the probability p_u to be a spam page by incorporating all of the label information of its adjacent queries. Similarly, we can calculate p_q for each query q from adjacent URLs. We can describe this procedure formally as follows. For each q/u, we denote its label as l_q/l_u, which is $\mathbf{P}$ for spam. Thus, every URL u in set L would obtain a label score, which formally as $P(l_u = \boldsymbol{P}) = 1$. Besides, every URL u in set $U - L$ would be initialized by $P(l_u = \boldsymbol{P}) = 0$. Then, we have

$$P(l_q = \mathbf{P}) = \sum_{u:(q,u)\in E} f_{qu} P(l_u = \mathbf{P}) \tag{1}$$

where

$$f_{qu} = \frac{n_{qu}}{\sum_{u':(q,u')\in E} n_{qu'}} \tag{2}$$

Query q may connect many URLs in the query-URL bipartite graph. f_{qu} can be interpreted as the transition probability from query q to a certain URL u, and n_{qu} is the times URL u clicked for query q. Thus, every query q could calculate a label score by all of its connected URLs from Eqs. 1 and 2. The bigger f_{qu} is, the more influencing its corresponding URL u has on determining the label score of query q.

Similarly, for each URL u in U, the probability $P(l_u = \boldsymbol{P})$ can be calculated as

$$P(l_u = \mathbf{P}) = \sum_{q:(q,u)\in E} f_{uq} P(l_q = \mathbf{P}) \tag{3}$$

Algorithm 2. Promoting Algorithm Utilize Position Bias

Input: labeled seed set L, click-through data $C(G)$
Output: $P(l_u{=}\boldsymbol{P})$ for all URLs in G
if promote by time hypothesis **then**
  calculate ω_u as $rank_u$
end if
if promote by click hypothesis **then**
  calculate ω_u as $1/click\_number_{rank_u}$
end if
if promote by probability hypothesis **then**
  calculate ω_u as $1/\prod_{i=1}^{rank_u-1}(1-click\_probability_i)$
end if
repeat
  for $u \in L$, set $P(l_u{=}\boldsymbol{P}){=}1$
  for all $q \in Q$, calculate $P(l_q{=}\boldsymbol{P})$ as $\sum_{u:(q,u)\in E}\omega_u f_{qu} P(l_u = \mathbf{P})$
  for all $u \in U$, calculate $P(l_u{=}\boldsymbol{P})$ as $\sum_{q:(q,u)\in E}\omega_u f_{uq} P(l_q = \mathbf{P})$
until Algorithm converges
Output $P(l_u{=}\boldsymbol{P})$ for all $u \in U$

where

$$f_{uq} = \frac{n_{qu}}{\sum_{q':(q',u)\in E} n_{q'u}} \tag{4}$$

URL u may have many adjacent queries in the query-URL bipartite graph. And f_{uq} present the transition probability from URL u to a certain query q.

Note that both f_{qu} and f_{uq} are positive number, and they should satisfy the formulas $\sum_{u:(q,u)\in E} f_{qu} = 1$ and $\sum_{q:(q,u)\in E} f_{uq} = 1$.

Using Equations mentioned above, we can calculate $P(l_q = \boldsymbol{P})$ and $P(l_u = \boldsymbol{P})$ recursively for all the queries and URLs in the click-through bipartite graph. If a query/URL has a high score after algorithm, the query/page is more likely a spam.

There is a sample portion of a bipartite graph extracted from search engine log, as shown in Fig. 1. When algorithm propagate from URLs to queries, the spam value of queries would be updated. With $q2$ taken as an example, the value would be calculated as *value(u2)*2/6+value(u4)*1/6+value(u5)*3/6*. Similarly, the value of URLs are calculated by the value of queries who click those URLs.

4 Promoting Utilize Position Bias

We use three methods to promote label propagation algorithm. We change the weight between query and URL which occur click, and we can find the influence by position bias.

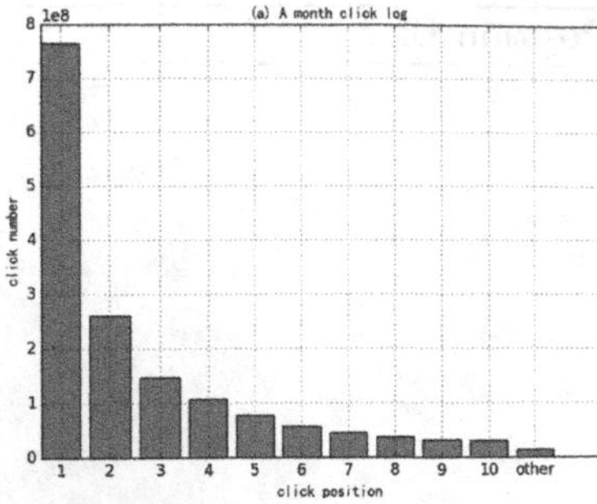

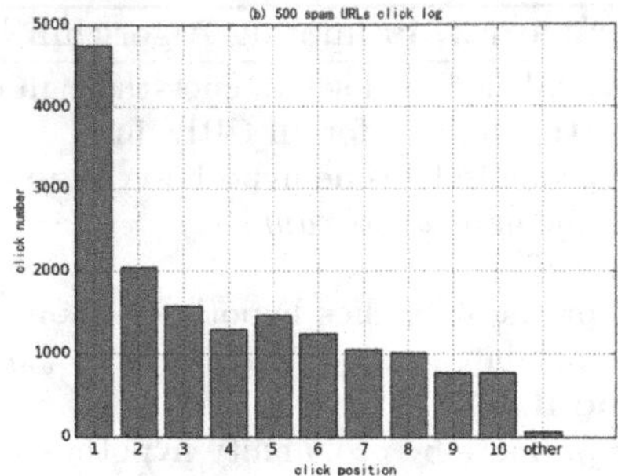

Fig. 2. Click distribution of different position. (a) contains all click data of a month search engine result, and (b) contains 500 spam URLs with 16, 021 times click.

4.1 Time Method

Time Hypothesis: An explanation for position bias is that the search result with lower position need more time to be checked. We assume that the user views search results from top to bottom, and the time spend for checking each result is equal. If the user need time t to check a search result, then he/she may spend time rt to reach to the result with rank r and decide to click it or not. Thus, we can use r to weigh the link between query and clicked URL.

Add weight r to f_{qu} and f_{uq} in Eqs. 2 and 4, and the transition probability can be calculated as

$$f_{qu} = \frac{r_{qu} n_{qu}}{\sum_{u':(q,u') \in E} r_{qu'} n_{qu'}} \tag{5}$$

and

$$f_{uq} = \frac{r_{qu} n_{qu}}{\sum_{q':(q',u) \in E} r_{qu'} n_{q'u}} \tag{6}$$

Note that the same URL may have different rank for different query. The time method only change the transition probability f_{qu} and f_{uq}, which means that the label propagation formulas of $P(l_q = \boldsymbol{P})$ and $P(l_u = \boldsymbol{P})$ are the same as label propagation algorithm in Sect. 3.

4.2 Click Method

Click Hypothesis: Another explanation for position bias is that the rank of search result with less click need more attention. We collect a month click log of search engine, and calculate click distribution of different position, which shown in Fig. 2(a). We can observe that click number descend rapidly with the increase of rank value. It is intuitive that the results with lower position are hard to be clicked, and they should be given a higher weight. Thus, we can use the reciprocal of click number in rank r to weigh the link between query and clicked URL.

We define that m_r is the click number in rank r, and r_{qu} is the rank of URL u in the link of query q and URL u. Add weight m_r to f_{qu} and f_{uq} in Eqs. 2 and 4, and the transition probability can be calculated as

$$f_{qu} = \frac{n_{qu}/m_{r_{qu}}}{\sum_{u':(q,u')\in E} n_{qu'}/m_{r_{qu'}}} \tag{7}$$

and

$$f_{uq} = \frac{n_{qu}/m_{r_{qu}}}{\sum_{q':(q',u)\in E} n_{q'u}/m_{r_{qu'}}} \tag{8}$$

Note that the click number in position more than 10 are too small to be use to calculate transition probability directly. Thus, we simplify the problem that use the click number in position 11 for all position more than 10.

4.3 Probability Method

Probability Hypothesis: Final explanation for position bias is that the lower the result's position is, the smaller probability the result is clicked. We assume that the user views search results from top to bottom, deciding whether to click each result before moving to the next, and stop search when click any result. We use click number m_r to calculate actual click probability pa_r in position r, which shown in Eq. 9.

$$pa_r = \frac{m_r}{\sum_{r'=1}^{\infty} m_{r'}} \tag{9}$$

We define pm_r as the theoretical probability in position r when there is not position bias, which can be formulated as

$$pm_r = pa_r \prod_{i=1}^{r-1} (1 - pm_i) \tag{10}$$

Thus, we can use the division value w between theoretical probability pm_r and actual probability pa_r to weigh the link of query and clicked URL, which can eliminate the influence of position bias. Add weight w to f_{qu} and f_{uq} in Eqs. 2 and 4, and the transition probability can be calculated as

$$f_{qu} = \frac{w_r n_{qu}}{\sum_{u':(q,u')\in E} w_{qu'} n_{qu'}} \tag{11}$$

and

$$f_{uq} = \frac{w_r n_{qu}}{\sum_{q':(q',u)\in E} w_{qu'} n_{q'u}} \tag{12}$$

where

$$w_r = \frac{pa_r}{pm_r} = \frac{1}{\prod_{i=1}^{r-1}(1 - pm_i)} \tag{13}$$

and

$$r = rank_{qu} \tag{14}$$

Note that we use the weight value w in position 11 for all position more than 10 like Sect. 4.2 because of the unreliably small data.

5 Experiments

We conducted experiments to test the performance of the proposed methods on spam URLs mining.

5.1 Data Sets

We collected a month click-through data from an enterprise search engine of an IT company, and structured a click-through bipartite graphs with the data. If there are more than one click between a query and a URL, we added a link between them. In other words, we discarded the links between queries and URLs whose click frequency is equal to 1. Finally, there are 30, 708, 979 queries and 40, 606, 253 URLs on the bipartite graph, which each query has on average 2.49 clicked URLs.

To decrease the effect of manual label, we considered only one type spam. Because it is easy to judge whether the page is a pornographic page or not, while other type spam may be hard to determine sometimes.

We deleted the healthy URLs whose out-degree is more than 2, 000, because those popular URLs could gain spam score from many spam queries and propagate it to other healthy queries and URLs, which may influence the performance of spam URLs mining. The amount of those URLs is 37, which contain many famous URLs such as www.sogou.com and weixin.qq.com, and all of them had been checked manually as healthy URLs.

We selected some URLs randomly from all of 40 million URLs, and labelled them as exact spam pages or not. Then, we collected 500 spam pages as seed URLs for LP algorithm, which will be labelled a spam value before the calculation of each round. In this work, we gave seed URLs the value 1.0, and the biggest URL score from LP algorithm would be not more than 1.0.

When the end of LP algorithm, each URL would obtain a spam score, which is from 0 to 1. We sorted those URLs by their spam score, and put them into 10 bins. Assume that sum of all URLs' score is N, then we put the first $n1$ URLs into the first bin whose total score is $N/10$. Similarly, we put the next $n2$ URLs into the second bin whose total score is $N/10$ too. The other URLs was put into the remaining 8 bins as the same way. Then we randomly selected 20 URLs from each bin in 4 methods, total 800 URLs, which is our test set in this work.

Table 1. AUC comparison of four methods.

	LP	LP-r	LP-m	LP-w
AUC	0.71997	0.72922	0.73231	0.72609
Impv	-	1.28%	1.71%	0.85%

We labelled the test URLs as 0 or 1. Value 0 means that the page of URL doesn't contains any pornographic materials, and value 1 means that the page contains some pornographic or sexual materials.

5.2 Experiment Setup

The baseline of our work is label propagation model mentioned in Sect. 3, which denoted as **LP**. And there are three experiment models described in Sect. 4: (1) Consider users' view behavior by time hypothesis, which denoted as **LP-r**; (2) Consider users' click behavior by click hypothesis, which denoted as **LP-m**; (3) Consider users' satisfying probability by probability hypothesis, which denoted as **LP-w**.

To decrease the influence of click number and present the actual effectiveness of position bias, we made $n_{qu} = 1$ for all the links of queries and clicked URLs for baseline and three experiment models. In other words, we only concerned whether there is a link between a certain query and URL. Note that we still deleted those edges which have only one click, because they are too unreliable to be used in this experiment.

To evaluate the performance of different methods in spam URLs mining, we employed AUC [1] as evaluation measures, which calculated by manual label result and algorithm score result of test set. We evaluated the quality of the partition of spam URLs and non-spam URLs. If a model has higher AUC, its algorithm may give spam URLs with a higher score and non-spam URLs with a lower score. Moreover, higher AUC may lead to higher precision in top result of test set.

5.3 Experimental Results

5.3.1 Appearance of Position Bias

With the development of search engine, most spam results may be filtered or put at very low position for common queries. In Fig. 2, we show the click distribution in different position. The data set of Fig. 2(a) is experimental click graph data, which contains a month click data collected from search engine. And Fig. 2(b) shows the distribution of 500 experimental spam seed URLs which occur 16, 021 times click.

Obviously, the click probability in various position is unequal, which represent the appearance of position bias in click graph. As the presentation of comparison results, we can observe that the click position of spam URLs are lower than normal URLs in general. It is helpful to demonstrate the performance improvement in label propagation model, because spam URLs with lower position may be gave a bigger edge weight and obtain a higher score in label propagation algorithm.

5.3.2 Performance Comparison of Various Methods

We used full seed set with 500 URLs for this experiment. In Table 1, we show the AUC results of all methods and the improvement of three modified methods compared with LP.

We can observe that the AUC results of four methods are all bigger than 0.7, which means that label propagation model represents pretty effectiveness on spam URLs identification. Moreover, LP-m performs best in three modified methods, which represents click distribution directly. Click graph expresses position bias by click number, and the reciprocal of click number may correct position bias very well. Besides, the result of LP-r is pretty good, because click number in low position represent linear decrease approximately, which conform to the content of time hypothesis. However, the effectiveness of LP-w is not very good, because of the simple users behavior definition. Users may click several result or come back to check higher position results before satisfaction. The model of LP-w is too fuzzy to restore real users behavior.

5.3.3 Algorithm Robustness

The selection of seed set size is very important in semi-supervised algorithms. Therefore, we conducted an experiment to see how robust our algorithm is. We randomly sorted the spam URLs seed set, and selected top n seed URLs for experiment, where n is multiples of 50. Note that the seed set with small size is the subset of the seed set with big size, which can explore the improvement of performance by adding some seed URLs. To decrease randomly sorting bias, we repeat 10 times experiment and calculate the average of results, which is shown in Fig. 3.

We can find that the value of AUC increase as seed URLs added, because more seed may lead to more correct algorithm score of URLs. Besides, LP-m shows the best result with different size of seed set, while LP-r represent average level and

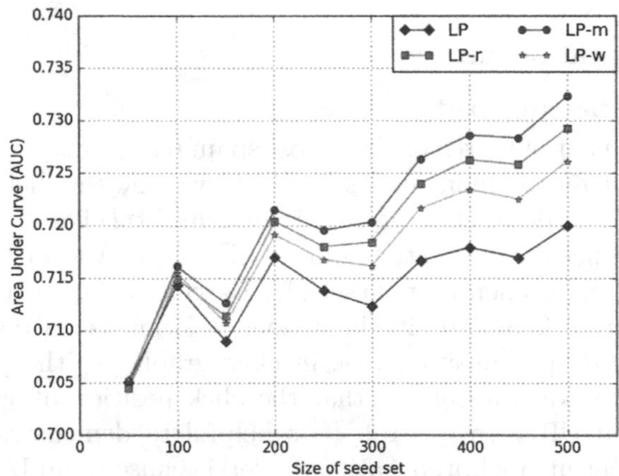

Fig. 3. AUC comparison with different size of seed set.

the result of LP-w is not very good, which conform to the result in Sect. 5.3.2. Moreover, three modified methods all perform better than baseline after only 100 seed URLs are used, which demonstrate that the appearance of position bias in click-through graph is obvious and our modified methods are very robust.

5.3.4 Some Cases of Non-spam URL with High Score

The AUC result can't reach to a very high value, because there are some non-spam URLs gained high score from label propagation algorithm. Therefore, we check those pages again to analyze the reasons.

A case is that it is a personal homepage whose name is the same as a porn star, while there is not any pornographic content. There is another case that this is a photo search result page or shopping site return page with query "old man", which usually be used as the key words of porn's title, while it is a healthy page absolutely.

Both of the two cases show that those pages may attract users with porn search intention to click it because of some ambiguous key words, which make them to obtain high score finally. It is noteworthy that those cases would decrease the value of AUC, but they don't influence the improvement of our methods compared with baseline.

6 Conclusion and Future Work

In this paper, we have studied the issue of incorporating position bias into click-through bipartite graph. We aim to eliminate position bias and promote performance of click graph models. We have proposed three hypotheses to explain and correct position bias, which represent pretty effective. Then we explore the robustness of our methods by analyzing the experimental results with different size of seed set. It is obvious that our methods always perform better than baseline. Finally, we analyze some cases of non-spam with high score from label propagation algorithm, which attract many clicks by matching some ambiguous key words.

As future work, we want to further enhance the efficiency of our methods and test their performance on larger data sets. To achieve the goal, we may need to structure edge weight vector and get the optimal solution based click graph data. In experiment section, we have to solve many problems of data set such as forbidden pages by search engine. We want to eliminate position bias maximumly, so that click-through bipartite graph models could represent best effectiveness.

References

1. Fawcett, T.: An introduction to ROC analysis. Pattern Recogn. Lett. **27**(8), 861–874 (2006)
2. Wu, W., Li, H., Xu, J.: Learning query and document similarities from click-through bipartite graph with metadata. In: Proceedings of the sixth ACM International Conference on Web Search and Data Mining, pp. 687–696. ACM (2013)

3. Wei, C., Liu, Y., Zhang, M., Ma, S., Ru, L., Zhang, K.: Fighting against web spam: a novel propagation method based on click-through data. In: Proceedings of the 35th International ACM SIGIR Conference on Research and Development in Information Retrieval, pp. 395–404. ACM (2012)
4. Luo, C., Liu, Y., Ma, S., Zhang, M., Ru, L., Zhang, K.: Pornography detection with the wisdom of crowds. In: Banchs, R.E., Silvestri, F., Liu, T.-Y., Zhang, M., Gao, S., Lang, J. (eds.) AIRS 2013. LNCS, vol. 8281, pp. 227–238. Springer, Heidelberg (2013). doi:10.1007/978-3-642-45068-6_20
5. Joachims, T., Granka, L., Pan, B., Hembrooke, H., Gay, G.: Accurately interpreting clickthrough data as implicit feedback. In: Proceedings of the 28th Annual International ACM SIGIR Conference on Research and Development in Information Retrieval, pp. 154–161. ACM (2005)
6. Craswell, N., Zoeter, O., Taylor, M., Ramsey, B.: An experimental comparison of click position-bias models. In: Proceedings of the 2008 International Conference on Web Search and Data Mining, pp. 87–94. ACM (2008)
7. Jiang, S., Hu, Y., Kang, C., Daly Jr., T., Yin, D., Chang, Y., Zhai, C.: Learning query and document relevance from a web-scale click graph. In: Proceedings of the 39th International ACM SIGIR Conference on Research and Development in Information Retrieval, pp. 185–194. ACM (2016)
8. Xue, G.R., Zeng, H.J., Chen, Z., Yu, Y., Ma, W.Y., Xi, W., Fan, W.: Optimizing web search using web click-through data. In: Proceedings of the Thirteenth ACM International Conference on Information and Knowledge Management, pp. 118–126. ACM (2004)
9. Li, X., Wang, Y.Y., Acero, A.: Learning query intent from regularized click graphs. In: Proceedings of the 31st Annual International ACM SIGIR Conference on Research and Development in Information Retrieval, pp. 339–346. ACM (2008)
10. Yi, J., Maghoul, F.: Query clustering using click-through graph. In: Proceedings of the 18th International Conference on World Wide Web, pp. 1055–1056. ACM (2009)
11. Richardson, M., Dominowska, E., Ragno, R.: Predicting clicks: estimating the click-through rate for new ads. In: Proceedings of the 16th International Conference on World Wide Web, pp. 521–530. ACM (2007)
12. Agichtein, E., Brill, E., Dumais, S., Ragno, R.: Learning user interaction models for predicting web search result preferences. In: Proceedings of the 29th Annual International ACM SIGIR Conference on Research and Development in Information Retrieval, pp. 3–10. ACM (2006)
13. Antonellis, I., Molina, H.G., Chang, C.C.: Simrank++: query rewriting through link analysis of the click graph. Proc. VLDB Endowment **1**(1), 408–421 (2008)
14. Jeh, G., Widom, J.: SimRank : a measure of structural-context similarity. In: Proceedings of the Eighth ACM SIGKDD International Conference on Knowledge Discovery and Data Mining, pp. 538–543. ACM (2002)
15. Cao, H., Jiang, D., Pei, J., He, Q., Liao, Z., Chen, E., Li, H.: Context-aware query suggestion by mining click-through and session data. In: Proceedings of the 14th ACM SIGKDD International Conference on Knowledge Discovery and Data Mining, pp. 875–883. ACM (2008)

A Study of User Image Search Behavior Based on Log Analysis

Zhijing Wu, Xiaohui Xie, Yiqun Liu(✉), Min Zhang, and Shaoping Ma

State Key Laboratory of Intelligent Technology and Systems,
Department of Computer Science and Technology,
Tsinghua University, Beijing, China
wzjingzai@163.com, xiexh_thu@163.com,
{yiqunliu,z-m,msp}@tsinghua.edu.cn

Abstract. Study of user behavior in Web search helps understand users' search intents and improve the ranking quality of search results. To better understand user's Web image search behavior in practical environment, we investigate user behavior by analyzing a query log collected in one week from a popular image search engine in China. We focus on individual query analyses, temporal distribution, click-through behavior on the search engine result pages (SERPs), and behaviors on preview pages. Compared to general Web search, image search users usually submit shorter query strings and their selections of query terms are more diverse. We find that there exists a huge difference among users in image search click-through behavior. Users are more likely to do exploratory search compared to that in general Web search. This finding may provide us some insights about users' behavior in the context of image search. Our findings may also benefit multiple perspectives of image search, such as UI design, effectiveness evaluation, ranking algorithms, and etc.

Keywords: Image search · User behavior · Log analysis · Search intent

1 Introduction

Understanding user behavior is one of the prime concerns of Web search related studies. It provides opportunities for advertising, suggestions on the design of user interface, and assessment indicators for result relevance.

One of the most common approaches to understand user behavior is analysis of query logs. Previous studies on general Web search focused on individual query, session analysis, and click position on the SERPs [3,8,10]. These findings help us gain better understanding in how users use search engine to get information. However, the way to show results in image search differs greatly from that of general Web search. Most image search results are organized in grids

This work is supported by Natural Science Foundation of China (Grant No. 61622208, 61532011, 61672311) and National Key Basic Research Program (2015CB358700).

J. Wen et al. (Eds.): CCIR 2017, LNCS 10390, pp. 69–80, 2017.
https://doi.org/10.1007/978-3-319-68699-8_6

instead of the linear result list in general Web search. Therefore, there exists a huge difference in user behavior between general Web search and image search [4,10]. Queries in image search have more zero hits and are more specific. What's more, preview pages are provided by image search engine. The image preview page is an enlarged preview of an image, which is usually shown after the image is clicked on the SERP. On the preview page, the enlarged picture is presented together with navigation buttons (e.g. Previous, Next), download button, and thumbnails for further exploration of the results (see Fig. 1). This interaction mechanism is significantly different from the landing page reading process of general Web search engines. If users want to view more results, they can jump to another search result directly instead of turning back to the SERP. However to the best of our knowledge, few existing works investigate its influence in user behavior.

Fig. 1. Image preview page (shown after the image is clicked on the SERP).

In this study, we analyze user behavior including individual query analyses, temporal distribution, click-through behavior on the SERPs, and behavior on preview page based on the logs collected from a popular image search engine.

To summarize, the main contributions of this work are as follow:

- We find a number of differences in search behavior between general Web search and image search. Compared to general Web search, image search users usually submit shorter query strings and their selections of query terms are more diverse.
- We show that clicked images for the same query vary greatly across users, which potentially indicate serious challenge for click models to perform as well as they do in general Web search.
- We analyze users' exploratory behavior on preview page. 61.4% of users view more than one image on a preview page. It provides further evidence of exploratory behavior in image search.

The paper is organized as follows. In the next section, we review related work. Section 3 introduces our dataset. We report the analysis results of user behavior in image search in Sects. 4–7. Finally, we discuss conclusions and future work.

2 Related Work

With the wide application of Web search engines, log analysis has become one of the most common approaches to understand user behavior. There are many previous studies in large-scale Web search studies [3,8,13,14], from which we can get a timeline of the evolution of general Web search. Intents behind queries can be classified into three categories: informational, navigational, and transactional [1]. Search engines need to deal with all three types because each type is best satisfied by very different results.

Several image search studies characterizing the general user behavior also have been performed in the past [2,4–6]. Their approach take different factors into consideration, such as query length, query reformulation patterns, and the search depth. Andre [4] made a large-scale analysis of query logs to characterize some of the differences between general Web search and image search. It derived four main characteristics that make image search unique from its Web search counterpart. Compared to general Web search, image search leads to shorter queries, tends to be more exploratory, and requires more interactions [10]. Another related work in understanding image search behavior was conducted by Goodrum and Spink [7] They found that image queries contained on average 3.74 words. They also reported a high percentage of unique terms. These studies help us better understand the general user behavior in image search.

What do users search for in image search? A query log analysis showed that more than 20% of image search queries are related to location, nature, and daily life categories [15]. Pu [10] classified the 1,000 most frequent image queries based on a proprietary subject-based categorization scheme. They found that the majority of the queries were in the entertainment domain. Most recently, Park, et al.[2] further examined the differences between head queries and long-tail queries. They looked at the query types which belong to the intersection of subject-based and facet-based categorizations to uncover more fine-grained categories that cover a significant amount of search requests. It sheds light on the importance of considering query categories to better understand user behavior on image search platforms.

Understanding interactive behavior with image search result pages is also of vital importance. Interactive behavior provides abundant implicit user feedback for image search engine. Smith [16] presented a study that compares click-through on image searches with what has been discovered for traditional text search. They also evaluated searches for different types of images. Since the presented results are quite different (image v.s. text), the findings of previous studies based on traditional text search interactions were not applicable to image search interactions.

Most of the above studies mainly focused on interactive behavior on image search result pages. The variance of click behavior between different users still lacks fine-gained analysis. Although Park et al. [2] showed several user behavior patterns across query types on preview pages, little attention has been paid to the general situation.

3 Dataset

We take 7 days' image search server logs on desktop collected in March 2017. We extract four types of behavior from the log: the SERP information, click events, download events, and preview page events. The details of the data are presented in Table 1. From *the SERP information*, we can get the rank position of all thumbnail result images. For *click events*, we have the id of the clicked image. For *download events*, we extract the entrance information including time and image id. For *preview page events*, we have the event type (e.g. mouse scroll, click the thumbnails below).

Table 1. The details of the data.

Data type	Data field
SERP information	id, user, query, time, resolution, page, images
Click events	id, user, query, time, image
Download events	user, query, time, entrance
Preview page events	user, query, time, event type, entrance

The dataset contains approximately 19.2M searches, 7.9M unique queries, 7.6M sessions, and 5.3M users (see Table 2). In this study, we also use another dataset collected in January 2017 from a popular Web search engine in China. It contains 287.5M searches and 67.9M unique queries.

Table 2. Dataset statistics.

	Image search	Web search
Number of searches	19,231,768	287,487,327
Number of unique queries	7,910,311	67,904,480
Number of sessions	7,573,304	#
Number of users	5,255,563	#

4 Query Distributions

In this section, we analyze the query distributions from different aspects, including query length, query frequency, and the usage of advanced search. We also make a comparison between image search and general Web search.

4.1 Query Length

As the first step of our analysis, we look at the number of words and characters per query. It should be noted that the number of words here refers to the number of words separated by spaces. As shown in Fig. 2, there are 5.69 characters in each query on average. Among all the queries, 6.3% contain only one word and each query contains only 1.05 words on average. It means that the average length of words used in image search is 5.69/1.05=5.42, which is much larger than the average length of Chinese words. Users usually put two or more Chinese words together as the query string, instead of using spaces to separate each word. The search engine should run an effective word segmentation program after receiving a query.

We also investigate that the average number of words per query is 6.64 in Web search. This reflects that users are lazy when they input a query string. Image search engine gets even less information about what users really want to search for. Therefore we need to pay more attention to analysis of user behavior in image search. It helps understand users' search intent and improve the quality of search results.

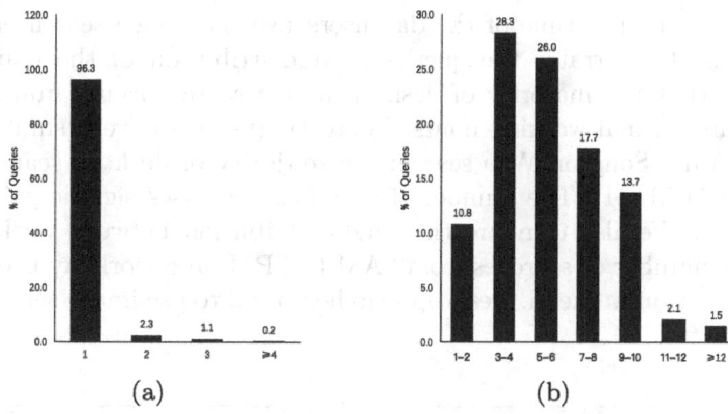

Fig. 2. The distribution of query length: (a)number of words per query, (b)number of characters per query.

4.2 Query Frequency

We also focus on the query frequency distribution. We get how many times each unique query was submitted in the 7 days.

As expected, a small number of queries account for a large part of all queries. At the head of the distribution, the top 0.17% of unique queries occur more than 100 times, which account for 26.12% of all queries and the top 3.5% of unique queries account for 50% of all queries, the top 23% of unique queries account for 70% of all queries. It also has a long tail, 78.14% of unique queries were submitted only once, and they account for 30.41% of all queries.

This is very similar to Jaimie Y. Park's report in English environment [2]: 75% of unique queries were issued only once, and they account for approximately 25% of the traffic in the sample; the other 25% of queries account for 75% of all traffic. Only the top 5.99% queries account for 70% of all queries in Web search. It seems that their selections of query terms are less diverse.

This goes to show that a large part of queries committed by users are repeated. A small number of queries account for a large part of users' needs. If the search engine can pay more attention to improving the ranking quality of those popular searches' results, users' satisfaction will improve significantly.

4.3 Usage of Advanced Search

Users can add some specific words and symbols (such as "and", "or", "not", '+') to query strings to use advanced search. The percentage of advanced search is only 0.73% reported in 2007 [3]. In our analysis, the percentage has grown substantially to 4.57% in image search, 6.06% in Web search.

5 Temporal Distribution

Next, we focus on what time of the day users use the image search engine on desktop. Figure 3 illustrates the queries as a distribution of the hour of the day. It shows that the majority of desktop image search occurs from 9AM to 5PM which are normal working hours. Interestingly, this is very similar to the statistics by Yang Song on Web search: the majority of desktop search occurs from 8AM to 5PM [11]. The number of searches decreases significantly during the lunch time. We also compare the time distribution between weekend and workday. The number of searches from 9AM to 5PM on a workday is obviously greater than that on weekend. Desktop searchers tend to use image search engine at work.

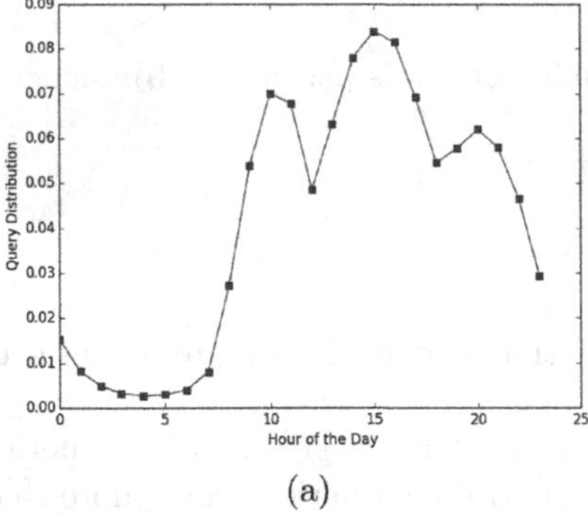

(a)

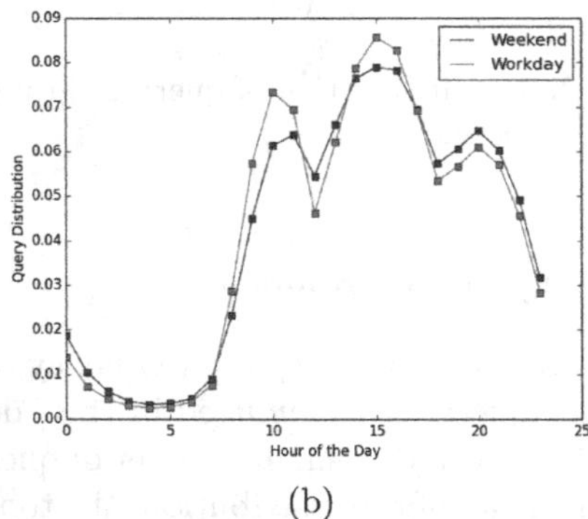

(b)

Fig. 3. Temporal distributions: (a)for hour of the day, (b)comparison between weekend and workday.

6 Session Characteristics

A session is "a series of queries by a single user made within a small range of time" as defined in Craig Silverstein's study [8]. In this section, we partition a user's actions into separate sessions when the time between consecutive actions exceeds 30 min [2]. We look at the distribution of queries per session and average number of sessions for a user in one day.

As is shown in Fig. 4, the average number of queries per session is 2.54. 56.6% of sessions contain only one query. 91% of sessions contain less than 6 queries. It is similar to the previously published result [2]: 2.95 queries per session.

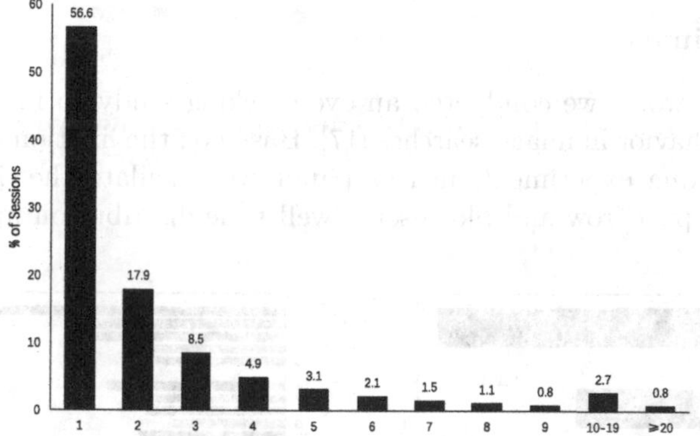

Fig. 4. Number of queries per session.

5,255,291 users ever used image search engine in the 7 days. The average number of queries per user is 3.66, sessions per user is 1.44. 78.9% of users used the image search engine only once in the 7 days.

7 Interaction with Search Results

Understanding users' interaction with search results helps make search intent explicit. Users' interaction behavior can be used by search engine as relevance feedback data. In this section, we analyze browsing depth, the position distribution of clicks and click entropy distribution.

7.1 Browsing Depth

There are 5 rows of images on each page in the data we use. As users scroll down the result page, images are loaded page after page automatically. We use how many pages of results user explores during one query to define browsing depth. From Table 3 we can see a clear distribution of browsing depth. We find that image search has deeper browsing depths than Web search. Experiments show that 85% of users view only the first page of results in Web search [3].

Table 3. Distribution of browsing depth.

Browsing depth (page)	1	2	3	4	⩾ 5
Queries (%)	3	21	17	11	48

Differ from the automatically loading of image search results, normal Web search engines have only 10 results each page. Users must click on the next page button if they want to browse more results. So rather than view more results, users tend to change query strings to get new results. This also indicates that image search is more exploratory than text search.

7.2 Dwell Time

In our previous work, we conducted an eye-tracking study to investigate users examination behavior in image searches [17]. Based on the fixation data collected in the eye-tracking experiment, in this paper we calculate the dwell time for images on each page/row and plot users dwell time distribution in Fig. 5.

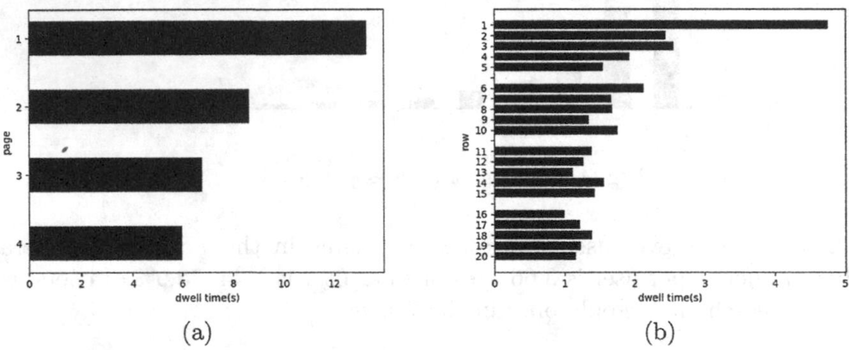

Fig. 5. Distribution of dwell time on each (a)page, (b)row.

As is shown in Fig. 5, the dwell time decreases with the page number and row number. The first page and first row have longer dwell time than the other positions. It shows that users pay more attention to images placed in the first page, especially in the first row.

7.3 Click Position Distribution

Next, we examine the position distribution of click. Most image search engines adopt two-dimensional result placement instead of linear result list in general Web search. As is stated in previous section, each result page contains 5 rows. Since each of the search engine result pages we collected contains 5 pages of results. Figure 6 shows the distributions of click counts on each row and page.

Clicks on the first page account for 57.68% of all clicks. A sharp decay is observed over the top 2 rows and top 2 pages. The top-ranked images have more clicks than those of lower ranked. It shows that position has more influence on click. On the other hand, this indicates that the first page have better relevance and users trust in search engine's ranking.

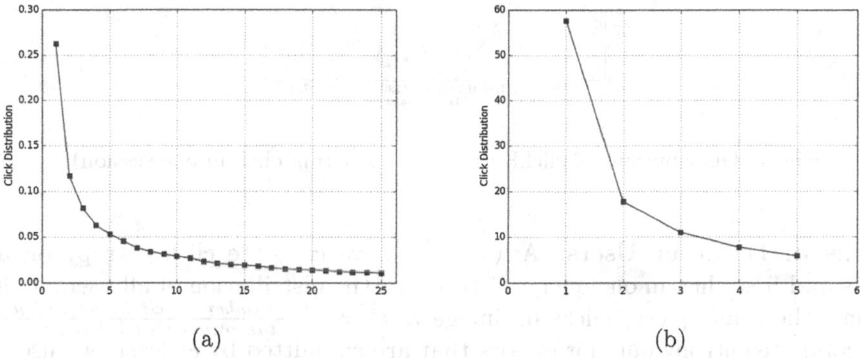

Fig. 6. Distribution of click position: % of Clicks on (a)N-th row, (b)N-th page.

7.4 Click Entropy Distribution

In this section, we use click entropy to explain the following questions.

- What's the distribution of clicks in one session?
- Is click behavior different between users in the same query?

The click entropy is calculated as follows:

$$ClickEntropy = \sum_{P_i} P_i log(P_i). \quad (1)$$

Clicks in One Query: In one query, P_i refers to the distribution of one user's clicks on image i. $P_i = \frac{the\ number\ of\ clicks\ on\ image\ i}{the\ number\ of\ all\ clicks}$. We compute entropy only for queries that have at least one click. Figure 7 shows that for 72.04% of queries there is only one image clicked (maybe multiple clicks on the same image). A few queries have clicks on two or more images. However, in general Web search clicks are more dispersive. For 32.2% of queries, there is only one doc clicked [12].

Possibly because there are few click behavior in image search (average number of clicks is 0.89 per query). Based on Rongwei Cen's findings [12], if users' clicks are definite, we tend to think users are satisfied with the search results.

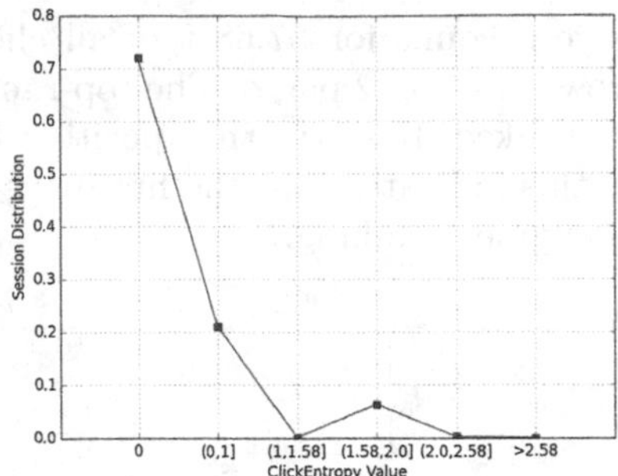

Fig. 7. Distributions of clickEntropy (considering click in one session).

Clicks of Different Users: Across users, we compute click entropy on all clicks and first click in one query. P_i refers to the distribution of all users' (who submit the same query) clicks on image i, $P_i = \frac{\textit{the number of clicks on image i}}{\textit{the number of all clicks}}$. We compute entropy only for queries that are submitted by at least two users.

Figure 8 shows that for 4.7% of queries there is only one image clicked. Clicked images for the same query vary greatly across users. It potentially indicates a large challenge for click models to perform as well as they do in general Web search.

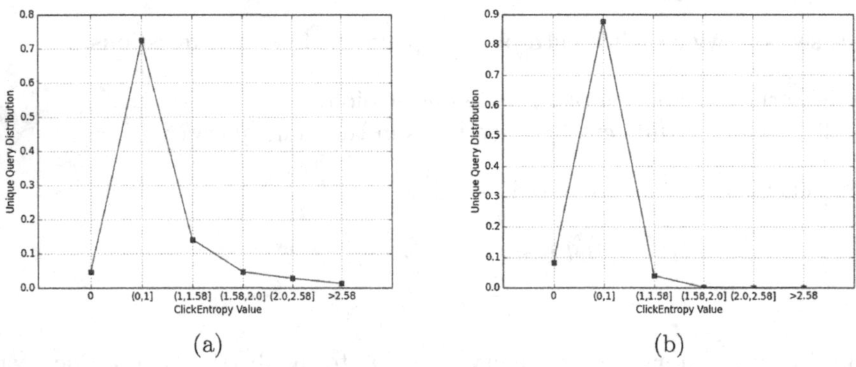

Fig. 8. Distributions of clickEntropy (considering click in one query between different users): (a)all clicks of different users in one query, (b)the first click of different users in one query.

7.5 Behavior on Preview Page

Most of the image search engines provide preview of an image for users after clicking on it. Users can click the thumbnail, previous and next button to view more images. They can also download the full-size image. We make an analysis on these interaction behaviors.

Table 4 shows that the average preview duration is 4.67 min. It illustrates that users spend long time on further exploration. There is a huge difference between the average number of clicks on previous button and next button. Few users ever look back on previous images. For 38.60% of result clicks, only one image is viewed on the preview page. In means that a user clicks on an image, views the image, and does not preview any other result images on the page. In other words, 61.4% of users view more than one images on a preview page. It provides further evidence of exploratory behavior in image search.

Table 4. Preview page characteristics.

Preview Duration (min)	4.67
Single Image Previews (%)	38.60
Average number of download	0.07
Average number of previous button clicked	1.20
Average number of next button clicked	20.15
Average number of mouse scroll	11.21
Average number of thumbnail clicked	2.18

8 Conclusions and Future Work

In this paper, we carry out an analysis of user behavior in image search based on logs. It includes individual query analyses, temporal distribution, click-through behavior on the result pages, and behavior on preview page search. We obtain three interesting findings. (1) A number of differences in search behavior between general Web search and image search is found. Compared to general Web search, image search users usually submit shorter query strings and their selections of query terms are more diverse. (2) We show that clicked images for the same query vary greatly across users, which potentially indicate serious challenge for click models to perform as well as they do in general Web search. (3) We analyze users' exploratory behavior on preview page. 61.4% of users view more than one image on a preview page. It provides further evidence of exploratory behavior in image search.

Our study makes a comprehensive analysis on user behavior in image search. Interesting directions for future work include the impact of image content, relevance, and diversity on the click. As mobile search is getting larger than desktop search, we also plan to investigate user behavior in image search on mobile devices and examine the difference between mobile devices and desktop.

References

1. Broder, A.: A taxonomy of web search. SIGIR FORUM. **36**(2), 3–10 (2002)
2. Park, J.Y., O'Hare, N., Schifanella, R., Jaimes, A., Chung, C.: A large-scale study of user image search behavior on the web. In: Proceedings of CHI (2015)

3. Huijia, Y., Liu, Y., Zhang, M., Liyun, R., Ma, S.: Research in search engine user behavior based on log analysis (in chinese). J. Chin. Inf. Process. **21**(1), 109–114 (2007)
4. André, P., Cutrell, E., Tan, D.S., Smith, G.: Designing novel image search interfaces by understanding unique characteristics and usage. In: Gross, T., Gulliksen, J., Kotzé, P., Oestreicher, L., Palanque, P., Prates, R.O., Winckler, M. (eds.) INTERACT 2009. LNCS, vol. 5727, pp. 340–353. Springer, Heidelberg (2009). doi:10.1007/978-3-642-03658-3_40
5. Goodrum, A., Spink, A.: Visual information seeking: a study of image eries on the world wide web. In: Proceedings of the ASIS Annual Meeting, vol. 36, pp. 665–674. ERIC (1999)
6. O'Hare, N., de Juan, P., Schifanella, R., He, Y., Yin, D., Chang, Y.: Leveraging user interaction signals for web image search. In: Proceedings of the 39th International ACM SIGIR conference on Research and Development in Information Retrieval, pp. 559–568. ACM (2016)
7. Goodrum, A., Spink, A.: Image searching on the excite web search engine. Inf. Process. Manage. **37**(2), 295–311 (2001)
8. Silverstein, C., Henzinger, M., Marais, H., et al.: Analysis of a very large web search engine query log [J]. SIGIR Forum **33**(1), 6212 (1999)
9. Kamvar, M., Baluja, S.: A Large Scale Study of Wireless Search Behavior: Google Mobile Search. In: Proceedings of SIGCHI Conference on Human Factors in Computing Systems, pp. 701–709. ACM Press (2006)
10. H-T, Pu.: A comparative analysis of web image and textual queries. OIR **29**(5), 457–467 (2005)
11. Song, Y., Ma, H., Wang, H., Wang, K.: Exploring and exploiting user search behavior on mobile and tablet devices to improve search relevance. In: Proceedings of 22nd International Conference on World Wide Web, pp. 1201–1212, Rio de Janeiro, ACM (2013)
12. Cen, R., Liu, Y., Zhang, M., Liyun, R., Ma, S.: Reliability analysis for the behavior of web retrieval users. J. Softw. **21**(5), 1055–1066 (2010)
13. Jansen, B.J., Spink, A., Bateman, J., Saracevic, T.: Real life information retrieval: a study of user queries on the web. SIGIR Forum **32**(1), 5–17 (1998)
14. Spink, A., Jansen, B., Wolfram, D., Saracevic, T.: From E-sex to E-commerce: web search changes. IEEE Comput. **35**(3), 107–110 (2002)
15. Zhang, L., Chen, L., Jing, F., Deng, K., Ma, W.: Enjoyphoto : a vertical image search engine for enjoying high-quality photos. In: MM 2006, pp. 367–376 (2006)
16. Smith, G., Brien, C., Ashman, H.: Evaluating implicit judgments from image search clickthrough data. JASIST **63**, 2451–2462 (2012)
17. Xie, X., Liu, Y., Wang, X., Wang, M., Wu, Z., Wu, Y., Zhang, M., Ma, S.: Investigating examination behavior of image search users. In: The 39th ACM SIGIR International Conference on Research and Development in Information Retrieval (2017)

User Preference Prediction in Mobile Search

Mengyang Liu, Cheng Luo, Yiqun Liu(✉), Min Zhang, and Shaoping Ma

State Key Laboratory of Intelligent Technology and Systems,
Department of Computer Science and Technology,
Tsinghua University, Beijing 100084, China
liumengyang13@gmail.com, luochengleo@gmail.com
{yiqunliu,z-m,msp}@tsinghua.edu.cn

Abstract. As search requests from mobile devices are growing very quickly, mobile search evaluation becomes one of the central concerns in mobile search studies. Beyond traditional Cranfield paradigm, side-by-side user preference between two ranked lists does not rely on user behavior assumptions and has been shown to produce more accurate results comparing to traditional evaluation methods based on "query-document" relevance. On the other hand, result list preference judgements have very high annotation cost. Previous studies attempted to assist human judges by automatically predicting preference. However, whether these models are effective in mobile search environment is still under investigation. In this paper, we proposed a machine learning model to predict user preference automatically in mobile search environment. We find that the relevance features can predict user preference very well, so we compare the agreement of evaluation metrics with side-by-side user preferences on our dataset. We get inspiration from the agreement comparison method and proposed new relevance features to build models. Experimental results show that our proposed model can predict user preference very effectively.

Keywords: Mobile search · Search evaluation · User preference prediction

1 Introduction

With the rapid development of mobile devices, people can address their information needs almost anywhere and anytime with their mobile devices. Since more and more search requests are from mobile devices, in particular smart phones, the massive shift of users makes the evaluation of mobile search become an essential issue.

In the field of search evaluation, the traditional Cranfield [1] evaluation method consists of corpus, queryset, relevance judgements, and metrics.

This work is supported by Natural Science Foundation of China (Grant No. 61622208, 61532011, 61672311) and National Key Basic Research Program (2015CB358700).

J. Wen et al. (Eds.): CCIR 2017, LNCS 10390, pp. 81–92, 2017.
https://doi.org/10.1007/978-3-319-68699-8_7

The Cranfield evaluation method calculates the evaluation metrics, such as Accuracy, Recall, etc., according to the document lists returned by the retrieval system. Moffat et al. studied the correlation between different evaluation metrics and user behavior [2]. They found that different evaluation metrics implied different user behavior assumptions and the evaluation metrics can usually be presented as a weighted sum of relevance scores for each result in the results list.

Recently, the *preference-based* evaluation of the search engine result pages (SERPs) has been widely concerned in the search evaluation. In user preference judgment, the assessors are presented with two SERPs in parallel. Then they are instructed to annotate their preferences based on their satisfaction with each search result list. This method of annotation naturally take more contextual information into consideration than the traditional Cranfield evaluation method. Sanderson et al. found that there was a strong correlation between users' preference and various evaluation metrics [3]. Although the preference evaluation may be more accurate, the cost of collecting user preference labels is relatively high. To evaluate N ranking systems, it is a necessity to collect user preferences between every two systems. That's to say, there will be $(N + 1) * N/2$ annotations for each query. Compared to user preference, traditional Cranfield evaluation methods are much easier to reuse: once the relevance judgment is done, the evaluation metrics can be calculated with little marginal effort. Thus, researchers want to investigate machine-assisted method to assist human judges thus reducing judgement cost.

Awadallah et al. studied the method of automatically predicting user preferences based on relevance features and click-through [4]. However, their study was carried out in traditional desktop search environment, where the search results are mainly presented in textual form ("*ten-blue-link*"). In contrast, modern SEPRs usually contains far richer information from heterogeneous information resource (News, Images, Videos, and etc.). These results may have a significant impact on users' interactions with SERP. On the other hand, mobile search is quite different from traditional desktop search from the perspective of interaction. Modern mobile devices are usually equipped with a touchable screen and users mainly interact with their mobile devices with touch gestures rather than mouse and keyboard. The visual area of mobile device is also much smaller comparing to desktop monitors. The impact of these differences between mobile search and desktop search on users' preference lacks investigation.

In this study, we focus on automatically predicting user preferences in the context of mobile search with a machine learning model. More specifically, we present a series of features based on the SERP and the search results relevance. We find that the most powerful features are relevance features. We calculate the agreement of a series of traditional desktop evaluation metrics with the users preference. We get inspiration from the agreement comparison method and proposed a new relevance feature. We prove that we can get good prediction effect based on this feature and propose a method to improve the feature. The results prove that our method does predict the preference very effectively.

To summarize, we make the following contributions in this study:

- We compare the difference between the relevance features and other indicators in terms of predicting user preference.
- We investigate the relationship between the evaluation metrics and user preference in mobile environment.
- We propose a model to predict the user preference between ranked lists to reduce human effort in making preference judgments.

The remaining of this paper is organized as follows. We first review related studies in Sect. 2. The dataset and features are given in Sect. 3. In Sect. 4, we describe the study methods and experimental results, and analyse the experimental results. Finally, Sect. 5 summarizes the conclusions and discusses the direction of future work.

2 Related Work

2.1 Search Evaluation

Retrieval system performance evaluation is one of the most important research issues in information retrieval (IR) related studies. The traditional search evaluation method, referred to as Cranfield paradigm, is mainly based on corpus, fixed queryset, relevance judgment of "query-document" pairs, and evaluation metrics. A lot of evaluation metrics have been proposed based on different insights about users' search behavior, such as NDCG [7], ERR [8], RBP [9], TBG [10] and etc. Moffat et al. concluded that different evaluation metrics were based on different user behavior models [2]. For example, *Reciprocal Rank* (RR) is a metric suitable for navigational intents and it assumes that users will be satisfied by the first relevant document in the list. Maskari et al. found that the evaluation metrics correlate well with users' satisfaction [11]. However, some researchers argue that the relevance annotation does not take into account the interaction between the results, which may lead to certain differences between the evaluation of the results and actual users' feelings. The advantage of Cranfield evaluation paradigm is that the relevance judgment can be easily reused. However, the fixed user behavior models behind metrics may lead to failure in reflecting actual users' experience during search.

The preference-based evaluation method tries to infer users' satisfaction about result lists based on users' implicit or explicit feedback. The preference-based evaluation method was first proposed by Joachims [12]. In their approach, a user's click can be interpreted as a preference vote to a specific result. Their study facilitated the emergence of the Interleaving and A/B Testing methods. The Interleaving method selects results from two retrieval systems alternately to form a new result list in a fair way. Then the users' click behavior will be collected to analyze the relative preference. In the A/B Testing method, different users will be served with different systems, then the performance of different system will be compared based on some metrics, e.g. Click Depth, Click-Through Rate (CTR), and etc.

Thomas et al. proposed an evaluation method, referred to as *User Preference Test*, which presents two search result lists to the users side-by-side [13]. The users are then instructed to choose the result list that they preferred. This method was used to compare the first and third page of search results from Google. Their experimental results show that users prefer the results of the first page. This observation also demonstrates that user preference can be used to predict the quality of search results effectively (Fig. 1).

Fig. 1. User preference annotation interface schematic diagram.

Awadallah et al. investigated the method of predicting user preferences automatically to assist human judges [14]. They proposed a series of features and used different feature combinations to train the automatic prediction model. The results showed that the relevance based features are the most effective. However, their experiments were carried out in the traditional desktop search environment. Traditional textual results, which usually contains a title, a URL and a short summary. This is quite different from the heterogeneous results returned by current retrieval system.

2.2 Search on Mobile Device

There are many differences between mobile search and traditional desktop search, for example: (1) The distribution of query topic is not the same, people prefer to search pictures, entertainment class information on mobile device [18]; (2) The way of interacting is different, rich gesture operations can be conducted on mobile device, such as scaling, sliding, etc. Guo et al. found that there is a significantly correlation between the interactive behavior and document relevance on mobile device [5]; (3) Screen size is not the same, the screen of mobile device is usually smaller and less results can be displayed so that more interaction costs are required to get the same amount of information than the desktop. Jones et al. found that small screen would make the search tasks to be more difficult to complete [6]. Due to these differences, traditional techniques on desktop are not necessarily adequate on mobile devices (Fig. 2).

Fig. 2. Traditional desktop SERP (left) and mobile SERP (right).

3 Dataset and Features

The evaluation dataset we used contains SERPs from four commercial search engines, referred to as A, B, C and D. There are 50 queries in the dataset. Thus there are 200 SERPs in total. For all SERPs, the relevance of the top 5 results is accessed by the professional assessors with six-level relevance criteria. The relevance value ranges from 0 to 5, indicating completely irrelevant to completely relevant respectively. The labeling procedure is similar to typical TREC [14] settings. For user preference assessment, we will show assessors two SERPs of the same query side-by-side. One SERP is from A and the other SERP is from B, C, or D. Each "SERP-SERP" pair is marked by seven assessors with five-level preference criteria (from −2 to +2, indicating that the left page is much better than the right page to the left page is much better than the right page). If there is a disagreement between assessors, following Zhou et al. [15], we adopt the majority annotation as the final user preference.

We use a series of features to compare the two SERPs. We first consider the relevance evaluation features. We think that the relevance of the search results may have the greatest impact on the user's preferences. In addition to the relevance features, we also consider other features that may affect user preferences. The first is textual features, since when users browse the search results, the *textual similarity* between query and result may affect users' perceived relevance. The second one is heterogeneous features. Compared to the traditional SERPs, the SERPs used in our experiment have far richer heterogeneous information (pictures, videos, news, etc.), which may affect user's preferences. The third one is height features, different height of the results may cost the user different browsing time, which may be a factor affecting the preferences.

3.1 Relevance Features

The researches about document relevance and user preference do not agree with each other all the time. Some studies find that there is no significant correlation between user preference and document relevance [16], but there are some studies show that there is a significant correlation between relevance and preference [13]. We adopt a hypothesis from the study by Awadallah et al. [4]: *user preference has correlation with document relevance, and the relevance features of the entire page is very important.* For relevance features, we calculate not only the features value of the two pages respectively, but also the differences between the two SERPs. More specifically, we selected the following features:

- **Relevance:** Relevance of result in each rank
- **Precision@5:** Proportion of relevant documents in top 5 results
- **Best@5:** The most relevant document in top 5 results
- **Worst@5:** The least relevant document in top 5 results
- **FirstRR:** The rank reciprocal of the first relevant document in result list

3.2 Other Features

Text Features: We mainly focus on the textual content of result title and snippet. For each search result, we calculated the length of the text, the number of words in the text and the similarity of the text and query, extract the maximum and average of all results as text features.

Heterogeneous Features: The SERPs used in our experiment contains more diversity information than the traditional ten-blue-link SERPs, and we hope to use the diversity information to assist us predict user preferences. We classify all the results into ordinary results and vertical results, vertical results include five categories: multimedia (such as video, audio, pictures, etc.), wikipedia forum (such as Encyclopedia, Post Bar, Knows, etc.), text reading (such as news, library, WeChat articles, etc.), life services (such as shopping, booking, express, etc.) and other vertical results.

Height Features: The mobile search results list contains many card-style search results. The geometric height of these cards in the user's screen is quite different, some cards even cover more than a screen. The user needs to spend more time to browse the higher results, which may have an influence on the user's preferences, so we extract the height of each result and the height of the landing page as features of the predict model.

4 Experiments and Results

Our goal is to predict the user's preference between two SERPs. First, we classify the user preference into three categories: the user preference of -1 and -2 are classified as preferring the page in the right side, denoted as -1; the user preference of +1 and +2 are classified as preferring the page in the left side, denoted as +1; the user preference of 0 means no preference, still denoted as 0. We train a classifier to predict the user preference based on the features mentioned in Sect. 3, all experiments used 5 fold cross validation. We have tried different learning models, such as Logistics Regression, Support Vector Machine, Random Forest and Gradient Boosting Decision Tree. But the difference of prediction effect of the best model and the worst model is not greater than 3%, so we only present the results of the Logistic Regression in this paper.

4.1 Preference Predicting

We train a classifier only with relevance evaluation feature first (Relevance Only), then we train a classifier without relevance evaluation feature (All but Relevance), finally we train classifier with all the features (All Features).

Table 1. Results for different features combination.

Model	Accuracy
Relevance Only	74.73%
All but Relevance	58.70%
All Features	75.37%

The results of this experiment are shown in Table 1. The table shows the Accuracy for each classifier. It can be seen that the Accuracy of the classifier without relevance evaluation features is only 58.70%, which means that the other type of features we used can not describe the user preference very well. The Accuracy of the classifier only with relevance evaluation features is 74.73%, this indicates that the relevance features can describe the user's preference well. And there is no significant improvement when we use all features to train the classifier, indicating that the other features can not improve the predict effect obviously under the existence of relevance evaluation features.

4.2 Agreement with Preference

For the relevance features can predict the user preference very well, so we try to use the evaluation metrics directly to predict the user preference. We first calculate the agreement of a series of traditional desktop evaluation metrics with the user's preference, and also calculate the HBG metric which is based on the characteristics of mobile interaction [17], the parameters of HBG are from the original paper, the results are shown in Table 2.

Compared with the prediction precision in Sect. 4.1, we can tell that our prediction precision is not worse than the traditional desktop search evaluation metrics, which indicates that our proposed classifier is indeed effective. However, compared with the HBG evaluation metric, there is still a gap, which indicates that HBG is a good evaluation metric in mobile search. On the other hand, we

Table 2. Agreement between evaluation metrics and user preference.

Metrics	Agreement
HBG	85%
ERR	75%
DCG	72%
nDCG	59%
AP	53%

think this is acceptable because the parameters of HBG is calculated based on real user behavior data, it is reasonable that HBG metric can better reflect the mobile search users' preferences.

4.3 New Feature

Because the agreement between the evaluation metrics and user preference is relatively high. We get inspiration from the agreement comparison method and proposed a new relevance feature. It is called the difference ratio of evaluation metrics which divides the max metric value by the difference value of metrics. We only use the difference ratio feature to train classifier to predict user preference, the Accuracy of the prediction can reach 74%.

Although using the evaluation metric difference ratio feature can get a satisfactory prediction effect, we still expect to be able to improve the difference ratio feature to get better prediction effect. So we have further analysis on this feature. As shown in Fig. 3, using ERR as an example, gives the distribution of difference ratio and the user preference (DCG has a similar distribution).

As shown in Fig. 3, for each class of user preference judgment, the difference ratio is distributed in a relatively concentrated area. This also proves that the evaluation metric difference ratio feature can reflect the user preference better.

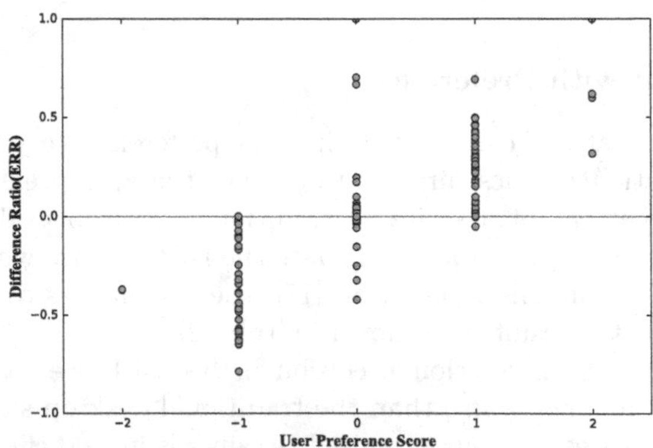

Fig. 3. Distribution of evaluation metric (ERR) difference ratio.

While analyzing the data that we can not give correct prediction, we find that if the two contrast pages are low quality pages, the difference ratio can not describe the user's preference very well. This is easy to understand for the difference ratio feature only reflects the relative differences but not the absolute differences between the contrast pages.

In order to solve the problem of low-quality pages, we start our analysis from the aspect of the distribution of evaluation metrics score. As shown in Fig. 4, we firstly plot the distribution of the minimum ERR between the two contrast pages. We find that the distribution of ERR can be roughly divided into two parts, the one is less than 0.1 and the other is greater than 0.2. So we consider that the page with ERR smaller than 0.1 is a low-quality page. Therefore, we propose a solution that combines the absolute value with the relative difference of evaluation metrics. If both pages belong to the low-quality pages at the same time, then the difference ratio is 0. Otherwise the difference ratio is calculated according to the original method. We use the modified difference ratio feature to train the new classifier to predict the user preference, the Accuracy rate can reach 78%. This indicates that our proposed solution does very effective. Which needs to be emphasized is that we just provide a general idea to solve the problem of low quality pages, rather than giving a perfect solution. We believe that we can find a more reasonable answer in a larger dataset, we would like to explore this in our future work.

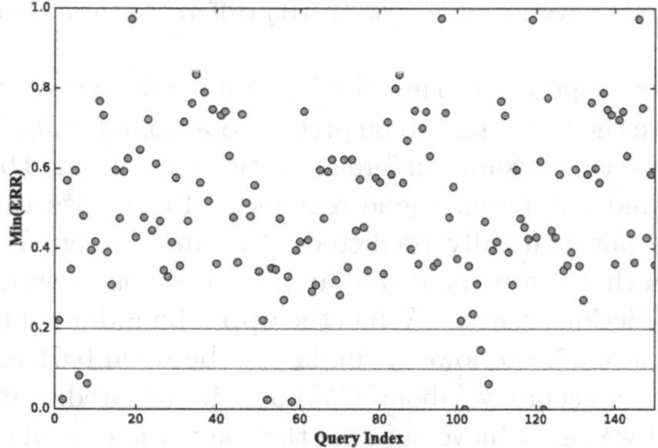

Fig. 4. Distribution of the minimum evaluation metric (ERR) of two SERPs. The two blue lines indicate that the metric value equals to 0.1 and 0.2. (Color figure online)

4.4 Assist Human Annotator

Since our goal is not to build a perfect user preference predictor, we just want to use the automatic predictor to assist human annotator to reduce the cost of preference evaluation. If we can pick out the data which is difficult to be predicted automatically, make this part annotated by human annotator, leaving the remaining part to be predicted with automatic predictor, then we can save the cost while ensuring the quality.

Looking back to Fig. 3, we find that there is an overlap when the absolute difference ratio is small, so we re-plot the distribution of the data which the absolute difference ratio is less than 0.1. As show in Fig. 5, there is a large amount of overlap in the preference distribution when the absolute value of the difference ratio (ERR for example) is around 0.05. We suspect that the reason may be that whether there exists preference or not is a very vague concept when the difference between two pages is not very obvious. So we think this part of user preference data is not easy to be predicted automatically.

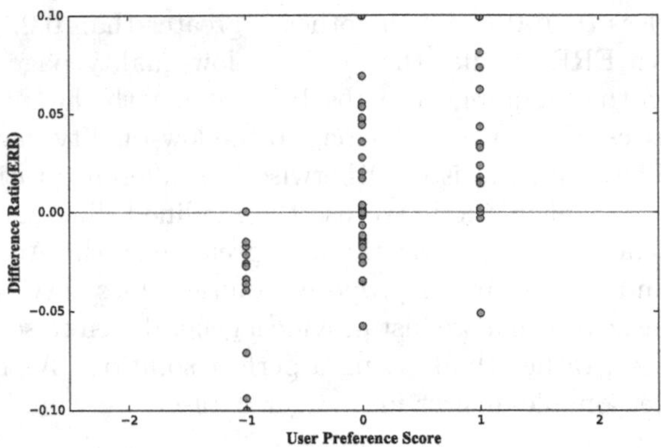

Fig. 5. Distribution of evaluation metric (ERR) difference ratio that in [−0.1, 0.1].

The solution we proposed is to manually annotate the data that difficult to be predicted and leave the remainder to be predicted automatically. We set different vague regions when the absolute difference ratio is at about 0.05, compare the prediction effect under different vague regions and calculate the proportion of data that can be automatically predicted. As can be seen from Table 3, the proportion of data that can be predicted automatically is showed in the followed parentheses of prediction accuracy. When the upper boundary of the vague region is between 0.06 and 0.07, the lower boundary is between 0.01 and 0.02, we can get higher prediction accuracy (about 85%) and it can predict most of the data (about 85%). And we can achieve our goal that save the cost of evaluation while ensuring the quality.

Table 3. Predicting effect in different vague regions.

min/max	0.06	0.07	0.08	0.09
0.01	88.4%(76%)	89.3%(74.7%)	89.0%(72.7%)	88.9%(72%)
0.02	**85.6%(83.3%)**	**86.2%(82%)**	85.8%(80%)	85.7%(79.3%)
0.03	**83.5%(88.7%)**	**84.0%(87.3%)**	83.6%(85.3%)	83.5%(84.7%)
0.04	80.8%(94%)	81.3%(92.7%)	80.9%(90.7%)	80.7%(90%)

5 Conclusions and Future Work

In conclusion, in this paper, we compare the differences between the relevance metrics and other metrics in user preference prediction in the mobile search environment. The results show that the relevance evaluation feature has a very good effect on the user preference prediction. We further examine the correlation between the evaluation metrics and the preference, and put forward the method of converting the search evaluation metrics of single search engine into the preference results of two search engines. That's to say, using the difference ratio of two evaluation metrics can effectively describe the user's preference.

Yet there are some limitations to our study: (1) The current dataset is not large enough, we would like to verify our conclusions on a larger dataset in the future. (2) Some of the threshold parameters used in our experiments are based on statistical analysis of current small-scale data, these parameters can be obtained on a larger scale dataset by the method of machine learning in the future.

References

1. Cleverdon, C.W., Keen, M.: Aslib Cranfield research project-factors determining the performance of indexing systems (1966)
2. Moffat, A., Thomas, P., Scholer, F.: Users versus models: what observation tells us about effectiveness metrics. In: Proceedings of the 22nd ACM international Conference on Information & Knowledge Management. ACM (2013)
3. Sanderson, M., et al.: Do user preferences and evaluation measures line up? In: Proceedings of the 33rd International ACM SIGIR Conference on Research and Development in Information Retrieval. ACM (2010)
4. Hassan Awadallah, A., Zitouni, I.: Machine-assisted search preference evaluation. In: Proceedings of the 23rd ACM International Conference on Information and Knowledge Management. ACM (2014)
5. Guo, Q., et al.: Mining touch interaction data on mobile devices to predict web search result relevance. In: Proceedings of the 36th international ACM SIGIR Conference on Research and Development in Information Retrieval. ACM (2013)
6. Jones, M., et al.: Improving web interaction on small displays. Comput. Networks **31**(11), 1129–1137 (1999)
7. Jrvelin, K., Keklinen, J.: Cumulated gain-based evaluation of IR techniques. ACM Trans. Inform. Syst. (TOIS) **20**(4), 422–446 (2002)
8. Chapelle, O., et al.: Expected reciprocal rank for graded relevance. In: Proceedings of the 18th ACM Conference on Information and Knowledge Management. ACM (2009)
9. Moffat, A., Zobel, J.: Rank-biased precision for measurement of retrieval effectiveness. ACM Trans. Inform. Syst. (TOIS) **27**(1), 2 (2008)
10. Smucker, M.D., Clarke, C.L.A.: Time-based calibration of effectiveness measures. In: Proceedings of the 35th International ACM SIGIR Conference on Research and Development in Information Retrieval. ACM (2012)
11. Al-Maskari, A., Sanderson, M., Clough, P.: The relationship between IR effectiveness measures and user satisfaction. In: Proceedings of the 30th Annual International ACM SIGIR Conference on Research and Development in Information Retrieval. ACM (2007)

12. Joachims, T.: Evaluating retrieval performance using clickthrough data, pp. 79–96 (2003)
13. Thomas, P., Hawking, D.: Evaluation by comparing result sets in context. In: Proceedings of the 15th ACM International Conference on Information and Knowledge Management. ACM (2006)
14. Voorhees, E.M., Harman, D.K.: Experiment and evaluation in information retrieval (2005)
15. Zhou, K., et al.: Evaluating aggregated search pages. In: Proceedings of the 35th International ACM SIGIR Conference on Research and Development in Information Retrieval. ACM (2012)
16. Hersh, W., et al.: Do batch and user evaluations give the same results? In: Proceedings of the 23rd Annual International ACM SIGIR Conference on Research and Development in Information Retrieval. ACM (2000)
17. Luo, C., et al.: Evaluating mobile search with height-biased gain. In: Proceedings of the 40th International ACM SIGIR Conference on Research and Development in Information Retrieval. ACM (2017)
18. Song, Y., et al.: Exploring and exploiting user search behavior on mobile and tablet devices to improve search relevance. In: Proceedings of the 22nd International Conference on World Wide Web. ACM (2013)

How Users Select Query Suggestions Under Different Satisfaction States?

Zhenguo Shang[1], Jingfei Li[1], Peng Zhang[1(✉)], Dawei Song[1,2(✉)], and Benyou Wang[1]

[1] School of Computer Science and Technology, Tianjin University, Tianjin, China
asdszg@163.com, {jingfeili,pzhang,dwsong}@tju.edu.cn, wabyking@163.com
[2] The Computing Department, The Open University, Milton Keynes, UK

Abstract. Query suggestion (or recommendation) has become an important technique in commercial search engines (e.g., Google, Bing and Baidu) in order to improve users' search experience. Most existing studies on query suggestion focus on formalizing various query suggestion models, while ignoring the study on investigating how users select query suggestions under different satisfaction states. Specifically, although a number of effective query suggestion models have been proposed, some basic problems have not been well investigated. For example, (i) how much the importance of query suggestion feature for users with respect to different queries; (ii) how user's satisfaction for current search results will influence the selection of query suggestions. In this paper, we conduct extensive user study with a search engine interface in order to investigate above problems. Through the user study, we gain a series of insightful findings which may benefit for the design of future search engine and query suggestion models.

Keywords: Query suggestion · User study · Eye tracking · User satisfaction · Novelty

1 Introduction

Query suggestion technology has been widely used in modern search engines (e.g., Google, Bing and Baidu) in order to help users type potentially intended queries and improve users' search experience. In this paper, we will further investigate the query suggestion problem in order to enhance our understanding on: how users select query suggestions under different satisfaction states.

A number of query recommendation models have been developed in order to suggest similar or relevant queries against original issued query for users. Existing query recommendation models mainly depend on the query related documents and global query logs. They usually extract related query terms from these data as the recommendation candidate set. Then, reorder the query terms in the candidate set based on some specific ranking rules and select the top n queries as the final query recommendations. Although a number of effective

J. Wen et al. (Eds.): CCIR 2017, LNCS 10390, pp. 93–105, 2017.
https://doi.org/10.1007/978-3-319-68699-8_8

query suggestion models have been proposed, some basic problems on query suggestion have not been well investigated. For example, (i) we still do not know how much the importance of query suggestion in search engine for users with respect to different queries; (ii) if user's satisfaction for current search results will influence the selection of query suggestions.

In order to investigate above problems, we conducted extensive user's study based on a simulated search engine interface which can record user's search behaviors, e.g., result rating, query recommendation selection and eye movement, etc. To this end, we first develop a simulated search engine interface which contains two main interaction areas, including search results area and query recommendation area. In the search results area, we display a series of web pages (each result contains a title and a snippet) with respect to a specific query. In the query recommendation area, we display some query recommendations for users with respect to the issued query. Furthermore, we design 15 queries corresponding to 3 types of search tasks, i.e., navigational search task, information search task and transactional search task. We ask participants to search for the 15 designed queries with the search engine interface, meanwhile rate the relevance for each displayed webpage and the satisfaction for overall search results, then select any number of query suggestions.

In order to deepen our understanding on aforementioned research problems, we conducted a series of data analyses from different perspectives. We first analyze user's interest degree for query recommendation functionality of the search engine by computing the correlation between user satisfaction and the number of selected recommendations. We find that users' interest degrees for query recommendation are different across different search tasks and under different user satisfactions. Similar findings are found through eye tracking data analysis. Then, we analyze the relationship between user satisfaction and the novelty of the selected recommendations. From the analysis results, we find that when users are satisfied with the current search results, they will tend to select novel query recommendations. For different search tasks, the trend will be different. More detailed findings will be introduced in the following sections.

The rest of the paper is organized as follows: Sect. 2 reviews the relative work briefly. Section 3 presents the research problems of this paper. Section 4 introduces the methods of the user study in detail. The analysis of experiment results are given in Sect. 5. Finally, we will conclude this paper and give some possible research directions in the future.

2 Related Work

There are two directions of work that are related to this paper, i.e., query recommendation and query classification.

Query recommendation techniques focus on providing similar (or relevant) query suggestions to users by analyzing users' potential query intention and information need. Classic query recommendation models can be generally grouped into two categories from the perspective of the data sources they depend on, they are document collection based and query log based query recommendation models [9]. The document collection based models recommend relevant queries by mining co-occurred terms from human-edited documents which include the issued query [1]. While, the log-based models recommend relevant queries by mining frequently issued historical queries from logs [2, 13]. For example, Guo et al. proposed a query recommendation approach based on social annotation data [6]. Dupret used the click-through data in query logs to generate query recommendation automatically [5]. Previous researchers have realized that the search engines should recommend different types of queries to users according to the different user satisfaction states [8]. They model users' satisfaction for current search results with click through data, then recommend queries based on historical query logs considering the real-time satisfaction. The query recommendation model is proposed based on an assumption that users will tend to select novel recommendation queries if they are satisfied with current search results, and vice versa. However, the assumption has not been well studied and verified. In this paper, we will investigate this assumption and aim to gain more insightful findings about users' behaviors on the selection of query suggestions.

In this paper, we aim at investigating query suggestion problems for different categories of queries. Therefore, query classification is also related to our work. Broder [3] propose for the first time a taxonomy for web search queries according to different search tasks. The proposed taxonomy classifies web queries into three types, i.e., navigational, informational and transactional. Navigational queries aim to get certain web sites. The informational query intent is to get some relevant information about the query topic, which may be uncertain. And the aim of users' transactional search tasks may be completing some types of interaction in some specific websites. For example, when the query is "book flight", users may want to finish the search task of booking flight ticket. On this basis of classification, Lee et al. explored the user's click features in different query types [7]. In this paper, we aim to explore users' behaviors in selecting recommendation queries for aforementioned three types of search tasks.

3 Research Questions

3.1 RQ1: How Much the Importance of Query Recommendation for Users with Respect to Different Queries?

As we know, most commercial search engines (e.g., Google, Bing and Baidu) contain the query recommendation functionality displayed in some specific areas (e.g., bottom and right side) of the search engine results page (SERP). However, we still do not know if the query recommendation functionality is important for all queries and how much the importance for users with respect to different queries. In this paper, we will explore the relationship between user satisfaction and the interest of users who want to use the query-recommendation functionality in search engine for different types of search tasks.

We explore this research problem motivated by the fact that traditional search engines usually response (query recommendation) users' input in the same way, while ignore the special cases where users may don't need the query recommendation functionality. This traditional one-fit-all mechanism may be not appropriate and even bring harm to users' search experience for some queries. If we know when users are interested in the query recommendation or not, search engines can adjust the query recommendation algorithms or adjust the deployment of the result page with respect to different types of queries in order to cater to users' different tastes.

3.2 RQ2: If Users' Satisfaction Will Influence Their Behaviors on Selecting Query Suggestions?

Most existing query recommendation models focus on mining the similarity and relevance between originally issued query and the candidate recommendation queries, and providing most similar (or relevant) query terms to users without considering users' satisfaction for current search results. This query recommendation mechanism may not be reasonable for some cases. Intuitively, if users have been satisfied with the current search results, search engines should provide more novel recommendation queries. On the contrary, for the cases where users are not satisfied with the current search results, the search engines should provide some relevant recommendation queries with increasing representation ability. In order to investigate this research question, we will explore the relationship between user satisfaction and the novelty of user-selected query recommendations for different search tasks.

We explore this research question motivated by an intuitive fact that users may prefer different recommendation queries under different satisfaction states. If we gain the knowledge about how users select query suggestions under different satisfaction states for specific search tasks, search engine can adjust the recommendation algorithms to provide suggestions with different novelty. In this way, the recommendation algorithm will be more wisdom.

4 User Study

In this section, we first introduce the search tasks (queries) and participants of user study in detail. Then, present the experimental environment of Tobii Pro TX300 eye tracker. Finally, introduce the process of user study. In the user study, we collected 450 (30×15) valid datum.

4.1 Search Tasks

We designed 15 queries corresponding to three different types of search tasks, including navigational, informational and transactional. Navigational task aims to reach a particular web site. Informational task's intent is to acquire information from some web pages. Transactional task is to perform some web-mediated activity. We ask each experiment participant to finish the designed 15 query tasks (there are five queries in each type of task). For each search task, users are randomly provided 3 documents as the search results. The designed queries in the user study are shown in Table 1. In order to make the experimental environment as practical as possible, we develop a simulated search engine interface to display the search results. The layout of search engine interface is shown in Fig. 2. Ten recommended candidate queries are displayed in the right of the interface. When users conduct the task, they will be asked to rate the relevance for each document and rate the overall satisfaction scores in the bottom of the interface. Users can select any number of recommendation queries by selecting the checkboxes. The process of experiment can be found in Fig. 1.

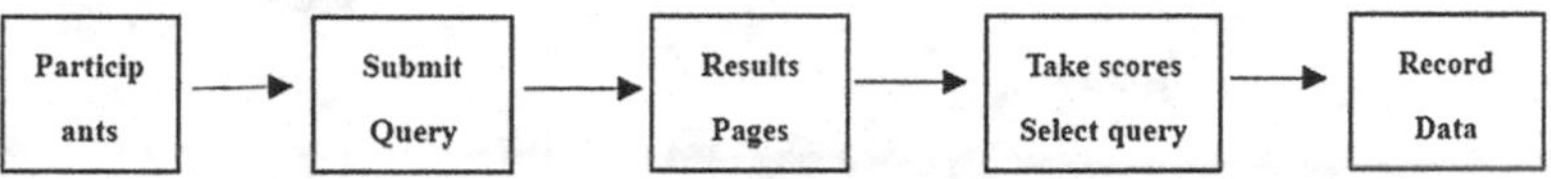

Fig. 1. The framework of user study.

Liu et al. has proved that the title and snippets of returned results are major judgement of relevant for users before they click the search results [10]. Xing has found that the user's attention will decrease with the decreasing of the result position for different types of query tasks [12]. We preselected 6 results from the real commercial search engine as the results candidates. Among the result candidates, 3 web pages are positive results which can meet the information needs and the others are negative results which can not meet the users' information need. In order to differ users' satisfaction as far as possible, the interface will randomly sample three results from the predefined candidate results and display on the search result pages for each query. In this way, the search results will be diversified and the satisfaction will become various across different users. This setting is to simulate the real search scenario in commercial search engine. Users give each result a relevance score and a overall satisfaction score, the full score are 5, 1 is defined as very dissatisfied, 2 defined dissatisfied, 3 defined as kind, 4 defined as satisfied, 5 defined as very satisfied.

For all recommended queries with respect to each original query, we manually divided them into 4 levels of novelty compared to original query. If the recommendation queries belong to the same topic, they are defined as level 1; if belong to similar topic, defined as level 2; if belong to novel topic, defined as level 3; if belong to completely novel topic, defined as level 4. We employ five participants who have related knowledge background to annotate the novelty data manually. Considering the diversity of recommend queries with different novelty, we select all candidate recommend queries from the commercial search engine.

4.2 Participants

The participants of our user study are recruited through university Advertisements. In total, 30 persons (including 13 female and 17 male) participated in

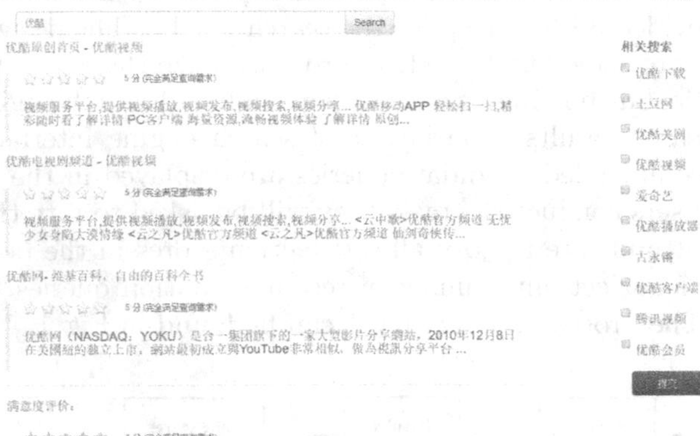

Fig. 2. The simulated search engine interface (take query "youku" as an example).

Table 1. The query lists of user study experiment.

Task Types	Query	Description
Navigational	youku	Search youku home page
	tmall	Search for youku home page
	12306	Search for tmall home page
	ifeng	Search for 12306 book ticket home page
	E-Banking Service of China Construction Bank	Search for China Construction Bank home page
Informational	what machine learning algorithms there are	Acquire the information about machine learning algorithms
	latest movie	Acquire latest movie information
	two sessions reports	Acquire two sessions reports information
	university rankings	Acquire university rankings information
	twelve zodiac	Acquire twelve zodiac information
Transactional	download QQ	Finish interaction download QQ
	inquiry express	Finish interaction inquiry express
	download Eclipse	Finish interaction download Eclips
	flight booking	Finish interaction booking ticket
	download cet-6 exams	Finish interaction download cet-6 exams

our study, all of them are between 20 and 26 years old (average 23.2). Each participant is asked to complete the designed 15 search tasks. They are payed $10 per person after completing all tasks. They come from the same university and different majors, including math, computer science, etc. Their eyesight are normal or correct normal (no achromatopsia or tritanope) to ensure their eye movement trace can be captured and recorded by eye tracker.

4.3 Environment

The user study is conducted on a Windows 7 system, running a IE Web browser and equipped with Tobii Pro TX300. The eye tracker with a 23" screen resolution of 1920 * 1080 pixels captures gaze positions at a frequency of 300 HZ and have an accuracy of 0.4°. The participants seated in front of the eye tracker 50–60 cm to ensure the user's eye-tracking data can be recorded at high precision by the Tobii studio.

4.4 Process

Before conducting the experiment, the participants have to make a 5 points calibration to make the eye movement data more precisely recorded. When the calibration is accepted, the participants are introduced the content and procedure of experiment and take a test task before they begin the formal tasks. After the test, each participants should finish 15 search tasks sequentially, the number of navigational, informational and transactional tasks are 5, respectively. For each task, the search query is predefined in the interface, the participants need just to click the search button, the 3 search results will be returned randomly. After each task, there is a break allowing users have a rest and the experiments have no time limitation.

For each task, the participant is asked to look at each of the three results and rate a relevance score. Then, according to the whole quality of the returned results, rate a overall satisfaction score. Finally, the participants select the recommended queries in the recommendation lists according to their own judgment. The number of selected recommended queries maybe one or more. If the participants feel there is no necessary to select any recommended queries, the number of selected queries can be zero. After the participants submit their interaction data, they will finish one task and they can continue to search until finishing the 15 query tasks.

5 Analysis and Discussion

From the user study, two kinds of data are collected, namely (1) users' interaction behavior information including the rating scores of each returned results, the overall satisfaction scores and the clicked recommend queries; (2) users' eye movement data including fixation point and duration, etc. In total, we collected 450 records (30 participants each finish 15 query tasks). We will analyze the collected data from different angles in order to investigate the aforementioned two research questions, i.e., RQ1 and RQ2.

5.1 Investigating RQ1

5.1.1 Correlation Analysis

In this subsection, we explore the correlation between user satisfaction and user-interest on query recommendation functionality in order to investigate how much the importance of query recommendation for user with respect to different types of queries. To this end, we obtain the average overall satisfaction scores for each queries (S_sat), the average of max relevance score of three returned results (Avg_max) and the average number of selected recommended queries (Avg_num) in each task, then plot them in one line chart and observe their change trend across different queries. S_sat and Avg_max can reflect users' satisfaction for current search results to some extent. The Avg_max is defined as follows:

$$Avg\_max = \frac{1}{|P|}\sum_{i=1}^{|P|} max(i) \tag{1}$$

where P is the participant sets, it represent the 30 recruited persons, $max(i)$ represents the max scores among the three returned results. The results are shown in Fig. 3.

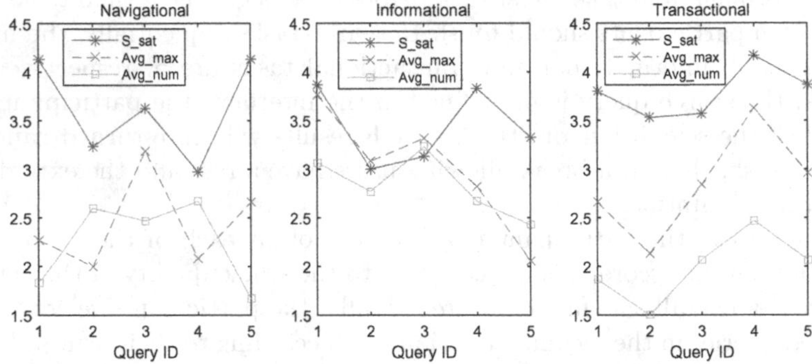

Fig. 3. The change trend for average overall satisfaction scores for each queries (S_sat), the average of max relevance score of three returned results with respect to a specific query (Avg_max) and the average number of selected recommended queries (Avg_num) in each task, across different queries.

From Fig. 3, we gain a series of meaningful findings. For navigational search tasks, there is strong negative correlations between user satisfaction and the average number of clicked query in navigational query tasks, which shows that when users are satisfied with current search results, their interests on the query recommendation functionality will decrease, on the contrary, they become more interested in the query recommendation functionality. For informational and transactional search tasks, we find that there is significant positive correlation between user satisfaction and average selected recommendation queries, which shows that if users are satisfied with current search engine, they will have more interests to continue to search for other information (do other transactions). Compared among three types of search tasks, we find that informational search tasks have the highest average number of selected recommended queries, which shows that users may have the most interest on query recommendation when they are searching for informational tasks.

5.2 Investigating RQ2

In this subsection, we will compute correlation between user satisfaction and the novelty of user-selected recommended queries in order to investigate the research question RQ2. To this end, we formalize some evaluation metrics for current search results, i.e., MRR [4] and NDCG [11]. MRR and NDCG can reflect the quality of the returned search results in an objective way. Moreover, the average satisfaction and overall satisfaction to reflect subject satisfaction. The overall satisfaction can be obtained directly. The average satisfaction is the average scores of the results for each query. MRR is defined as follows:

$$MRR = \frac{1}{|Q|} \sum_{i=1}^{|Q|} \frac{1}{rank_i} \tag{2}$$

where Q is the result sets, $rank_i$ is the rank position of the ith relevant search results. MRR metric reflects the impact of rank position to search experience. We define the results whose overall satisfaction scores are more than 3 as relevant results, others as irrelevant results. NDCG is defined as follows:

$$DCG@n = \sum_{i=1}^{n} \frac{2^{rel(j)} - 1}{log(1+j)} \tag{3}$$

In this equation, n is rank position in the search results. $rel(j)$ defined as the correlation of the ith results, in this experiment it represented by the score of the ith satisfaction. $log(1+j)$ is the position discount factor. The normalized DCG defined as:

$$nDCG@n = \frac{DCG@n}{idealDCG@n} \tag{4}$$

where $idealDCG@n$ is the DCG in the condition of ideal ranking results.

We firstly explore the consistency of subjective metrics and objective metrics, in order to investigate the research question RQ2 in different angles. To this end, we calculate the Pearson correlation coefficient between user overall satisfaction (S_sat) and user average satisfaction (S_avg), MRR and NDCG@3 as shown in Table 2. The results show that there is significant positive correlation between them and the evaluation metrics MRR and average satisfaction could reflect the overall satisfaction more accuracy. It's also reflect the fact that the user satisfaction is influenced by the rank position and number of relevant results. In the results, MRR has more strong positive correlation than NDCG@3, which illustrates that user satisfaction are more influenced by the results rank position.

Table 2. Pearson correlation coefficient between user overall satisfaction and user average satisfaction, MRR and NDCG@3

	NDCG@3	MRR	S_avg
S_sat	0.49	0.83	0.87

Because the search results are only three in this experiment. The average satisfaction have the strongest positive correlation means the user's intuitive feeling are consistency with judgment.

After that, we explore the correlation between user satisfaction and the novelty of user-selected query recommendations. The results are illustrated in Fig. 4, which shows that there are positive correlative between them in three types tasks. That is to say, when users are satisfied with the current search results they prefer to select more novel recommended queries.

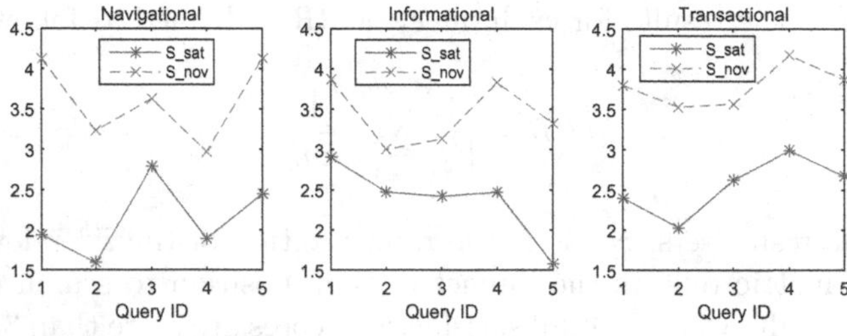

Fig. 4. The scores of satisfaction (S_sat) and novelty (S_nov) in different types tasks.

There are weak positive correlation in the navigational tasks for that users are usually have specific search targets. The logical relations are weak between the current query and the next novel query. At the most time, when users are satisfied, they want to stay in the current results or close search engine page rather than click a novelty query. For the informational task, there is strong positive correlation. When they get enough relevant information, they will be interested in the novel recommended queries. Because they prefer to achieve more information. For the transactional tasks, there are strong positive correlative. Most of the transactional tasks have potential search intent, when they finish the current query, they will naturally begin a novel query tasks. For example, after the user booked airplane, usually they will search for the hotel or airport pick up services.

The Fig. 5 illustrate the relationship between MRR, NDCG@3, average satisfaction and the novelty of user selected queries. NDCG evaluation metrics usually represent the quality of current results. The results shows that strong positive correlation in the navigational tasks, medium negative correlation in the informational tasks and medium positive correlation in tractional tasks. There are clear or specific search needs in the navigational and tractional tasks, so the user can easily get satisfied if the special sites or information are contained in the

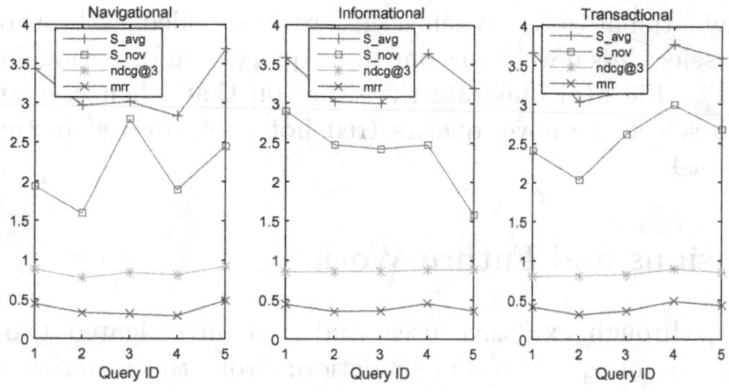

Fig. 5. The scores of novelty (S_nov),average satisfaction(S_avg), MRR and NDCG@3 in different types tasks.

returned results. For other tasks, there are no fixed sites or specific information need, users' satisfaction is cumulative. Only the information is accumulated to a certain amount, will they feel satisfied and then stop the current query or start a new novel search task.

MRR evaluation metrics usually reflect the impact of results position to user satisfaction. The results show that strong positive correlative in informational and transactional tasks and weak negative correlative in navigational tasks. The results of average satisfaction evaluation are similar with MRR. Different evaluation metrics reflect different relationship between the user satisfaction and the novelty of user selected query. In the navigational tasks, there are strong positive correlation reflected by MRR, NDCG@3, S_avg and overall satisfaction.

5.3 Further Study with Heat Map of User Eye Movement

In this subsection, we further investigate the research questions through analyzing the eye movement data. Specifically, we plot the superposed heat map and choose the typical user heat maps for each types tasks. The results show that weak relationship in navigational tasks while strong in informational and transactional tasks. The Fig. 6(a) reflects that users are more focus on relative

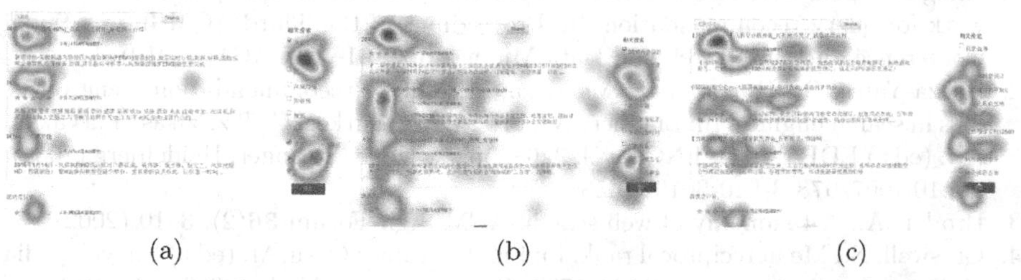

Fig. 6. (a) Heat map of a navigational task; (b) Heat map of a informational task; (c) Heat map of a transactional task.

queries in navigational tasks. When users are unsatisfied with current results, they prefer to select less novel and same topic query in informational tasks shows in Fig. 6(b). The Fig. 6(c) illustrate the situation that when user are satisfied, they prefer to select the novel queries (red hot spot covered in the figure) in transactional tasks.

6 Conclusions and Future Work

In this paper, through extensive user study, we investigated two important research questions on query recommendation. From the experimental results, we can gain a series of meaningful findings. For example, when users are satisfied with the current search results, they are interested in query recommendation functionality in the informational and transactional query tasks, but not interested in navigational query tasks. When users are searching for navigational tasks, they will not prefer to select the recommended queries. Moreover, when users are satisfied with current search results, they will prefer novel recommended queries especially in transactional query tasks. The findings of this paper may benefit for the design of future search engine, especially the design of query recommendation functionality.

In the future, this work can be extended to more general recommendation systems, e.g., commodity recommendation and news recommendation. Practical recommendation algorithms or products which consider current user satisfaction could be developed.

Acknowledgements. This work is supported in part by the Chinese National Program on Key Basic Research Project (973 Program, grant No. 2014CB744604, 2013CB329304), the Chinese 863 Program (grant No. 2015AA015403), the Natural Science Foundation of China (grant No. U1636203, 61272265, 61402324), the Tianjin Research Program of Application Foundation and Advanced Technology (grant no. 15JCQNJC41700), and the European Union's Horizon 2020 research and innovation programme under the Marie Skłodowska-Curie grant agreement No. 721321.1qQ.

References

1. Anagnostopoulos, A., Becchetti, L., Castillo, C., Gionis, A.: An optimization framework for query recommendation. In: Proceedings of the Third ACM International Conference on Web Search and Data Mining, pp. 161–170. ACM (2010)
2. Baeza-Yates, R., Hurtado, C., Mendoza, M.: Query recommendation using query logs in search engines. In: Lindner, W., Mesiti, M., Türker, C., Tzitzikas, Y., Vakali, A.I. (eds.) EDBT 2004. LNCS, vol. 3268, pp. 588–596. Springer, Heidelberg (2004). doi:10.1007/978-3-540-30192-9_58
3. Broder, A.: A taxonomy of web search. ACM Sigir Forum **36**(2), 3–10 (2002)
4. Craswell, N.: Mean reciprocal rank. In: Liu, L., Tamer Özsu, M. (eds.) Encyclopedia of Database Systems, pp. 1703–1703. Springer, New York (2009). doi:10.1007/978-0-387-39940-9_488

5. Dupret, G., Mendoza, M.: Automatic query recommendation using click-through data. In: Debenham, J. (ed.) Professional Practice in Artificial Intelligence. IFIP, vol. 218, pp. 303–312. Springer, Boston, MA (2006). doi:10.1007/978-0-387-34749-3_32
6. Guo, J., Cheng, X., Xu, G., Shen, H.: A structured approach to query recommendation with social annotation data. In: Proceedings of the 19th ACM International Conference on Information and Knowledge Management, pp. 619–628. ACM (2010)
7. Lee, U., Liu, Z., Cho, J.: Automatic identification of user goals in web search. In: International Conference on World Wide Web, WWW 2005, Chiba, Japan, pp. 391–400, May 2005
8. Li, J., Shang, Z., Zhang, P., Song, D.: An auto-adaptive query recommendation model based on users search satisfaction state. J. Frontiers Comput. Sci. Technol. (2015)
9. Li, Y., Wang, B., Li, J.: A survey of query suggestion in search engine. J. Chin. Inform. Process. **24**(6), 75–84 (2010)
10. Liu, Y., Miao, J., Zhang, M., Ma, S., Ru, L.: How do users describe their information need: query recommendation based on snippet click model. Expert Syst. Appl. **38**(11), 13847–13856 (2011)
11. Rvelin, K., Jaana, I.: Cumulated gain-based evaluation of IR techniques. ACM Trans. Inform. Syst. **20**(4), 422–446 (2002)
12. Xing, Q.: A study of click models in web search. PhD thesis, Tsinghua University (2014)
13. Zhang, Z., Nasraoui, O.: Mining search engine query logs for query recommendation. In: Proceedings of the 15th International Conference on World Wide Web, pp. 1039–1040. ACM (2006)

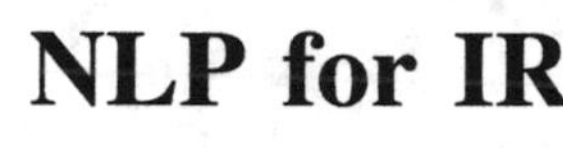

NLP for IR

Tripartite-Replicated Softmax Model for Document Representations

Bo Xu[1], Hongfei Lin[1(✉)], Lin Wang[1], Yuan Lin[2], Kan Xu[1], Xiaocong Wei[1], and Dong Huang[1]

[1] School of Computer Science and Technology, Dalian University of Technology, Dalian, China
{xubo2011, nilgnaw}@mail.dlut.edu.cn, {hflin, xukan}@dlut.edu.cn

[2] WISE lab, Dalian University of Technology, Dalian, China
zhlin@dlut.edu.cn

Abstract. Text mining tasks based on machine learning require inputs to be represented as fixed-length vectors, and effective vectors of words, phrases, sentences and even documents may greatly improve the performance of these tasks. Recently, distributed word representations based on neural networks have been demonstrated powerful in many tasks by encoding abundant semantic and linguistic information. However, it remains a great challenge for document representations because of the complex semantic structures in different documents. To meet the challenge, we propose two novel tripartite graphical models for document representations by incorporating word representations into the Replicated Softmax model, and we name the models as Tripartite-Replicated Softmax model (TRPS) and directed Tripartite-Replicated Softmax model (d-TRPS), respectively. We also introduce some optimization strategies for training the proposed models to learn better document representations. The proposed models can capture linear relationships among words and latent semantic information within documents simultaneously, thus learning both linear and nonlinear document representations. We examine the learned document representations in a document classification task and a document retrieval task. Experimental results show that the learned representations by our models outperform the state-of-the-art models in improving the performance of these two tasks.

Keywords: Document representations · Replicated softmax model · Text mining

1 Introduction

Text mining tasks usually require their inputs to be fixed-length vectors for machine learning algorithms to deal with. With the number of documents increasing rapidly on the Internet, there is an urgent need to represent documents effectively to improve the performance of these tasks. Research on document representations has been studied for years, and various models have been developed [1–6]

Vector space model, as a traditional method, is based on the bag-of-words assumption, and has been widely used in information retrieval, which represents

J. Wen et al. (Eds.): CCIR 2017, LNCS 10390, pp. 109–121, 2017.
https://doi.org/10.1007/978-3-319-68699-8_9

documents with feature vectors. In these vectors, each dimension stands for a word in the vocabulary, which is weighted based on weighting schemes, such as term frequency inverse document frequency (tf-idf). However, there are two weaknesses of these models. For one thing, the word order may be lost by assuming the independency among words, and as a result representations for different documents may be the same when the documents contain the same words. For another, these models have little sense about the semantics embedded in documents for representing the documents, and the dimensionality of representations by these models is proportional to the size of vocabularies, which limits their generalization ability in different tasks due to the curse of dimensionality. Therefore, it poses a great challenge for effective document representations.

Recently, neural network-based models exhibit powerful capabilities in representing words, phrases and even documents. As one of the most effective models, Restricted Boltzmann Machines (RBMs) [7, 8] have been used for document modeling in the Replicated Softmax model (RPS) [9], which outperforms LDA [10] in different text mining tasks. The RPS model is constituted with an ensemble of RBMs with shared parameters, and outputs the values of hidden units as replacements to the original documents, thus learning fix-length features for document representations. To enhance the RPS model, Srivastava et al. [11] introduces another hidden layer into the original RPS model to learn document representations more effectively and more efficiently. However, these models still are subject to the bag-of-words assumption, which may lose semantic information about words when learning the document representations. Therefore, we may further enhance these models by taking more semantic information of words into consideration to learn the representations of documents.

Inspired by the successful use of distributed word representations based on neural network language model [12, 13], some studies attempt to integrate distributed word representations into topic models. For example, Niu et al. [14] proposed a method to train LDA-based topic models using a neural network language model [13], which achieves better performance and outperforms original LDA. Nguyen et al. [15] replaced the probability of generating words from topic distribution with the probability of co-occurrence between a topic vector and a word vector, and introduce a Bernoulli distribution to optimize the learning process. In some other works, distributed word representations have been widely used to generate presentations of phases, sentences and paragraphs [13]. These studies indicate that word representations may enhance document modeling with abundant semantic information.

Based on this idea, we propose two Tripartite-Replicated Softmax models (TRPS) to learn better document representations. The proposed models integrate distributed word representations into the Replicated Softmax model to encode more semantic information about words into document representations. The learned models, as tripartite graphical belief networks, can learn both linear representations and nonlinear representations for documents simultaneously. We conduct extensive experiments on publicly available datasets in a text classification task and an information retrieval task to examine the performance of the proposed models, and investigate the effectiveness of the learned representations for documents with different lengths. Experimental results show that the proposed models outperform state-of-the-art models, and achieve impressive improvements over the Replicated Softmax model and the Over-Replicated Softmax model.

2 Replicated Softmax Model

Before introducing the proposed models, we first introduce the undirected graphical model, Replicated Softmax model (RPS), which can automatically extract low-dimensional latent semantic representations of documents. The RPS model comprises of a family of RBMs with shared parameters, and each RBM has Softmax visible variables that can have one of the different states. The RPS model takes document vectors as inputs and low-dimensional representations of documents as outputs, where the input vectors are generated based on term frequencies in each document.

Specifically, let K be the size of vocabulary, N be the length of a document (the total number of words in a document), F be the number of units in the hidden layer (the dimensionality of the outputted representations). The values of hidden units are defined as binary stochastic hidden topic features $h \in \{0,1\}^F$. Let V be an $N \times K$ observed binary matrix with $v_{ik} = 1$ if visible unit i takes on the kth value. The energy of the RPS model is defined as:

$$E(\mathbf{V},\mathbf{H};\theta) = -\sum_{i=1}^{N}\sum_{j=1}^{F}\sum_{k=1}^{K} W_{ijk}h_j v_{ik} - \sum_{i=1}^{N}\sum_{k=1}^{K} v_{ik}b_{ik} - N\sum_{j=1}^{F} h_j a_j \tag{1}$$

where $\theta = \{\mathbf{W}, \mathbf{a}, \mathbf{b}\}$ are parameters of the model, W_{ijk} is a symmetric interaction term with value k between the visible unit i and the hidden unit j, b_{ik} is the bias of visible unit i with value k, and a_j is the bias of hidden unit j. The probability that the model assigns to a visible binary matrix $\mathbf{V}$ is defined as follows.

$$P(\mathbf{V};\theta) = \frac{1}{Z(\theta,N)}\sum_{\mathbf{h}} \exp(-E(\mathbf{V},\mathbf{h};\theta)) \tag{2}$$

$$Z(\theta,N) = \sum_{\mathbf{V}'}\sum_{\mathbf{h}'} \exp(-E(\mathbf{V}',\mathbf{h}';\theta)) \tag{3}$$

where $Z(\theta, N)$ is a normalization factor. The RPS model creates a separate RBM for each document with as many Softmax units as the number of words in each document. All of these Softmax units share the same weights without consideration of the order of words, connecting them to binary hidden units. Therefore, the energy of the state $\{\mathbf{V}, \mathbf{h}\}$ for a document that contains N words can also be interpreted as follows.

$$E(\mathbf{V},\mathbf{H};\theta) = -\sum_{j=1}^{F}\sum_{k=1}^{K} W_{jk}h_j\hat{v}_k - \sum_{k=1}^{K}\hat{v}_k b_k - N\sum_{j=1}^{F} h_j a_j \tag{4}$$

where $\hat{v}_k = \sum_{i=1}^{N} v_{ik}$ refers to the number of occurrences of the kth word. The bias terms of the hidden variables are scaled up by the length of the document, which is important for the hidden units to behave sensibly when dealing with documents of different lengths. The conditional distributions are given by Softmax and logistic functions as follows.

$$P(v_{ik} = 1 | \mathbf{h}) = \frac{\exp(\mathbf{hW_{\bullet k}} + b_k)}{\sum_{k'}^{K} \exp(\mathbf{hW_{\bullet k'}} + b_{k'})} \tag{5}$$

$$P(h_j = 1 | \mathbf{V}) = \sigma(\hat{\mathbf{v}}\mathbf{W}_{\mathbf{j}\bullet}^{\mathbf{T}} + Na_j) \tag{6}$$

Exact maximum likelihood learning of RPS is intractable as computing the derivatives of the partition function takes time exponentially proportional to $min\{D, F\}$. In practice, the learning can be approximated by contrastive divergence (CD) [16].

3 Tripartite-Replicated Softmax Model

The RPS model takes the outputs of the binary hidden layer as document representations, which can be considered as nonlinear mappings from inputs to low-dimensional latent representations of documents. However, some semantic information in documents may be ignored because of the bag-of-words assumption. To capture more semantic information in the learned representations, we attempt to introduce distributed word representations into the RPS-based model in consideration of much semantic information embedded in word representations. It has been demonstrated that the learned distributed representations of words have powerful capabilities to embed rich

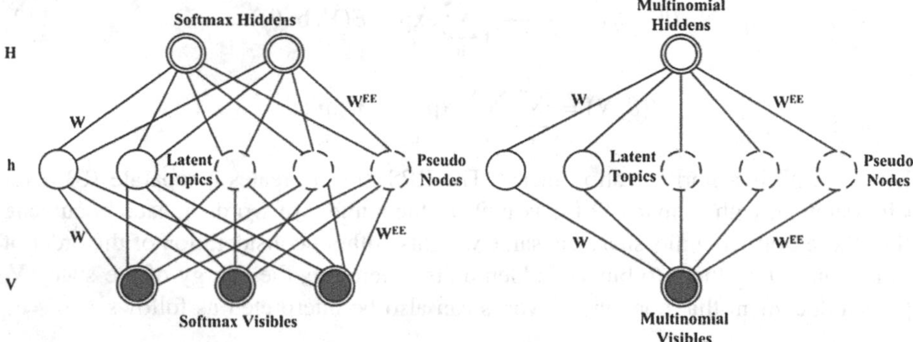

Fig. 1. The Tripartite-Replicated Softmax model. The bottom layer represents softmax visible units V. The middle layer represents binary latent topics h. The top layer represents softmax hidden units H. All visible and hidden softmax units share the same set of weights, connecting them to binary hidden units, and all visible and second hidden softmax units are connected with distributed word representation-based weights. The dotted nodes are pseudo nodes only used for model outputs. Left: The model for a document containing N = 3 words with M = 2 softmax hidden units. Right: A different interpretation of the model, in which N softmax units with identical weights are replaced by a single multinomial unit which is sampled N times and the M softmax hidden units are replaced by a multinomial unit sampled M times. (Color figure online)

semantics within the context. From this point of view, we introduce the word representations to learn meaningful linear representations of documents together with the nonlinear representations in the proposed models. We name the new model as Tripartite-Replicated Softmax model (TRPS), whose structure is illustrated in Fig. 1.

Before introducing distributed word representations, we first add another hidden layer **H** to the RPS model to give a complementary prior over the hidden units, which has been proved effective to provide more flexibility in defining the prior [11]. As a result, the model's prior over the latent topics **h** can be viewed as the geometric mean of two probability distributions: one is defined by an RBM composed of **V** and **h**, and the other is defined by an RBM composed of **h** and **H**.

On this basis, we fully connect visible units and units in the second hidden layer by extra edges (the blue edges in Fig. 1). We then define weights on these edges based on the distributed word representations. The dotted circles on the blue edges stand for a set of pseudo nodes, which can output linear representations of the documents. We call these nodes as pseudo nodes because no activations on these nodes are executed during the model training, and we only use the outputs of the models when the final model is learned. The connections between visible units and units in the second hidden layer convey much semantic information from word representations during the model training, and may generate more useful document representations. Like the RPS model, the TRPS model can also be interpreted as an RBM-based model that uses a single visible multinomial unit with support {1,…, K}, which is sampled N times.

Specifically, let L be the dimensionality of the word representations, K be the vocabulary size, and **E** be the word representation matrix in $K \times L$. We define the weights on the edges between the visible layer and the second hidden layer as $\mathbf{W}^{\mathbf{EE}} = \mathbf{W}^{\mathbf{E}}(\mathbf{W}^{\mathbf{E}})^{\mathrm{T}}$ where $w_{kl}^{E} = c_k * E_{kl}$. **c** is a vector in K, and can be tuned during the model training. $(\mathbf{W}^{\mathbf{E}})^{\mathbf{T}}$ is the transpose of the matrix $\mathbf{W}^{\mathbf{E}}$. The values of the elements in **c** are forced to be non-negative, standing for the weights on each word in the vocabulary. We define the outputs for pseudo nodes as $\mathbf{V}\mathbf{W}^{\mathbf{E}}$, which can be taken as a linear conversion of the original inputs by weighting on each word, and can be considered as linear document representations learned from the proposed model. It is worth noting that the number of pseudo nodes equals to the dimensionality of the word representations based on the above setting. The energy of the proposed model can be formalized as follows.

$$E(\mathbf{V},\mathbf{h},\mathbf{H};\theta) = -\sum_{j=1}^{F}\sum_{k=1}^{K} W_{jk}h_j(\hat{v}_k + \hat{h}'_k) - \sum_{k=1}^{K}(\hat{v}_k + \hat{h}'_k)b_k - (M+N)\sum_{j=1}^{F} h_j a_j - \sum_{k}^{K}\sum_{k'}^{K}\sum_{l}^{L} W_{kl}^{E}W_{k'l}^{E}\hat{v}_k\hat{h}'_{k'} \tag{7}$$

where $\theta = \{\mathbf{W}, \mathbf{a}, \mathbf{b}, \mathbf{c}\}$ are parameters of the TRPS model. **E** is the matrix based on the pre-trained distributed word representations. $\hat{v}_k = \sum_{i=1}^{N} v_{ik}$ refers to the number of occurrences for the kth word in the document, and $\hat{h}'_k = \sum_{i=1}^{M} \hat{h'}_i$ refers to the number of occurrences for the kth words in the second hidden layer **H**. Based on the energy definition, we define the joint probability distribution of our model as follows.

$$P(\mathbf{V};\theta) = \frac{1}{Z(\theta,N)} \sum_{\mathbf{H}} \sum_{\mathbf{h}} \exp(-E(\mathbf{V},\mathbf{h},\mathbf{H};\theta)) \quad (8)$$

$$Z(\theta,N) = \sum_{\mathbf{V}'} \sum_{\mathbf{h}'} \sum_{\mathbf{H}'} \exp(-E(\mathbf{V}',\mathbf{h}',\mathbf{H}';\theta)) \quad (9)$$

Similar to the RPS model, the complexity of the TRPS model is mainly caused by $Z(\theta,N)$ and $\sum_{\mathbf{H}} \sum_{\mathbf{h}} \exp(-E(\mathbf{V},\mathbf{h},\mathbf{H};\theta))$, we can adopt Contrastive Divergence (CD) algorithm [16] or Persistent Contrastive Divergence (PCD) [17] algorithm to estimate these two probability. The corresponding conditional probability can be derived as follows.

$$P(v_{ik} = 1|\, \mathbf{h},\mathbf{H}) = \frac{\exp(\mathbf{h}\mathbf{W}_{\bullet\mathbf{k}} + b_k + \hat{\mathbf{h}}'\mathbf{W}^{\mathbf{EE}}_{\bullet\mathbf{k}})}{\sum_{k'=1}^{K} \exp(\mathbf{h}\mathbf{W}_{\bullet\mathbf{k}'} + b_{k'} + \hat{\mathbf{h}}'\mathbf{W}^{\mathbf{EE}}_{\bullet\mathbf{k}'})} \quad (10)$$

$$P(\hat{h}'_{ik} = 1|\, \mathbf{V},\mathbf{h}) = \frac{\exp(\mathbf{h}\mathbf{W}_{\bullet\mathbf{k}} + b_k + \hat{\mathbf{v}}\mathbf{W}^{\mathbf{EE}}_{\bullet\mathbf{k}})}{\sum_{k'}^{K} \exp(\mathbf{h}\mathbf{W}_{\bullet\mathbf{k}'} + b_{k'} + \hat{\mathbf{v}}\mathbf{W}^{\mathbf{EE}}_{\bullet\mathbf{k}'})} \quad (11)$$

$$P(h_j = 1|\, \mathbf{V},\mathbf{H}) = \sigma((\hat{\mathbf{v}} + \hat{\mathbf{h}}')\mathbf{W}^{\mathbf{T}}_{\mathbf{j}\bullet} + (M+N)a_j) \quad (12)$$

From the probabilities, we can see that the main difference of our model compared with the RPS model lies in the conditional probability $P(v_{ik} = 1|\mathbf{h},\mathbf{H})$. In TRPS, the probability is not only determined by the binary hidden layer, but also impacted by the second layer. The information flow between the visible layer and the second layer conveys semantic information based on distributed word representations. As a result, the document representations by our model learn both original document information from the visible layer and semantic information from the second hidden layer. The learned weight **c** balances the importance of different words in the document, and help to give a linear representation of documents.

Since final outputs of our model are the values of units in binary hidden layer, we further modify the TRPS model by converting the edges from the visible units to the units in second hidden layer as directed edges. This modification restricts that the conditional probability for units in the visible layer is only impacted by the latent topic units to encode as much document information as possible in the model outputs, namely $P(v_{ik} = 1|\mathbf{h},\mathbf{H}) = P(v_{ik} = 1|\mathbf{h})$. We refer to the new model as directed Tripartite Replicated softmax model (d-TRPS).

The d-TRPS model is similar to the TRPS model except that the energy representation $E(\mathbf{V},\mathbf{h},\mathbf{H};\theta)$ is different, which produces different conditional probability $P(v_{ik} = 1|\mathbf{h})$ and $q(h'_i = k \mid \mathbf{V})$. In our experiments, we will examine the effectiveness of both of the proposed models.

3.1 Model Training

Since introducing the high-level hidden layers makes it difficult to calculate the posterior probability for the TRPS model and the d-TRPS model, we resort to some methods to estimate the posterior probability to obtain better local optimum to approximate the global optimum. We give detailed explanations of the training process of our model as follows. We first give the probability distribution in its free energy form.

$$\mathrm{P}(\mathbf{V};\theta) = \frac{\exp(-FE(\mathbf{V};\theta))}{\sum_{\mathbf{V}} -FE(\mathbf{V};\theta)} \tag{13}$$

where $FE(\mathbf{V};\theta)$ refers to the free energy of our model. To calculate the maximum likelihood for $\mathrm{P}(\mathbf{V};\theta)$, we give the derivative of parameter θ as follows.

$$-\frac{\partial \log(p(\mathbf{V};\theta))}{\partial\theta} = \frac{\partial FE(\mathbf{V};\theta)}{\partial\theta} - \sum_{\mathbf{V}'} P(\mathbf{V}')\frac{\partial FE(\mathbf{V}';\theta)}{\partial\theta} \tag{14}$$

The first item in Eq. (14) is related to the data distribution, and can be estimated using mean field variational approach. The estimation can be formalized as follows to approximate the posterior probability $\mathrm{P}(\mathbf{h},\mathbf{H}|\mathbf{V};\theta)$.

$$Q^{MF}(\mathbf{h},\mathbf{H}\,|\,\mathbf{V};\mu) = \prod_{j=1}^{F} q(h_j|\mathbf{V}) \prod_{i=1}^{M} q(h_i^{'}\,|\,\mathbf{V}) \tag{15}$$

where $\mu = \{\boldsymbol{\mu}^{(1)}, \boldsymbol{\mu}^{(2)}\}$ are the parameters of the distribution, and can be estimated iteratively as follows.

$$\mu_{\mathrm{j}}^{(1)} \leftarrow \sigma((\hat{\mathbf{v}} + M\boldsymbol{\mu}^{(2)})\mathbf{W}_{\mathbf{j}\bullet}^{\mathbf{T}} + (N+M)a_j) \tag{16}$$

$$\mu_{\mathrm{k}}^{(2)} \leftarrow \frac{\exp(\boldsymbol{\mu}^{(1)}\mathbf{W}_{\bullet\mathbf{k}} + b_k + \hat{\mathbf{v}}\mathbf{W}_{\bullet\mathbf{k}}^{\mathbf{EE}})}{\sum_{k'=1}^{K}\exp(\boldsymbol{\mu}^{(1)}\mathbf{W}_{\bullet\mathbf{k}'} + b_{k'} + \hat{\mathbf{v}}\mathbf{W}_{\bullet\mathbf{k}'}^{\mathbf{EE}})} \tag{17}$$

The second item in Eq. (14) is related to the distribution by the model, and can be estimated using Persistent Contrastive Divergence (PCD) [17] algorithm. Specifically, let $x_t = \{\mathbf{V_t},\mathbf{h_t},\mathbf{H_t}\}$ and θ_t be the state and parameters at current time. To obtain a new state x_{t+1}, we update x_t using alternating Gibbs sampling, and obtain a new parameter θ_{t+1} by making a gradient step. After iterations, we obtain a new state $\mathbf{V_{t'}}$, which can be taken as the estimation for the item. Besides, we normalize and scale the trained word representations as the initial weight, and tune the parameter $\mathbf{c}$ in the training process.

3.2 Pre-training

As mentioned above, due to the complexity of computation, the global optimum is intractable. Therefore, we adopt some estimation method to obtain a local optimum to accelerate the model training. To obtain better initial values for parameters of our models, we use a pre-training phase to make the model have good initial values. The pre-training for the proposed models are based on the RPS model except that we replace the posterior probability $P(h_j = 1|\mathbf{V})$ and the energy $E(\mathbf{V},\mathbf{h};\theta)$ as follows.

$$P(h_j = 1 \mid \mathbf{V}) = \sigma((1 + \frac{M}{N})\hat{\mathbf{v}}\mathbf{W}_{\mathbf{j\bullet}}^{\mathbf{T}} + (M+N)a_j) \tag{18}$$

$$E(\mathbf{V},\mathbf{h};\theta) = -\sum_{j=1}^{F}\sum_{k=1}^{K} W_{jk}h_j(\hat{v}_k * \frac{M+N}{N}) - \sum_{k=1}^{K}(\hat{v}_k * \frac{M+N}{N})b_k - (M+N)\sum_{j=1}^{F} h_j a_j \tag{19}$$

We conduct the pre-training to achieve two goals: one is to make the initial values of units in the second layer to have a nearer distribution as the visible layer, since the latent topic layer units are determined by the other two layers; the other is since the numbers of units in the input layer and units in the second hidden layer are different, we adjust the values of units on the second hidden layer with a normalization factor *M*/*N*. Since the TRPS model and the d-TRPS model have similar structures, we take the same pre-training method to optimize the two models.

4 Experiments

4.1 Experimental Settings

We use two publicly available datasets, the 20-newsgroups dataset and the Tagmynews dataset, to examine the effectiveness of our models in document classification task and information retrieval task. We preprocess documents from both datasets by removing non-letter characters and stopwords, and lowercasing all the letters. We also remove the words not appearing in pretrained distributed word representations in advance. Finally, we keep top-3000 most frequent words for training our models. For the 20-newsgroups dataset, we use the existed division of the training set and the testing set. For the Tagmynews dataset, we perform 10 fold cross validations, each fold uses 90% of the data as the training set and 10% as the testing set, and the reported results are averaged over all the folds. We give detailed statistics of these datasets in Table 1.

Table 1. Statistics about the datasets

Datasets	#documents (train + test)	# words	Average length of docs
20-newsgroups	11314 + 7532	1,115,342	99.58
20-newssingle	11298 + 7516	344,188	30.46
Tagmynews	29344 + 3260	487,152	14.94

To examine the performance of the proposed models on documents with different lengths, we take the 20-newsgroups dataset as long documents and the Tagmynews dataset as short documents. We also extract the last paragraph of each document in the 20-newsgroups dataset as a new short document to construct a new dataset with medium length, denoted as 20-newssingle. We refer to the 20-newsgroups, 20-newssingle and

TagmyNews datasets as 20-NG, 20-NS, and TMN, respectively. We compare our models with the Replicated Softmax model [9] and the Over-Replicated Softmax model (ORPS) [11]. The ORPS model is a modified RPS model with a second hidden layer.

4.2 Training Details

In our experiment, we use the word representations provided by Google to train the TRPS model and the d-TRPS model. A validation set is held out from the training set for hyper-parameter selection. Using the validation set, we choose the dimensionality of the word representations as 300, which equals to the number of pseudo nodes. We choose the value of M and the number of hidden units over a grid search using the validation set, and finally set M for the 20-newsgroups dataset and $M = 20$ for the other two datasets. We also find that a large M cannot contribute much to the performance in the proposed models. We set the number of units in the binary hidden layer as $F = 200$, since more nodes on the binary hidden layer may cause overfitting. We use the Ada-delta [19] in our training to avoid manual tuning of the learning rate. We transform the raw term frequencies using equation $\log(1 + w_i)$, and round the values to improve the performance.

4.3 Document Classification

To evaluate our models on document classification task, we first learn all the models on the training sets to represent documents as feature vectors, and then use logistic regression as the classifier to examine the effectiveness of the learned document representations. We give the experimental results in Table 2. For the proposed models, we

Table 2. Performance of models for document classification task on three datasets

Dataset	Methods	Nonlinear Dim. = 200	Linear Dim. = 300	Combination Dim. = 500
20-newsgroups	RPS	79.01	–	79.34
	ORPS	80.21	–	81.43
	TRPS	77.55	73.29	89.80
	d-TRPS	**81.65**	**73.45**	**92.81**
20-newssingle	RPS	46.19	–	43.09
	ORPS	48.78	–	49.59
	TRPS	48.36	**37.23**	**51.22**
	d-TRPS	**50.01**	32.96	50.94
Tagmynews	RPS	77.94	–	76.93
	ORPS	79.29	–	**80.28**
	TRPS	77.64	**61.32**	79.42
	d-TRPS	**79.88**	57.94	80.25

show the classification accuracies with 200-dimensional nonlinear representations, 300 dimensional linear representations and their combinations in a 500 dimensional vector. For other models, since they only output nonlinear representations, we show their results in 200 and 500 dimensional vectors to facilitate the comparisons.

From the table, we can find that the TRPS model achieves comparable performance with the baseline models, especially when combined with linear features, and the d-TRPS model further enhances the performance, which shows transforming the edges from the inputs to the hidden units as direct connections is effective to build more effective models.

4.4 Document Retrieval

Similar to the settings in [11], we take documents in testing sets as queries, and take documents in training sets as documents. We label a document as relevant when the document and the query are in the same class, and label a document as irrelevant when the document and the query are in different classes. We measure the distances between a query representation and its corresponding document representations by cosine similarity, and rank the documents based on similarity scores. Since our model can learn both the linear and the nonlinear representations for documents, we compute their similarity with query representation respectively, and combine them by the equation $\lambda s_1 + (1 - \lambda)s_2$, where s_1 is the similarity score with nonlinear features, and s_2 is the score with linear features. We tune the parameter λ on the validation set, and find that the best performance can be achieved when set $\lambda = 0.9$. We evaluate the retrieval performance in terms of precision-recall curves, shown in Fig. 2.

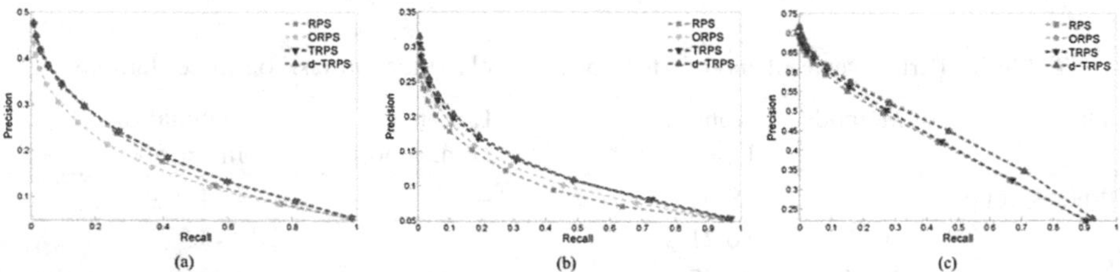

Fig. 2. Precision-Recall curves for (a) 20-newsgroups, (b) 20-newssingle and (c) Tagmynews datasets when a query document from the test set is used to retrieve similar documents from the training corpus. Results are averaged over all queries of each dataset.

From the results, we find that in the information retrieval task the TRPS model outperforms the baseline models, and the d-TRPS model achieves the best performance on all the datasets. We also find that the proposed models achieve better results especially when modeling long documents. A possible explanation for this may be that the TRPS model connects the input layer and the second hidden layer to make them compete for the contribution to the binary hidden layer and benefit from the word representations. Furthermore, d-TRPS model replaces the undirected connections with directed connections to reduce the influence of the competition.

4.5 Discussion

For documents with different lengths, the TRPS model is especially suitable for modeling long documents according to the experimental results on both of the tasks, compared with the RPS-based models. This may attribute to the introduction of distributed word representations while learning the model. Distributed word representations are learned mainly based on the co-occurrence of words within a certain window size in a large corpus, thus capturing much semantic information of words, while long documents contains more complicated semantic structures compared with short documents. Therefore, the TRPS model learns better document representations for long documents. Meanwhile, compared with the RPS model, the conditional probability for the visible units in the TRPS model is not only impacted by the latent topic layer, but also impacted by the second hidden layer, which may miss some document information while learning the document representations.

The d-TRPS model further enhances the TRPS model by transforming the connection from the visible units to the second hidden units as directed edges, namely P $(v_{ik} = 1|\mathbf{h},\mathbf{H}) = P(v_{ik} = 1|\mathbf{h})$. This modification restricts the conditional probability for the units in the visible layer is only impacted by the latent topic units to encode as much document information as possible in the final document representations, thus achieving better performance.

As to the complexity of the proposed models, our models only require an extra parameter **c** as the weights on word representations, whose dimensionality equals to the dimensionality of the word representations. Therefore, the complexity of our models is comparable with the RPS-based models. Besides, since our model is general, the learned document representations can also be applied in other text mining tasks, and the learned weight vector **c** for words can be applied for weighting words in different tasks. When setting the weight vectors as zero vectors and setting $M = 0$, our model reduces to the RPS model. Therefore, the proposed models can be considered as a generalization of the Replicated Softmax model.

5 Conclusion and Future Work

In the paper, we propose two tripartite graphical models, the Tripartite-Replicated Softmax model and the directed Tripartite-Replicated Softmax model, to represent documents as fixed-length representations. The proposed models introduce distributed word representations based on neural network to encode much semantic information of words into the document representations, and learn the linear and the nonlinear representations simultaneously. Experimental results show that the proposed models, especially the d-TRPS model, outperform the state-of-the-art models for document representations in document classification task and document retrieval task, which indicates the effective of the learned document representations.

We will carry out our future work in two directions: one is to investigate more powerful word representations in our framework to further enhance the learned document representations, and the other is to explore effective ways to make the most of the linear and the nonlinear document representations, together with the learned weights on words, in consideration of characteristics of different tasks.

Acknowledgements. This work is partially supported by grant from the Natural Science Foundation of China (No. 61632011, 61572102, 61402075, 61602078, 61562080), State Education Ministry and The Research Fund for the Doctoral Program of Higher Education (No. 20090041110002), the Fundamental Research Funds for the Central Universities.

References

1. Grefenstette, E., Dinu, G., Zhang, Y.Z., et al.: Multi-step regression learning for compositional distributional semantics. arXiv preprint arXiv:1301.6939 (2013)
2. Mikolov, T., Le, Q.V., Sutskever, I.: Exploiting similarities among languages for machine translation. arXiv preprint arXiv:1309.4168 (2013)
3. Mitchell, J., Lapata, M.: Composition in distributional models of semantics. Cogn. Sci. **34** (8), 1388–1429 (2010)
4. Nam, J., Mencía, E.L., Fürnkranz, J.: All-in text: learning document, label, and word representations jointly. In: Thirtieth AAAI Conference on Artificial Intelligence (2016)
5. Yessenalina, A., Cardie, C.: Compositional matrix-space models for sentiment analysis. In: Proceedings of the Conference on Empirical Methods in Natural Language Processing, pp. 172–182. Association for Computational Linguistics (2011)
6. Zanzotto, F.M., Korkontzelos, I., Fallucchi, F., et al.: Estimating linear models for compositional distributional semantics. In: Proceedings of the 23rd International Conference on Computational Linguistics, pp. 1263–1271. Association for Computational Linguistics (2010)
7. Gehler, P.V., Holub, A.D., Welling, M.: The rate adapting Poisson model for information retrieval and object recognition. In: Proceedings of the 23rd International Conference on Machine Learning, pp. 337–344. ACM (2006)
8. Xing, E.P., Yan, R., Hauptmann, A.G.: Mining associated text and images with dual-wing harmoniums. arXiv preprint arXiv:1207.1423 (2012)
9. Hinton, G.E., Salakhutdinov, R.R.: Replicated softmax: an undirected topic model. In: Advances in Neural Information Processing Systems, pp. 1607–1614 (2009)
10. Blei, D.M., Ng, A.Y., Jordan, M.I.: Latent Dirichlet allocation. J. Mach. Learn. Res. **3**, 993–1022 (2003)
11. Srivastava, N., Salakhutdinov, R.R., Hinton, G.E.: Modeling documents with deep Boltzmann machines. arXiv preprint arXiv:1309.6865 (2013)
12. Mikolov, T., Chen, K., Corrado, G., et al.: Efficient estimation of word representations in vector space. arXiv preprint arXiv:1301.3781 (2013)
13. Mikolov, T., Sutskever, I., Chen, K., Corrado, G.S., Dean, J.: Distributed representations of words and phrases and their compositionality. In: Advances in Neural Information Processing Systems, pp. 3111–3119 (2013)
14. Niu, L.Q., Dai, X.Y.: Topic2Vec: learning distributed representations of topics. arXiv preprint arXiv:1506.08422 (2015)
15. Nguyen, D.Q., Billingsley, R., Du, L., et al.: Improving topic models with latent feature word representations. Trans. Assoc. Comput. Linguist. **3**, 299–313 (2015)
16. Hinton, G.E.: Training products of experts by minimizing contrastive divergence. Neural Comput. **14**(8), 1771–1800 (2002)

17. Tieleman, T.: Training restricted Boltzmann machines using approximations to the likelihood gradient. In: Proceedings of the 25th International Conference on Machine Learning, pp. 1064–1071. ACM (2008)
18. Salakhutdinov, R., Hinton, G.E.: Deep Boltzmann machines. In: International Conference on Artificial Intelligence and Statistics, pp. 448–455 (2009)
19. Zeiler, M.D.: ADADELTA: an adaptive learning rate method. arXiv preprint arXiv:1212. 5701 (2012)

A Normalized Framework Based on Multiple Relationships for Document Re-ranking

Wenyu Zhao and Dong Zhou(✉)

School of Computer Science and Engineering, Hunan University of Science and Technology, Xiangtan 411201, Hunan, China
719727262@qq.com, dongzhou1979@hotmail.com

Abstract. Document re-ranking has been widely adopted in Information Retrieval as a way of improving precision of top documents based on the first round retrieval results. There are methods that use semi-supervised learning based on graphs constructed based on similarities between documents. However, most of them only consider relationships between documents. In this paper, we propose an approach to take the relationships between documents, between words in documents, as well as between documents and words into consideration. We develop a novel generative model which integrates neural language model with latent semantic model, then we incorporate the relationships between documents and words into a normalized framework to re-rank documents based on the initial retrieval results. Experimental results show that the method show significant improvements in comparison with other baseline methods.

Keywords: Document re-ranking · Word Embeddings · Latent semantic model

1 Introduction

The goal of Information Retrieval (IR) is to generate a list of retrieval results that meet user's' information needs (referred to as queries). Results ranked by the correlation between queries and documents in corpus. Previous research has shown that few people would browse results after two pages when using the Google[1] search engine [1]. Researchers have found that only part of the first round retrieval results meet user's' information needs because of ambiguity among queries and texts. To tackle this problem, there are two methods can improve accuracy of retrieval results to a large extent without personal intervention. One is document re-ranking [2, 3], another is query expansion [4–6]. Query expansion needs a second retrieval or extra resources. Therefore, it is attractive to re-rank the initial results from a practical perspective. How to effectively improve the accuracy of top documents is an important issue in IR. From the document re-ranking perspective, inter-document relationships are favored by previous studies. For instance, Lee et al. [3] clustered texts to re-rank the retrieval results according to the combined similarities between documents and cluster. Yang et al. [7] considered the relationships of documents and incorporated these relationships into a normalized

1 http://www.google.com.

J. Wen et al. (Eds.): CCIR 2017, LNCS 10390, pp. 122–135, 2017.
https://doi.org/10.1007/978-3-319-68699-8_10

framework for document re-ranking. Zhou et al. applied Latent Dirichlet Allocation (LDA) model to re-rank initial results, by exploiting the latent structure of documents and queries [8]. As pointed out by other researchers [9], the LDA model tends to describe the statistical relationships of occurrence rather than real semantic information embedded in words [11]. Instead, recent advances in neural language models like Word Embeddings [12] are playing an increasingly vital role in capturing the semantics of context words.

There are some limitations in these approaches. One of the drawbacks is that those approaches only focused on the relationships between documents, by ignoring multiple relationships between documents and words in documents. In this paper, we propose a novel generative model to capture the semantics of words in corpus for computing the similarities between documents, between words in documents, as well as between documents and words. Then we incorporate the relationships between documents and words into a normalized framework to re-rank documents based on the initial results. To illustrate the effectiveness of our methodology, we apply the approach to a public dataset. Experimental results show that our proposed method is consistent and promising compared with the state-of-the-art techniques.

The rest of this paper is organized as follows. In Sect. 2, we briefly present related work on document re-ranking. We summarize the problem definition and demonstrate our approach in Sect. 3. Then, Sect. 4 describes the details about experiments and results. Finally, Sect. 5 concludes our paper and discusses the future works.

2 Related Work

To improve the accuracy of top documents, there are a large amount of studies of document re-ranking have been proposed in the IR domain. These studies can be broadly classified into four categories according to different resources used by researchers [7].

The first category performs re-ranking by using inter-document relationships. For example, Baliński et al. [2] re-ranked documents by using different inter-document distances based on text or hyperlinks to modify their relevance weights. Lee et al. [3] proposed their approach to re-rank documents according to the combined similarities between documents similarities and query-cluster similarities. Authors in [13] re-ranked documents by utilizing the document classification, where the documents with low classification scores would be irrelevant to the query. The second category of work uses various external resources for document re-ranking purpose. Qu et al. [14] computed similarities between document vectors and query vectors produced by manually built thesaurus for document re-ranking purpose. Kamps et al. [15] re-ranked documents based on the controlled vocabularies assigned to documents. The third category of work re-ranks documents by using specific information extracted from documents or queries. Authors in [16] used structural information (such as document title and query title) to re-rank documents. Xu et al. [17] made use of global and local information to re-rank documents through local context analysis from the top-ranked documents. Raviv et al. [18] utilized the information about single terms in the query and documents as well as term sequences marked as entities to retrieve documents. The fourth category explores the intrinsic structure of documents over graphs for document re-ranking purpose. Researchers in [19–21] proposed a structural re-ranking approach using asymmetric relationships among documents induced by language models scores, through a weighted version of PageRank algorithm and HITS-style cluster-based approaches. Diaz et al. [22] used score regularization to adjust ad-hoc retrieval scores from an initial retrieval so that topically related documents received similar scores. Deng et al. [23] built a latent space graph based on content and explicit links information for further developing this method. Authors in [24] proposed an affinity graph to re-rank documents by optimizing their diversity and information richness. Yang et al. [7] re-ranked documents by considering the relationships between documents. Zhou et al. applied LDA model to re-rank

Table 1. Basic notations used in the paper

Names	Description	Names	Description
w	a word	K	number of topics
d	a document	α	the prior parameter of model
q	a query	β	the prior parameter of model
n	number of documents in corpus	θ	multinomial distribution of topics
D	corpus consisted of n documents	φ	multinomial distribution of words
b	top documents extracted from first retrieval results	$w_{j,i}$	i-th word in the document d_j
m	number of words in top b documents	$z_{j,i}$	topic associated with the i-th word in the document d_j
f	the re-ranking results	dim	dimensionality of word embeddings
g	ranking scores of words in documents	N_{d_j}	number of words in documents d_j
y	a list of documents in the first retrieval results	$\vec{w}$	pivot word representation or pivot word embedding of w
l	distance between a query q and words in documents	$f^e_{j,i}$	dimension e of the embedding of word $w_{j,i}$
A	an $n \times n$ matrix representing graph between documents.	μ_z	mean of normal distribution of retrieval scores for topic z
M	an $m \times m$ matrix representing graph between words in documents	σ_z	deviation of normal distribution of retrieval scores for topic z
R	an n × m matrix representing graph between documents and words in documents	$n_{j,k}$	the number of times that topic sampled from document d_j
D_A	an diagonal matrix reflecting relationships between documents	$v_{k,w_{j,i}}$	the number of times $w_{j,i}$ generated by topic k
D_M	an diagonal matrix reflecting relationships between words in top b documents	x_i	a vector representing d_i in WE-LDA model
D_R	an diagonal matrix reflecting relationships between documents and words in top b documents	c_i	a vector representing w_i in WE-LDA model

initial results and combined LDA model with Explicit Semantic Analysis (ESA) models for producing global consistency across the semantic space to achieve the goal [25]. Ermakova et al. introduced an approach for document re-ranking based on topic structure of texts [26]. Tu et al. proposed a new term weighting approach to calculate the degree of semantic coherence between the query and documents for document ranking [27].

However, most of these researches focused on the relationships between documents, ignoring multiple relationships between documents and words in documents. In this paper, we propose a novel generative model to capture the semantics of words in corpus for computing the similarities between documents, between words in documents as well as, between documents and words. We further incorporate these relationships into a normalized framework to re-rank documents based on the initial results.

3 Framework of Document Re-ranking

3.1 Problem Definition

Given a query q, a set of documents $D = \{d_1, d_2, \ldots, d_n\}$ is retrieved by a standard IR model, that returns the first retrieval results $y \in D$. Normally, users are interested in the top retrieval results. However, the first retrieval results tend to be imperfect. Therefore, the goal of our document re-ranking is to reorder the initial retrieval documents set y by using effective algorithms and models. It can improve precision at the top retrieval documents and user satisfaction. In Table 1, we list the definition of basic notations used in our paper.

3.2 Affinity Graph Construction

In this section, we propose a novel generative model to construct affinity graphs for reflecting relationships between documents, between words in documents, as well as between documents and words in documents. As we known, LDA model and Word Embeddings can both capture the semantic of words in the text. But there exits differences between the LDA model and Word Embeddings [9, 10]. LDA model is based on the statistic relationships of words in the corpus, while Word Embeddings take advantage of the semantics of context words. In this paper, we make the most use of the merits of each other. Thus, we integrate the first retrieval results y and a query q (here a query viewed as a document) into a generative probabilistic topic model named Word Embeddings based on Topic Model (WE-LDA).

To jointly model words and word embeddings produced by Word Embeddings. WE-LDA model can learn latent topic to generate words in documents of the first retrieval results and corresponding Word Embeddings. We use the Skip-Gram model [12] to learn the Word Embeddings. Skip-gram model is usually used to predict context words of a target word in s sliding window. Each target word w will be associated with a vector $\vec{w} \in \mathbb{R}^{dim}$, where dim is the dimensionality of Word Embeddings. We regard the target word vector as a feature to predict the context words. We employ a normal distribution for Word Embeddings to infer latent topics with the documents and words. With Word

Embeddings trained by the Skip-Gram model, the generation process of the WE-LDA model can be summarized as follows (see Algorithm 1).

Algorithm 1 Generative process for WE-LDA model

Require: a query q
Require: documents of the first retrieval results y
Require: word embeddings calculated by Skip-Gram for all words in $q \cup y$
for each topic $k \in [1, K]$ **do**
  sample a distribution over words $\boldsymbol{\varphi} \sim \boldsymbol{Dirichlet(\beta)}$
end for
for each document $d_j \in (q \cup y)$ **do**
  draw a vector of topic proportions for this document $\boldsymbol{\theta}_j \sim \boldsymbol{Dirichlet(\alpha)}$
  for each word w_i in document d_j **do**
    conditional on $\boldsymbol{\theta}_j$ choose a topic $z_{j,i} \sim \boldsymbol{Mult}(\boldsymbol{\theta}_{d_j})$
    conditional on $z_{j,i}$ choose a word $w_{j,i} \sim \boldsymbol{Mult}(\varphi_{z_{j,i}})$
    for each dimension of the embedding of $w_{j,i}$, draw $f_{j,i}^{e} \sim \mathcal{N}(\mu_{z_{j,i}}^{e}, \sigma_{z_{j,i}}^{e})$
  end for
end for

In Algorithm 1, μ and σ are the mean and deviation of the normal distribution. $f_{j,i}^{e}$ is Word Embeddings. α and β are the Dirichlet prior parameters. $\boldsymbol{\theta}_j$ and $\boldsymbol{\varphi}$ are the posterior distributions of latent variables. In this model, the posterior distributions of latent variable and a particular topic $z_{j,i}$ can be approximated by Gibbs sampling method [28]. In the sampling procedure, for each word the topic is sampled by using latent topic information of words and Word Embeddings, update rule is as follows:

$$P(z_{j,i} = k) \propto \frac{n_{j,k,\neg i} + \alpha}{n_{j,\cdot,\neg i} + K\alpha} \times \frac{v_{k,w_{j,i},\neg} + \beta}{v_{k,\cdot,\neg} + V\beta} \times \prod_{e=1}^{dim} \frac{1}{\sqrt{2\pi}\sigma_{z_{j,i}}} exp\left(-\frac{(f_{j,i}^{e} - \mu_{z_{j,i}})^2}{2\sigma_{z_{j,i}}^2}\right) \tag{1}$$

In this formula, for each sampled document d_j, $n_{j,k,\neg i}$ represents the number of times that topic k has been sampled from the multinomial distribution specific to documents d_j with the current $z_{j,i}$ not counted. Another counter variable $v_{k,w_{j,i},\neg}$ denotes the number of times word $w_{j,i}$ has been generated by topic, but not counting the current word $w_{j,i}$. After that we can calculate the posterior estimate of $\boldsymbol{\theta}$ and $\boldsymbol{\varphi}$. Figure 1 shows a graphical representation of our generative model.

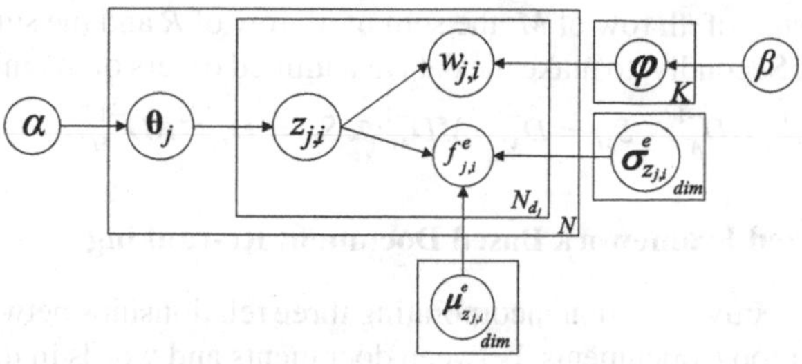

Fig. 1. Plate Notion of WE-LDA model

To construct matrices which can reflect relationships between documents, between words, as well as between documents and words, we use our generative model to compute the similarities between semantics of documents and words. In order to raise the effectiveness of document re-ranking, we choose words from top *b* documents in the first round retrieval results. Like the LDA model, we can obtain the topic vectors representing documents and words in documents after training the WE-LDA model. We compute the semantic similarities between documents, then we have

$$sim(d_i, d_j) = \frac{x_i.x_j}{||x_i||.||x_j||} \tag{2}$$

where x_i and x_j are the topic vector representing the document d_i and d_j in our generative model respectively. Thus, an matrix *A* representing the relationships between documents can be defined as $A_{ij} = sim(d_i, d_j)$. Likewise, we can get the semantic similarities between words in top *b* documents.

$$sim(w_i, w_j) = \frac{c_i.c_j}{||c_i||.||c_j||} \tag{3}$$

where c_i and c_j are the topic vector representing the word w_i and w_j in our generative model respectively. An matrix *M* denoting the relationships between words in top *b* documents can be defined as $M_{ij} = sim(w_i, w_j)$. Moreover, we can get the semantic similarities between documents and words in top *b* documents.

$$sim(d_i, w_j) = \frac{x_i.c_j}{||x_i||.||c_j||} \tag{4}$$

where x_i and c_j are the topic vector representing the document d_i and word w_j in our generative model respectively. An matrix *R* representing the relationships between documents and words in top *b* documents can be defined as $R_{ij} = sim(d_i, w_j)$. In order to show how to model these relationships into a normalized framework for document re-ranking, we define four diagonal matrices D_A, D_M, D_{RA}, D_{RM} for the ease of representation. The *i*th row of the diagonal element of D_A, D_M, D_{RA}, D_{RM} are equal to the sum of

ith row of A, the sum of ith row of M, the sum of ith row of R and the sum of ith column of R respectively. Secondly, to make the data in a unified orders of magnitude, we define matrices $S_A = D_A^{-1/2} A D_A^{-1/2}$, $S_M = D_M^{-1/2} M D_M^{-1/2}$, $S_R = D_{RA}^{-1/2} R D_{RM}^{-1/2}$.

3.3 A Normalized Framework Based Document Re-ranking

We propose an objective function incorporating three relationships between documents, between words in top b documents, between documents and words in top b documents. This normalized framework where the learned ranking f and g should be as consistent as possible with the given information M, A, R, y, l and we aim to minimize it.

$$
\begin{aligned}
Q(f,g) = {} & \frac{1}{2}\lambda \sum_{i,j=1}^{p} A_{ij}\left(\frac{1}{\sqrt{D_{A_{ii}}}} f_i - \frac{1}{\sqrt{D_{A_{jj}}}} f_j\right)^2 + \frac{1}{2}\delta \sum_{i,j=1}^{m} M_{ij}\left(\frac{1}{\sqrt{D_{M_{ii}}}} g_i - \frac{1}{\sqrt{D_{M_{jj}}}} g_j\right)^2 \\
& + \gamma \sum_{i=1}^{p}\sum_{j=1}^{m} R_{ij}\left(\frac{1}{\sqrt{D_{RA_{ii}}}} f_i - \frac{1}{\sqrt{D_{RM_{jj}}}} g_j\right)^2 + \rho \sum_{i=1}^{p} (f_i - y_i)^2 \\
& + \eta \sum_{j=1}^{m} (g_j - l_j)^2
\end{aligned}
\tag{5}
$$

where y_i is the vector representing the initial result of ith document, f_i is the vector representing the re-ranking result of ith document. In the objective function, the first term captures the relationships between documents. It means that a good ranking of documents should assign similar ranking score to similar documents. The second term captures the relationships between words in documents, meaning that similar words more likely to belonging to the same topic, which can enhance the relationships of documents where word come from. That can increase relevance between documents for document re-ranking. The third term captures the relationships between documents and words in documents. The fourth and fifth terms minimize the difference between the ranking scores and the given training data. The tradeoff among those terms are controlled by the regularization parameters $\lambda, \delta, \gamma, \rho, \eta$, $0 < \lambda, \delta, \gamma, \rho, \eta < 1$.

We can rewrite the objective function in the equivalent matrix-vector form:

$$
\begin{aligned}
Q(f,g) = {} & \lambda f^T (I - S_M) f + \delta g^T (I - S_A) g + \gamma \left(f^T f + g^T g - 2 f^T S_R g\right) \\
& + \rho (f - y)^T (f - y) + \eta (g - l)^T (g - l)
\end{aligned}
\tag{6}
$$

We can minimize the objective function with respect to f and g by differentiating it and set the corresponding derivatives to 0.

Differentiating $Q(f,g)$ with respect f and g respectively, we have

$$
\frac{\partial Q}{\partial f} = [(1 - \delta - \eta) I - \lambda S_A] f - \gamma S_R g - \rho y = 0 \tag{7}
$$

$$\frac{\partial Q}{\partial g} = \left[(1 - \lambda - \rho)I - \delta S_M\right]g - \gamma S_R^T f - \eta l = 0 \tag{8}$$

Denoting $F = \left[(1 - \delta - \eta)I - \lambda S_A\right], G = \left[(1 - \lambda - \rho)I - \delta S_M\right]$, we can first solve for g using Eq. (8) as

$$g = G^{-1}(\gamma S_R^T f + \eta l) \tag{9}$$

Then substitute g into Eq. (7), we obtain the closed form solution of final re-ranking results f

$$f = \left(F - \gamma^2 S_R G^{-1} S_R^T\right)^{-1}(\gamma\eta S_R G^{-1} l + \rho y) \tag{10}$$

Firstly, the framework is depended on the initial results. Secondly, we compute similarities between documents, between words in documents, between documents and words in documents and queries and words in documents by using our generative model. Then, we construct matrices which reflect those relationships and incorporate them together. Finally, we re-rank the initial results by using the normalized framework.

4 Evaluation

In this section, we present the experimental settings in details, then describe the compared state-of -the-art techniques and the evaluation methodology. At last, we present and analyze the results.

4.1 Experimental Setup

We perform our experiments on two public available text corpora which made up from the CLEF-2008 and CLEF-2009 of the European Library (TEL) Collections[2]. The primary source of data for this study consists of British Library Data written in English. For preparation, all documents written in English in the corpus were pre- processed by using English analyzer, Porter's stemmer and a stopword list.

In addition, the corpora were indexed by using Terrier[3] toolkit. Table 2 shows the statistics of the test collection.

We choose the following evaluation metrics to measure our approach and other baselines: normalized discounted cumulative gain (NDCG@1), mean reciprocal rank (MRR) and Precision@5 (P@5). Statistically significant differences were determined using a paired t-test at a confidence level of 95%.

[2] http://www.clef-campaign.org.
[3] http://www.terrier.org.

Table 2. Statistics of test collections

Collection	Contents	Language	Num of documents	Size	Queries
CLEF-2008	British Library Data	English (Main)	1000,100	1.2 GB	50
CLEF-2009	British Library Data	English (Main)	1000,100	1.2 GB	50

4.2 Experiment Runs

We compared our approach with several state-of-the-art methods for document re-ranking, including:

InitialResult. We use the quite popular TF-IDF model as the retrieval model to get the initial results.

Aff. The method re-ranked the results based on an affinity graph which depended on a linear combination of results from text search and authority ranking [24].

Structline. Kurland et al. performed re-ranking based on the centrality of graph by language model [19].

LDA. Zhou et al. proposed a generative model which used *LDA* model to re-rank initial results [8].

Yang. This is a method used in Yang et al.'s paper [7]. They considered incorporating the relationships of documents into a method to re-rank documents.

WELDAMFL. From our proposed method, we use the WE-LDA model to construct the affinity graphs of relationships between documents, between words in document, as well as between documents and words, then utilize the normalized framework to re-rank documents.

We describe the impact of eight parameters on document re-ranking performance. These parameters control relative importance of different types of information sources. Specifically, λ controls the importance of the information between documents, δ controls the importance of the information between words in documents, γ controls the importance of information between documents and words in documents. ρ controls the importance of information source, and η controls the importance of document re-ranking results. There exit another three parameters, b controls the number of top documents in the lists of result. K controls the number of topic in our generative model, ς controls the percentage of re-ranking results and initial results. To enhance efficiency of results re-ranking, b sets to be from 10 to 50, K ranges from 5 to 50, the rest of parameters range from 0.1 to 0.9. In this experiment, parameters-tuning are based on one corpus and applied to the rest of test collection. After several trails, results show that when K and ς is equal to 5 and 0.9 respectively, b sets to 20, λ, ρ, η are set 0.3 or 0.4, δ, γ are equal to 0.5 or 0.6, our algorithm achieves the best results.

Table 3. Evaluation results of various methods. Statistically significant differences between our method and the best baseline are indicated by *.

	CLEF-2008			CLEF-2009		
	P@5	NDCG@1	MRR	P@5	NDCG@1	MRR
InitialResult	0.492	0.68	0.7413	0.476	0.54	0.6824
Aff	0.504	0.68	0.7412	0.432	0.5	0.6237
Structline	0.504	0.69	0.744	0.484	0.56	0.7003
LDA	0.508	0.72	0.7575	0.496	0.58	0.7026
Yang	0.496	0.7	0.7512	0.488	0.54	0.681
WELDAMFL	0.52*	0.74*	0.7661*	0.508*	0.6*	0.7169*

4.3 Results

Table 3 shows the document re-ranking performance of baselines and our proposed method for different corpora. Statistically significant differences between our method and the best baseline are indicated by *. As illustrated by the results, the initial result is the lowest performance for all evaluation metrics. With the help of an affinity graph, *Aff* works consistently better than the *InitialResult* baseline. Moreover, *Structline* outperforms the *InitialResult* baseline. This demonstrates the power of the methods based on graphs. Taking the relationships of documents into consideration, *LDA* and *Yang* show a better performance of document re-ranking. We are able to see that our method effectively boosts document re-ranking performance compared with the baselines for different corpora for all evaluation metrics. The much better performance of our proposed method is due to incorporating more relationships relevant to documents, it can enhance relationships between documents and increase relevance between documents.

As seen from Table 3, several conclusions can be drawn. First, our proposed method outperforms all baselines in all metrics. Moreover, the difference between our proposed method and baseline runs is always statistically significant. We believe that the strong performance of our method is due to incorporating more relationships relevant to documents. Secondly, we integrate the Word Embeddings with LDA model, we can make the most use of merits between each other. It can accurately capture semantics of the context words for better embedded presentations. Thus, we can get better semantic similarities between documents, between words in documents. Finally, we use a normalized framework to regularize the smoothness of the initial ranking scores over graphs, which can refine the results by leveraging the global consistency over three affinity graphs.

4.3.1 Performance with Different Number of Topics

In this sub-section, we describe the impact of number of topic in our model on document re-ranking performance. We adjust the number of topics from 5 to 50. We evaluate the performance by the P@5, NDCG@1, MRR. The results are shown in Fig. 2. As illustrated by the results, when the number of topics sets to 5, the performance of our approach achieves good results in terms of P@5 and MRR.

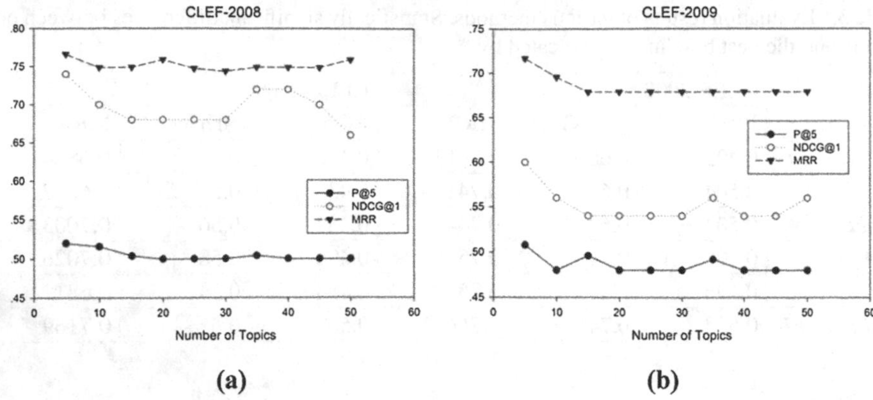

Fig. 2. Performance with different number of Topics

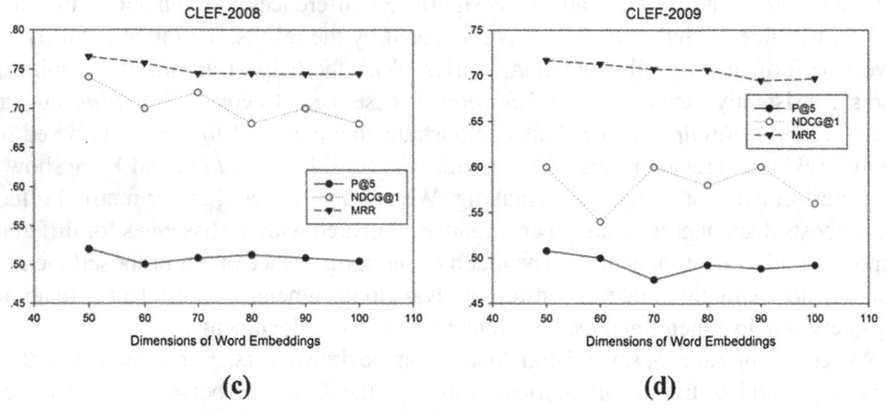

Fig. 3. Performance with different dimensions of Word Embeddings

4.3.2 Performance with Different Number of Dimensions of Word Embeddings

This sub-section examines the performance of our model on different dimensions of Word Embeddings. We set the dimensions of Word Embeddings from 50 to 100. We also use P@5, NDCG@1, MRR to evaluate the performance with different size of Word Embeddings. The results are shown in Fig. 3. As can be seen from the figure, the highest performance of dimensions of Word Embeddings is 50 in terms of P@5, NDCG@1 and MRR.

5 Conclusion

Document re-ranking can improve accuracy of top documents of the first round retrieval by using effective algorithms to adjust their positions. In this paper, we propose a novel generative model which integrate neural language model with latent semantic model to make the most use of the merits of each other to capture the semantics of words for calculating similarities between documents. Then we use a normalized framework to incorporate the relationships between documents, between words in top documents, as well as between documents and words in top documents for document re-ranking purpose. The proposed method performed well on a public dataset. Experimental results show that our method can effectively boost re-ranking performance. We will further improve the performance of document re-ranking by combining other effective algorithms and incorporating more detailed information in documents.

Acknowledgement. The work described in this paper was supported by National Natural Science Foundation of China under Project No. 61300129, Scientific Research Fund of Hunan Provincial Education Department of China under Project No. 16K030, Hunan Provincial Natural Science Foundation of China under Project No. 2017JJ2101, Hunan Provincial Innovation Foundation For Postgraduate under Project No. CX2016B575.

References

1. Zhang, Y., Jansen, B.J., Spink, A.: Time series analysis of a Web search engine transaction log. Inf. Process. Manage. **45**(2), 230–245 (2009)
2. Baliński, J., Daniłowicz, C.: Re-ranking method based on inter-document distances. Inf. Process. Manage. **41**(4), 759–775 (2005)
3. Lee, K.S., Park, Y.C., Choi, K.S.: Re-ranking model based on document clusters. Inf. Process. Manage. **37**(1), 1–14 (2001)
4. Zhou, D., Lawless, S., Wade, V.: Improving search via personalized query expansion using social media. Inf. Retrieval **15**(3–4), 218–242 (2012)
5. Zhou, D., Lawless, S., Wu, X., et al.: Enhanced personalized search using social data. In: Proceedings of the 2016 Conference on Empirical Methods in Natural Language Processing, pp. 700–710 (2016)
6. Diaz, F., Mitra, B., Craswell, N.: Query expansion with locally-trained word embeddings. In: Proceedings of the 2016 Conference on the Association for Computational Linguistics (2016)
7. Yang, L., Ji, D., Zhou, G., Nie, Y., Xiao, G.: Document re-ranking using cluster validation and label propagation. In: Proceedings of the 15th ACM International Conference on Information and Knowledge Management Arlington, Virginia, USA, pp. 690–697. ACM (2006)
8. Zhou, D., Wade, V.: Latent document re-ranking. In: Proceedings of the 2009 Conference on Empirical Methods in Natural Language Processing, vol. 3, Singapore, pp. 1571–1580. Association for Computational Linguistics (2009)
9. Vulić, I., Moens, M.-F.: Monolingual and cross-lingual information retrieval models based on (Bilingual) word embeddings. In: Proceedings of the 38th International ACM SIGIR Conference on Research and Development in Information Retrieval, Santiago, Chile, pp. 363–372 (2015)

10. Ai, Q., Yang, L., Guo, J., et al.: Improving language estimation with the paragraph vector model for ad-hoc retrieval. In: Proceedings of the 39th International ACM SIGIR Conference on Research and Development in Information Retrieval, pp. 869–872. ACM (2016)
11. Blei, D.M., Ng, A.Y., Jordan, M.I.: Latent dirichlet allocation. J. Mach. Learn. Res. **3**, 993–1022 (2003)
12. Mikolov, T., Sutskever, I., Chen, K., Corrado, G.S., Dean, J.: Distributed representations of words and phrases and their compositionality. In: Advances in Neural Information Processing Systems, pp. 3111–3119 (2013)
13. Plansangket, S., Gan, J.Q.: Re-ranking Google search returned web documents using document classification scores. Artif. Intell. Res. **6**(1), 59 (2016)
14. Qu, Y., Xu, G., Wang, J.: Rerank method based on individual thesaurus. In: Proceedings of the Second NTCIR Workshop on Research in Chinese & Japanese Text Retrieval and Text Summarization Tokyo, Japan, National Institute of Informatics (2001)
15. Kamps, J.: Improving retrieval effectiveness by reranking documents based on controlled vocabulary. In: McDonald, S., Tait, J. (eds.) ECIR 2004. LNCS, vol. 2997, pp. 283–295. Springer, Heidelberg (2004). doi:10.1007/978-3-540-24752-4_21
16. Luk, R.W.P., Wong, K.F.: Pseudo-relevance feedback and title re-ranking for Chinese information Retrieval. In: Proceedings of the Working Notes of the Fourth NTCIR Workshop Meeting Tokyo, Japan, National Institute of Informatics (2004)
17. Xu, J., Croft, W.B.: Improving the effectiveness of information retrieval with local context analysis. ACM Trans. Inform. Syst. (TOIS) **18**(1), 79–112 (2000)
18. Raviv, H., Kurland, O., Carmel, D.: Document retrieval using entity-based language models. In: Proceedings of the 39th International ACM SIGIR Conference on Research and Development in Information Retrieval, pp. 65–74. ACM (2016)
19. Kurland, O., Lee, L.: PageRank without hyperlinks: structural re-ranking using links induced by language models. In: Proceedings of the 28th Annual International ACM SIGIR Conference on Research and Development in Information Retrieval Salvador, Brazil, pp. 306–313. ACM (2005)
20. Kurland, O., Lee, L.: Respect my authority!: HITS without hyperlinks, utilizing cluster-based language models. In: Proceedings of the 29th Annual International ACM SIGIR Conference on Research and Development in Information Retrieval Seattle, Washington, USA, pp. 83–90. ACM (2006)
21. Kurland, O., Krikon, E.: The opposite of smoothing: a language model approach to ranking query specific document clusters. J. Artif. Intell. Res. (JAIR) **41**, 367–395 (2011)
22. Diaz, F.: Regularizing ad hoc retrieval scores. In: Proceedings of the 14th ACM International Conference on Information and Knowledge Management Bremen, Germany, pp. 672–679. ACM (2005)
23. Deng, H., Lyu, M.R., King, I.: Effective latent space graph-based re-ranking model with global consistency. In: Proceedings of the Second ACM International Conference on Web Search and Data Mining Barcelona, Spain, pp. 212–221. ACM (2009)
24. Zhang, B., Li, H., Liu, Y., Ji, L., Xi, W., Fan, W., Chen, Z., Ma, W.Y.: Improving web search results using affinity graph. In: Proceedings of the 28th Annual International ACM SIGIR Conference on Research and Development in Information Retrieval, Salvador, Brazil, pp. 504–511. ACM (2005)
25. Zhou, D., Lawless, S., Min, J., Wade, V.: Dual-space re-ranking model for document retrieval. In: Proceedings of the 23rd International Conference on Computational Linguistics: Posters, Beijing, China, pp. 1524–1532. Association for Computational Linguistics (2010)

26. Ermakova, L., Mothe, J.: Document re-ranking based on topic-comment structure. In: 2016 IEEE Tenth International Conference on Research Challenges in Information Science (RCIS), pp. 1–10. IEEE (2016)
27. Tu, X, Huang, J.X., Luo, J., et al.: Exploiting semantic coherence features for information retrieval. In: Proceedings of the 39th International ACM SIGIR Conference on Research and Development in Information Retrieval, pp. 837–840. ACM (2016)
28. Heinrich, G.: Parameter estimation for text analysis. University of Leipzig, Technical report (2008)

Constructing Semantic Hierarchies via Fusion Learning Architecture

Tianwen Jiang, Ming Liu, Bing Qin(✉), and Ting Liu

School of Computer Science and Technology,
Harbin Institute of Technology, Harbin 150001, Heilongjiang, China
{twjiang,mliu,qinb,tliu}@ir.hit.edu.cn

Abstract. Semantic hierarchies construction means to build structure of concepts linked by hypernym-hyponym ("is-a") relations. A major challenge for this task is the automatic discovery of hypernym-hyponym ("is-a") relations. We propose a fusion learning architecture based on word embeddings for constructing semantic hierarchies, composed of discriminative generative fusion architecture and a very simple lexical structure rule for assisting, getting an F1-score of 74.20% with 91.60% precision-value, outperforming the state-of-the-art methods on a manually labeled test dataset. Subsequently, combining our method with manually-built hierarchies can further improve F1-score to 82.01%. Besides, the fusion learning architecture is language-independent.

Keywords: Semantic hierarchies · Hypernym-hyponym relation · Fusion learning architecture

1 Introduction

Ontologies and semantic thesauri [16,22] are significant for many natural language processing applications. The main components of ontologies and semantic thesauri are semantic hierarchies. In the WordNet, semantic hierarchies are organized in the form of "is-a" relations. For instance, the words "`dog`" and "`canine`" have such relation, and we call "`canine`" is a hypernym of "`dog`". Conversely, "`dog`" is a hyponym of "`canine`". The hypernym-hyponym ("is-a") relation is the main relationship in semantic hierarchies. However, such manual semantic hierarchies construction as WordNet [16] and YAGO [22], the primary problem of them is the trade-off between coverage scope and human labor. A number of papers have proposed some approaches to extract semantic hierarchies automatically.

Hypernym-hyponym relation discovery is the key point of semantic hierarchies construction, also the major challenge. The usage of the context is a bottleneck in improving performance of hypernym-hyponym relation discovery. Several works focus on designing or learning lexical patterns [8,21] via observing context of hypernym-hyponym relation, which suffer from covering a small

T. Jiang—Ph.D Student.

J. Wen et al. (Eds.): CCIR 2017, LNCS 10390, pp. 136–148, 2017.
https://doi.org/10.1007/978-3-319-68699-8_11

proportion of complex linguistic circumstances. Besides, distributional inclusion hypothesis, which states that hypernyms tend to occur in a superset of contexts in which their hyponyms are found. In other words, hypernyms are semantically broader terms than their hyponyms [11,12]. However, it is not always rational. To acquire more contexts of words, Fu [6] applies a web mining method to discover the hypernyms of Chinese entities from multiple sources, assuming that the hypernyms of an entity co-occur with it frequently. The method works well for named entities. But for class names with wider range of meanings, this assumption may fail.

Word embedding is a kind of low-dimensional and dense real-valued vector encoding context information. Inspired by Mikolov [15] who founded that word embeddings can capture a considerable amount of syntactic/semantic relations, and found that hypernym-hyponym relations are complicated and a uniform linear projection cannot fit all of the hypernym-hyponym word pairs, Fu [5] proposed an architecture for learning semantic hierarchies via word embeddings with clustered hypernym-hyponym relation word pairs in advanced. But the method just focuses on linear transformation of word embeddings, using shallow level semantic of the representation. Besides, the method needs clustering for hypernym-hyponym relation word pairs in advanced.

Since word embeddings can capture a considerable amount of syntactic/semantic relations [15], we considered constructing a uniform architecture for semantic hierarchies learning based on nonlinear transformation of word embeddings. Inspired by advantages of discriminative model and generative model (see in Sect. 3), we fuse the two kind of models into one architecture. Considering word embeddings encode context information but ignore the lexical structures which contain some degree of semantic information, we integrate a very simple lexical structure rule into the previous fusion architecture aiming at building semantic hierarchies construction (see in Sect. 3.4).

For evaluation, the experimental results show that our method achieves an F1-score of 74.20% which outperforms the previous state-of-the-art methods. Moreover, and gets a much higher precision-value of 91.60%. Combining our method with the manually-built hierarchy can further improve F-score to 82.01%. The main contributions of our work are as follows:

- We present a uniform fusion architecture which can learn semantic hierarchies via word embeddings without any background knowledge.
- The method we proposed outperforms the state-of-the-art methods on a manually labeled test dataset especially with a good enough precision-value for application.
- The fusion learning architecture is language-independent which can be easily expanded to be suitable for other languages.

2 Related Work

During the early phase of semantic hierarchies study, some focused on building manually-built semantic resources, WordNet [16] is a representative thesauri among them. Such manually-built hierarchies have exact structure and high accuracy, but their coverage is limited, especially for fine-grained concepts and entities. Some researchers presented automatic approaches for supplementing manually-built semantic resources. Suchanek et al. [22] linked the categories in Wikipedia onto WordNet in construction of YAGO. However, the coverage is still limited by the scope of Wikipedia.

The major challenge for building semantic hierarchies is the discovery of hypernym-hyponym relations automatically. The usage of the context is a bottleneck in improving performance of discovery of hypernym-hyponym relations. Some researchers proposed method based on lexical pattern abstracted from context manually or automatically to mine hypernym-hyponym relations. Hearst [8] pointed out that certain lexical constructions linking two different noun phases (NPs) often imply hypernym-hyponym relation. A representative example is "`such NP1 as NP2`". Considering time-consuming of manually-built lexical patterns, Snow et al. [21] proposed an automatic method extracting large numbers of lexico-syntactic patterns to detect hypernym relations from a large newswire corpus. But the method suffers from semantic drift of auto-extracted patterns. Generally speaking, these pattern-based methods often suffer from low recall-value or precision-value because of the coverage and quality of extracted patterns.

Some measures rely on the assumption that hypernyms are semantically broader terms than their hyponyms. The assumption is a variation of the Distributional Inclusion Hypothesis [7,24]. The pioneer work by Kotlerman et al. [11] designed a directional distributional measure to infer hypernym-hyponym relations. Differently from Kotlerman et al. [11], Lenci and Benotto [12] focus on applying directional, asymmetric similarity measures to identify hypernyms. However the broader semantics hypothesis may not always infer broader contexts [5].

Considering the behavior of a person exploring the meaning of an unknown entity, Fu [6] applies a web mining method to discover the hypernyms of Chinese entities from multiple sources. The assumption is that the hypernyms will co-occur with its hyponym entity frequently. But the assumption maybe failed when involved with concept words which have boarder semantic compared with entities. Inspired by the fact [15] that word embeddings can capture a considerable amount of syntactic/semantic relations (e.g. $v(\texttt{king}) - v(\texttt{queen}) \approx v(\texttt{man}) - v(\texttt{woman})$, where $v(w)$ is the word embedding of the word w), Fu [5] present an approach to learn semantic hierarchies with clustered hypernym-hyponym relation word embedding pairs. However the method just focuses on linear transformation of word embeddings, using shallow level semantic of the representation. Besides, the method needs clustering for hypernym-hyponym relation word pairs in advanced and the precision-value on test data is not good enough for practical application. Shwartz et al. [19] included additional linguistic information for LSTM-based learning, but the method has co-occurrence requirements for hyponym-hypernym pairs in corpus.

Enlightened from good properties of word embeddings for capturing semantic relationship between words in work of Fu [5], we further explore capacity of word embedding for semantic hierarchies using neural networks based on a fusion learning architecture. Above all, the method we proposed do not need clustering for hypernym-hyponym relation word pairs in advanced.

3 Method

Given the hypernyms list of a word, our goal is building a semantic hierarchies construction of these hypernyms and the given word, following Fu [5], the process is presented in Fig. 1.

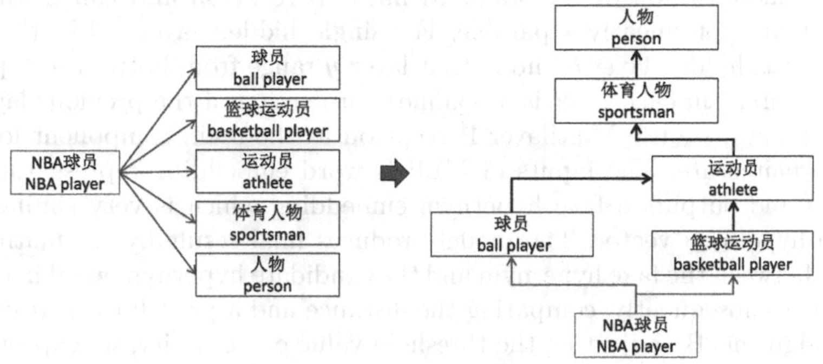

Fig. 1. An example of learning semantic hierarchies

Word embeddings representation is the only information we have to figure out whether there exists a hypernym-hyponym relation for the word pair. There are two major kind of architectures for such problem in general, one is discriminative and the other is generative. **Discriminative Architecture:** Discriminative architecture regards the discovery of hypernym-hypernym relation as a classification task, the process of learning semantic hierarchies equates to classifying word pair into **yes** or **no** for whether exists hypernym-hypernym relation. **Generative Architecture:** Generative architecture focus on generating the hypernym of a given hyponym directly. Direct generation is usual impractical, so generative method produces a fake target which is very similar with the true one.

We consider fusing these two models to discovery hypernym-hyponym relation much more precisely. We use Multilayer Perceptron (MLP) [18] to achieve a generative model and Recurrent Neural Network (RNN) [14] to implement a discriminative model.

3.1 Word Embedding Training

Various methods for learning word embeddings have been proposed in the recent years, such as neural net language models [1,15,17] and spectral models [3]. More recently, Mikolov et al. [13] propose two log-linear models, namely the Skip-gram and CBOW model, for inducing word embeddings efficiently on a large-scale corpus because of their low time complexity. Additionally, their experiment results have shown that the Skip-gram model performs best in identifying semantic relationship among words. For this reason, we employ the Skip-gram model for estimating word embeddings in this study.

3.2 Generative Architecture Based on MLP

Multilayer Perceptron (MLP) [18] is a feedforward artificial neural network which maps inputs onto a set of appropriate outputs. An MLP consists of multiple layers connecting each layer fully connected to the next one. Except for the input nodes, each node is a neuron with a nonlinear activation function (e.g. sigmoid). MLP is a modification of the standard linear perceptron and can distinguish data that are not linearly separable. For single hidden layer MLP, there are input layer x, hidden layer h and output layer y range from bottom to top. The value of neurons in each layer is a nonlinear projection of the previous layer.

In our work, we use Multilayer Perceptron as the main component for generative architecture. The inputs of MLP is word embedding representation of hyponym and outputs a fake hypernym embedding which is very similar with the true hypernym vector. The model produces final result by calculating the distance between the fake hypernym and the candidate hypernym word in continuous space, subsequently, comparing the distance and a predefined threshold to give a judgment. By adjusting the threshold value of similarity, we expect MLP model obtain much higher precision compared with the discriminative one.

3.3 Discriminative Architecture Based on RNN

Recurrent Neural Network (RNN) [14] is a kind of artificial neural network in which connections between neuron form a directed cycle, creating an internal state of the network which allows it to exhibit dynamic temporal behavior according to the history information. Unlike feedforward neural networks (e.g. MLP), RNNs can use their internal memory to process future inputs. The features of RNN makes them applicable to tasks such as unsegmented connected handwriting or speech recognition. There are some variants of original RNN, the most representative one of them is Long Short Term Memory (LSTM) [9] which is capable of learning long-term dependencies. In this paper we use the original simple recurrent networks (SRN). There are two major classes of SRN, know as Elman networks [4] and Jordan networks [10]. In this paper, we use Elman networks as the main component for discriminative architecture.

In Elman networks, the main architecture is composed of three classes layers, namely, input layer x, hidden layer h, and output layer y. Different from feedforward neural networks, the hidden layer h_t is depended on the previous time step hidden layer h_{t-1} and the current time step input layer x_{t-1}. And the output layer y_t is updated by the current time step hidden layer h_t. Formalization denotation in formulas are as follows:

$$h_t = f_h(Ux_t + Wh_{t-1} + b_h) \tag{1}$$

$$y_t = f_y(Vh_t + b_y) \tag{2}$$

Where U, V, W are matrices for linear projection. And f_h, f_y are nonlinear activation functions for nonlinear transformation.

The inputs of RNN is word embeddings representation of hyponym and candidate hypernym sequence. We regard the hyponym and candidate hypernym as a sequence, because that the judgment of candidate hypernym is depended on hyponym in discrimination process. Ignoring the outputs during recurrent process, we take the final output of the last input in the sequence as the result of discrimination.

3.4 Fusion Learning Architecture Combined with a Simple Lexical Structure Rule

Generative architecture can get a very high precision by adjusting the threshold value of similarity, but will pay a high price for low recall-value. Compared with generative method, discriminative architecture can obtain a higher recall-value with low guarantee for precision-value.

The feature of discriminative architecture indicates that if it determines a candidate hypernym-hyponym relation word pair as negative, then the word pair will have high probability for negative. We can use discriminative architecture to help the generative one to get rid of some false positive instance. For this reason, we fuse the generative and discriminative architectures together by applying Boolean operator "AND" to the results outputted by the two architectures. Excepting a much higher precision-value than the precious two models and almost the same recall-value as the generative one.

By combining discriminative and generative architectures, the fusion architecture can discover hypernym-hyponym much more precisely but becomes only focusing deep level semantic and ignoring the lexical structure information which is very useful for discovery of hypernym-hyponym relationship, especially for compound nouns (CNs), for instance, "洲际弹道导弹(Intercontinental Ballistic Missile)". The root word of CN often indicates a hypernym relation, like the word "导弹(Missile)" is the hypernyms of the precious CN. Root word of a CN can be obtained via using syntax dependency parsing or semantic dependency parsing of CN. Due to the word formation rule of Chinese, the root word is usually the last word in CN segmentation result. To supplement the capacity of learning semantic hierarchy from lexical structure, we use the simple lexical structure rule to assist previous fusion model.

The final fusion learning architecture (showed in Fig. 2) is composed of three parts, namely generative architecture, discriminative architecture and lexical structure rule module.

4 Experiments

In the experimental stage, we implement our fusion architecture for learning semantic hierarchies. To the end of this, we first introduce the preparation of

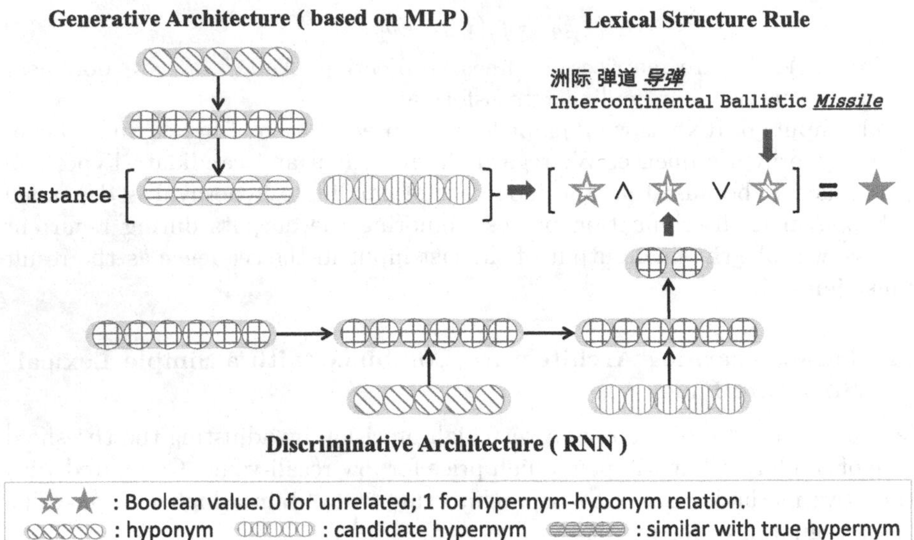

Fig. 2. Fusion learning architecture.

experimental setup. Next, we report the performance of fusion architecture and its components. Subsequently, we compare the performance of our method to those of several previous methods in different aspects and give an example for construction of semantic hierarchies.

4.1 Experimental Setup

Pre-trained Word Embeddings: We use a Chinese encyclopedia corpus named Baidubaike[1] to learn word embeddings, which contains about 30 million sentences (about 780 million words). The Chinese segmentation technology is provided by the open-source Chinese language processing platform LTP[2] [2]. Then, we employ the Skip-gram method (Sect. 3.1) to train word embeddings for the further experiment. We obtain the embedding vectors of 0.56 million words in total.

Dataset and Evaluation Metrics: The training data for learning semantic hierarchies is collected from CilinE[3] which contains 100,093 Chinese words and organized as a hierarchy of five levels, in which the words are linked by hypernym-hyponym relations. Finally, we obtain 15,242 word pairs of hypernym-hyponym relation for positive instances and constructed 15,242 negative instances for training.

[1] Baidubaike (https://baike.baidu.com/) is one of the largest Chinese encyclopedias.
[2] http://www.ltp-cloud.com/demo/.
[3] http://www.ltp-cloud.com/download/.

Table 1. The experimental data.

Relation	# of word pairs	
	Training	Test
Hypernym-hyponym	15,242	1,079
Hyponym-hypernym	7,621	1,079
Unrelated	7,621	3,250
Total	30,484	5,408

For comparability we use the same test dataset as Fu et al. [5] in evaluation stage. They obtain the hypernyms for 418 entities, which are selected randomly from Baidubaike, following their previous work [6]. The final data set was manually labeled and measured the inter-annotator agreement by using the kappa coefficient [20]. The kappa value is 0.96, which indicates a good strength of agreement. Training data and test data are showed in Table 1. We use precision-value, recall-value, and F1-score as metrics to evaluate the performances of the methods. Since the discovery of hypernym-hyponym relation is a binary classification task, we only report the performance of the positive instances recognition in the experiments.

Parameter Settings and Training: In our fusion architecture, there are MLP for generation (see in Sect. 3.2) and RNN for discrimination (see in Sect. 3.3) need to be trained. We experimentally study the effects of several hyper-parameters on this two neural networks: the number of neutrons in hidden layer, the selection of activation function. Table 2 shows all parameters used in the experiments. We use Adadelta [23] in the update procedure, which relies on two main parameters, ρ and ε, which do not significantly affect the performance. Following Zeiler [23], we choose 0.95 and $1e^{-6}$, respectively, as the values of these two parameters.

Table 2. Parameters used in our experiments.

Word dimension	# neutrons in hidden layer(RNN)	# neutrons in hidden layer (MLP)	Batch size	Adadelta parameter
$d_w = 300$	$n_h = 800$	$n_h = 500$	$b = 20$	$\rho = 0.95, \varepsilon = 1e^{-6}$

In the training stage, we train the discriminative architecture and generative architecture respectively. For training discriminative architecture based on RNN, we use the whole training data. But for training generative architecture based on MLP, we only use the positive instances in training data as Fu [5] for generating positive hypernym vectors.

4.2 Comparison with Previous Work

In this section, we compare the proposed method with previous methods, including pairwise hypernym-hyponym relation extraction based on patterns, word distributions, web mining, and based on word embeddings (see in Sect. 2). Results are shown in Table 3.

Table 3. Comparison of the proposed method with existing methods in the test set.

Method	P (%)	R (%)	F1 (%)
M_{Pttern}	97.47	21.41	35.11
M_{Snow}	60.88	25.67	36.11
$M_{balApinc}$	54.96	53.38	54.16
M_{invCL}	49.63	62.84	55.46
M_{Web}	87.40	48.19	62.13
M_{Emb}	80.54	67.99	73.74
$M_{lexicalRule}$	100.0	16.88	28.88
$M_{MLP_{gen.}}$	77.96	53.51	63.46
$M_{RNN_{dis.}}$	55.97	77.40	64.96
$M_{MLP+RNN}$	90.00	51.48	65.50
M_{Fusion}	91.60	62.36	**74.20**

Overall Comparison: $M_{Pattern}$ refers to the pattern-based method [8]. The method uses the Chinese Hearst-style patterns [6]. The result shows that only a small part of the hypernyms can be extracted based on these patterns because only a few hypernym relations are expressed in the fixed patterns, and most of them are expressed in highly flexible manners. M_{Snow} originally proposed by Snow et al. [21], this method relies on an accurate syntactic parser, and the quality of the automatically extracted patterns is difficult to be guaranteed. There are two previous distributional methods $M_{balApinc}$ [11] and M_{invCL} [12]. Each word is represented as a feature vector in which each dimension is the point-wise mutual information (PMI) value [7,24] of the word and its context words (see in Sect. 2). M_{Web} refers to a web mining method proposed by Fu et al. [6] which mines hypernyms of a given word w from multiple sources returning a ranked list of the hypernyms. M_{Emb} [5] refers to a novel method based on word embeddings achieving the best F1-value among previous methods.

$M_{lexicalRule}$ refers to using lexical structure rule to discover hypernym-hyponym relations (see in Sect. 3.4). The 100% precision-value on test data indicates the lexical rule is correct for most compound nouns (CNs). However the rule only takes effect for CNs which are minority. $M_{MLP_{gen.}}$ represents the generative architecture based on MLP neural network (see in Sect. 3.2), the method gets a higher F1-score than most of previous semantic hierarchy discovery method except M_{Emb}. $M_{RNN_{dis.}}$ represents the discriminative architecture based on RNN neural network (see in Sect. 3.3) which obtains the highest recall-value in comparison of the proposed methods. $M_{MLP+RNN}$ combines these two architectures based on MLP and RNN (see in Sect. 3.4), getting a much higher precision-value than any components and a comparable recall-value with $M_{MLP_{gen.}}$. M_{Fusion} refers to a fusion learning architecture composed of discriminative and generative architectures and assisted with lexical structure rule (see in Sect. 3.4). Assisted with the simple lexical structure rule, the fusion learning architecture get a better F1-score than all of the previous methods do and significantly improves the precision-value over the state-of-the-art method M_{Emb}.

Table 4. Out-of-CilinE data.

Method	P (%)	R (%)	F1 (%)
$M_{Wiki+CilinE}$	80.39	19.29	31.12
M_{Emb}	65.85	44.47	53.09
M_{Fusion}	79.92	44.94	**57.53**

Table 5. Combining manually-built hierarchies.

Method	P (%)	R (%)	F1 (%)
M_{Emb}	80.54	67.99	73.74
M_{Fusion}	91.60	62.36	**74.20**
$M_{Emd+CilinE}$	80.59	72.42	76.29
$M_{Fusion+CilinE}$	91.64	70.76	**79.85**
$M_{Emd+Wiki+CilinE}$	79.78	80.81	80.29
$M_{Fusion+Wiki+CilinE}$	91.00	74.63	**82.01**

Comparison on the Out-of-CilinE Data: Since the training data is extracted from CilinE, we are greatly interested in the performance of our method on the hypernym-hyponym relations outside of CilinE. We assume that as long as there is one word in the pair not existing in CilinE, the word pair is outside of CilinE. In our test data, about 61% word pairs are outside of CilinE. Table 4 shows the performances of the baseline method $M_{Wiki+CilinE}$ which means the existing manually-built hierarchies in Wikipedia and CilinE, previous state-of-the-art method M_{Emb} and our method M_{Fusion} on the out-of-CilinE data. In comparison, $M_{Wiki+CilinE}$ has the highest precision-value but has a lowest recall-value, M_{Emb} significantly improves recall-value and F1-score. By contrast, our method M_{Fusion} can discover a little bit more hypernym-hyponym relations than M_{Emb} with achieving a more than 14% precision-value improvement. And our method can get an F1-score of 57.53%, which is a new state-of-the-art result on the Out-of-CilinE Data.

Combined with Manually-Built Hierarchies: For further exploration, we combine our method M_{Fusion} with the existing manually-built hierarchies in Wikipedia and CilinE. The combination strategy is to simply merge all positive results from the two methods together, and then to infer new relations based on the transitivity of hypernym-hyponym relations. The same manner is allied to precious method M_{Emb} to be compared. The comparison is showed in Table 5. Combining our fusion method M_{Fusion} with manually-built hierarchies Wikipedia and CilinE can further improve F1-score to 82.01%, getting an about 1.7% improvement compared with the same manners on M_{Emb}.

4.3 Example of Learning Semantic Hierarchies

In Fig. 3, there is an example of learning semantic hierarchies based on our fusion architecture (M_{Fusion}) and combined method using manually-built hierarchies ($M_{Fusion+Wiki+CilinE}$). From the results, we can see that our method can actually learn the semantic hierarchies for a given word and its hypernyms list relatively precisely. The dashed line frames in Fig. 3(a) refers to the losing hypernym-hyponym relations words. For instance, our method fail to learn the two semantic hierarchies, namely, between "球员(ball player)" and "运动员(athlete)", and between "运动员(athlete)" and "体育人物(sportsman)". The reason maybe that their semantic similarity effects representations close to each other in the embedding space and our method can not find suitable projection for these pairs. Though failing to learn the two hierarchies, our method doesn't learn any wrong relations for this case. By combining our method with manually-built hierarchies, we can improve the capacity of learning semantic hierarchies. In this case, the combined method can build the semantic hierarchies correctly (see in Fig. 3(b)).

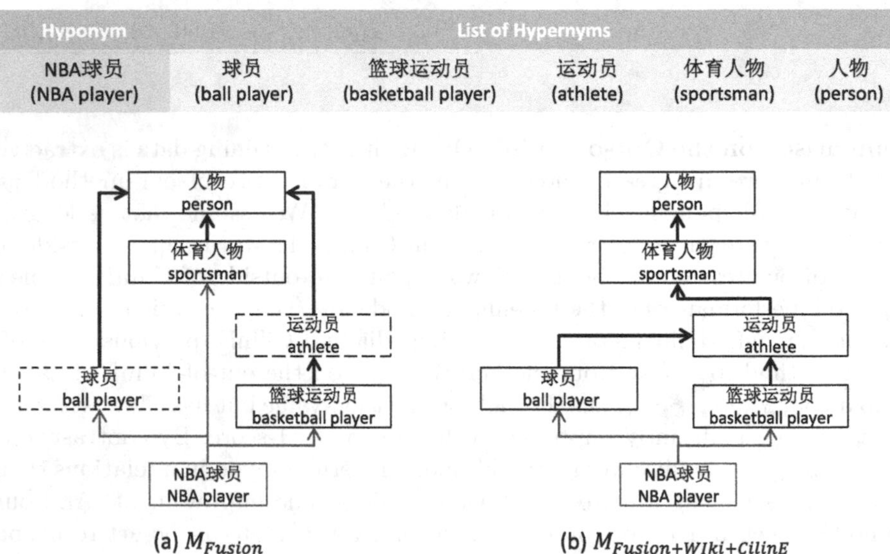

Hyponym	List of Hypernyms				
NBA球员 (NBA player)	球员 (ball player)	篮球运动员 (basketball player)	运动员 (athlete)	体育人物 (sportsman)	人物 (person)

Fig. 3. Example of learning semantic hierarchies.

5 Conclusion

This paper proposes a novel method for learning semantic hierarchies based on discriminative generative fusion architecture combined with a very simple lexical structure rule. The fusion architecture method can be easily expanded to be suitable for other languages. In experiments, the proposed method achieves the best F1-score of 74.20% on a manually labeled test dataset outperforming state-of-the-art methods with a much higher precision-value of 91.60% for application. Further experiments show that our method is complementary with some manually-built hierarchies to learn semantic hierarchy construction more precisely.

Fundings. The research in this paper is supported by National Natural Science Foundation of China (No. 61632011, No. 61772156), National High-tech R&D Program (863 Program) (No. 2015AA015407).

References

1. Bengio, Y., Ducharme, R., Vincent, P., Jauvin, C.: A neural probabilistic language model. J. Mach. Learn. Res. **3**, 1137–1155 (2003)
2. Che, W., Li, Z., Liu, T.: LTP: a Chinese language technology platform. In: Proceedings of the 23rd International Conference on Computational Linguistics: Demonstrations, pp. 13–16. Association for Computational Linguistics (2010)
3. Dhillon, P., Foster, D.P., Ungar, L.H.: Multi-view learning of word embeddings via CCA. In: Advances in Neural Information Processing Systems, pp. 199–207 (2011)
4. Elman, J.L.: Finding structure in time. Cognit. Sci. **14**(2), 179–211 (1990)
5. Fu, R., Guo, J., Qin, B., Che, W., Wang, H., Liu, T.: Learning semantic hierarchies via word embeddings. In: ACL, vol. 1 pp. 1199–1209 (2014)
6. Fu, R., Qin, B., Liu, T.: Exploiting multiple sources for open-domain hypernym discovery. In: EMNLP, pp. 1224–1234 (2013)
7. Geffet, M., Dagan, I.: The distributional inclusion hypotheses and lexical entailment. In: Proceedings of the 43rd Annual Meeting on Association for Computational Linguistics, pp. 107–114. Association for Computational Linguistics (2005)
8. Hearst, M.A.: Automatic acquisition of hyponyms from large text corpora. In: Proceedings of the 14th Conference on Computational Linguistics, vol. 2, pp. 539–545. Association for Computational Linguistics (1992)
9. Hochreiter, S., Schmidhuber, J.: Long short-term memory. Neural Comput. **9**(8), 1735–1780 (1997)
10. Jordan, M.I.: Serial order: a parallel distributed processing approach. Adv. Psychol. **121**, 471–495 (1997)
11. Kotlerman, L., Dagan, I., Szpektor, I., Zhitomirsky-Geffet, M.: Directional distributional similarity for lexical inference. Natural Lang. Eng. **16**(04), 359–389 (2010)
12. Lenci, A., Benotto, G.: Identifying hypernyms in distributional semantic spaces. In: Proceedings of the First Joint Conference on Lexical and Computational Semantics, vol. 1 - Proceedings of the Main Conference and the Shared Task, and vol. 2 - Proceedings of the Sixth International Workshop on Semantic Evaluation, pp. 75–79. Association for Computational Linguistics (2012)
13. Mikolov, T., Chen, K., Corrado, G., Dean, J.: Efficient estimation of word representations in vector space. arXiv preprint (2013). arXiv:1301.3781
14. Mikolov, T., Karafiát, M., Burget, L., Cernockỳ, J., Khudanpur, S.: Recurrent neural network based language model. In: Interspeech. vol. 2, p. 3 (2010)
15. Mikolov, T., Yih, W.T., Zweig, G.: Linguistic regularities in continuous space word representations. In: HLT-NAACL, vol. 13, pp. 746–751 (2013)
16. Miller, G.A.: Wordnet: a lexical database for English. Commun. ACM **38**(11), 39–41 (1995)

17. Mnih, A., Hinton, G.E.: A scalable hierarchical distributed language model. In: Advances in Neural Information Processing Systems, pp. 1081–1088 (2009)
18. Rosenblatt, F.: Principles of neurodynamics: perceptrons and the theory of brain mechanisms. Technical report, DTIC Document (1961)
19. Shwartz, V., Goldberg, Y., Dagan, I.: Improving hypernymy detection with an integrated path-based and distributional method. arXiv preprint (2016). arXiv:1603.06076
20. Siegel, S., Castellan Jr., N.J.: Nonparametric Statistics for the Behavioral Sciences, 2nd edn. McGraw-HiU Book Company, New York (1988)
21. Snow, R., Jurafsky, D., Ng, A.Y.: Learning syntactic patterns for automatic hypernym discovery. In: Advances in Neural Information Processing Systems, vol. 17 (2004)
22. Suchanek, F.M., Kasneci, G., Weikum, G.: Yago: a core of semantic knowledge. In: Proceedings of the 16th International Conference on World Wide Web, pp. 697–706. ACM (2007)
23. Zeiler, M.D.: Adadelta: an adaptive learning rate method. arXiv preprint (2012). arXiv:1212.5701
24. Zhitomirsky-Geffet, M., Dagan, I.: Bootstrapping distributional feature vector quality. Comput. Linguist. **35**(3), 435–461 (2009)

Jointly Learning Bilingual Sentiment and Semantic Representations for Cross-Language Sentiment Classification

Huiwei Zhou(✉), Yunlong Yang, Zhuang Liu, Yingyu Lin, Pengfei Zhu, and Degen Huang

School of Computer Science and Technology, Dalian University of Technology, Dalian 116024, Liaoning, China
{zhouhuiwei,huangdg}@dlut.edu.cn,
{SDyyl_1949,zhuangliu1992,cdzpf}@mail.dlut.edu.cn,
lyydut@sina.com

Abstract. Cross-language sentiment classification (CLSC) aims at leveraging the semantic and sentiment knowledge in a resource-abundant language (source language) for sentiment classification in a resource-scarce language (target language). This paper proposes an approach to jointly learning bilingual semantic and sentiment representations (BSSR) for English-Chinese CLSC. First, two neural networks are adopted to learn sentence-level sentiment representations in English and Chinese views respectively, which are attached to all word semantic representations in the corresponding sentence to express the words in the certain sentiment context. Then, another two neural networks in two views are designed to jointly learn BSSR of the document from word representations concatenated with their sentence-level sentiment representations. The proposed approach could capture rich sentiment and semantic information in BSSR learning process. Experiments on NLP&CC 2013 CLSC dataset show that our approach is competitive with the state-of-the-art results.

Keywords: Jointly learning · Cross-language · Sentiment classification

1 Introduction

With the exponential growth of online reviews, sentiment classification has attracted much attention in the field of natural language processing (NLP). Lots of previous researches [1–3] focus on the sentiment classification in English, the most commonly studied language in NLP. Such researches produce a great quantity of high-quality labelled corpora and sentiment knowledge in English. However, sentiment resources are imbalanced in different languages. To leverage resources in a resource-rich language (such as English) for sentiment classification in a resource-scarce language (such as Chinese), cross-language sentiment classification (CLSC) approaches have been investigated.

Existing CLSC approaches concentrate on bridging the gap between the source and the target languages. Machine translation (MT) services are usually employed to connect the two languages by translating one language (view) to another language.

J. Wen et al. (Eds.): CCIR 2017, LNCS 10390, pp. 149–160, 2017.
https://doi.org/10.1007/978-3-319-68699-8_12

Then training and testing are performed in one view [4] or in two independent views respectively [5, 6]. However, learning language-specific classifiers in each view fails to capture the common sentiment and semantic information between two languages during training process. Besides, MT services unavoidably introduce translation errors, which may even change the sentiment polarities of translated texts [6].

With the revival of interest in deep learning, common semantic representations [7–9] for the source and the target languages are also used to eliminate the gap between the two languages. Common semantic representations project data from the two languages into a common space, where training and testing are performed. However, high-quality common semantic representations require a large-scale parallel corpus, which is not always readily available in many real-world scenarios. Meanwhile, though common semantic representations capture cross-language semantics between the two languages, sentiment knowledge is hard to be exploited. Recently, some researches incorporate sentiment knowledge into common semantic representations to enhance the ability of sentiment expression [10–13]. However, these approaches learn semantics and sentiment separately, and fail to adjust the complex relations between sentiment and semantics.

This paper proposes an approach to jointly learning bilingual semantic and sentiment representations (BSSR) for CLSC. First, MT services are used to translate English training data into Chinese and translate Chinese test data into English. Then, two convolutional neural networks are applied to learn the sentiment representation of each sentiment fragment in the two views, based on the polarities of sentiment words, respectively. The learned fragment sentiment representations of all sentiment words in a sentence are averaged up as the sentence-level sentiment representations (SLSR). Finally, two Long Short-Term Memory (LSTM) in two views are designed to learn BSSR of the documents from the concatenations of word semantic and sentence-level sentiment representations, since LSTM has shown powerful ability on modeling long sequences [3]. Our approach is motivated by the intuition that sentence-level sentiment representations are composed by fragment sentiment representations in the sentence, and the words in different sentences express different sentiment though their word semantic representations (WSR) are the same.

The proposed BSSR, could capture sentiment and semantic information in cross-language document representations without relying on extra parallel corpora. Experiments on NLP&CC 2013 CLSC dataset show that our approach gets competitive performance compared with the state-of-the-art CLSC systems.

The major contributions of this work can be summarized as follows:

- We propose an approach to jointly learning BSSR of documents for CLSC, with only English training data and their translations. In the BSSR learning process, the sentiment information and cross-language semantic information can be adjusted to optimize CLSC performance.
- Cross-language semantic information is captured by minimizing the differences between the original and translated languages. It is proved that cross-language semantic information could bridge the gap between two languages effectively.

- We simply learn document representations from word semantic representations concatenated with sentence-level sentiment representations, which considers the words in the certain sentiment context. To the best of our knowledge, this representation approach has not yet been used for CLSC.

2 Related Work

In this section, we review the literature related to this paper from two perspectives: cross-language sentiment classification and sentiment classification.

2.1 Cross-Language Sentiment Classification (CLSC)

The goal of CLSC is to bridge the gap between the source language and target language. Wan [5] translates both the training and testing data to train different models in both source and target languages, which could learn language-specific classifiers in each view. However it is hard to capture the common sentiment and semantic knowledge of the two languages during training process.

Recently, there has been an interest in common semantic representations for two languages. Some approaches to common representation learning [7, 12, 13] require parallel data. These works use an encoder-decoder architecture with a shared hidden layer to learn common semantic representations. Zhou et al. [14] propose a cross-lingual representation learning model which simultaneously learns both the word and document representations in both languages. However, large-scale task-related parallel corpora may be a scarce resource for the English-Chinese sentiment classification task.

Some other works simply utilize the labelled corpora and their translations to learn common semantic representations. Zhou et al. [11] learn bilingual sentiment word embeddings (BSWE) with denoising autoencoders. BSWE is hard to adjust the complex relations between sentiment and semantics to optimize the CLSC performance. Zhou et al. [15] use a hierarchical attention model which is jointly trained with the bidirectional LSTM network to learn the document representations.

2.2 Sentiment Classification

Traditional sentiment classification approaches usually use a sentiment lexicon. The lexicon-based approaches compute the sentiment polarity for each text based on sentiment words and negatives [1]. Machine learning-based approaches employ sentiment words as important features to construct sentiment classifiers.

Tang et al. [2] incorporate sentiment polarity of the text to learn sentiment-specific word embeddings. Tang et al. [3] connect the target word representations with its context word representations to learn the semantic representations related to a target by LSTM models. Without requiring parser, CNN or LSTM could capture the fragment sentiment representations in a sentence by leveraging contexts of the sentiment words [16].

Inspired by Tang et al. [3], we attach contextual sentiment representations to the word semantic representations. Such concatenation representations are exploited for jointly learning BSSR in CLSC task.

3 Bilingual Semantic and Sentiment Representations for CLSC

Sentiment is expressed by phrases or sentences rather than by words [17]. The same words in different contexts could express different sentiment. To incorporate rich contextual sentiment information into BSSR learning process, we first extract fixed-length sentiment fragments centered at each sentiment word, and use CNN to learn the fragment sentiment representations based on the polarities of sentiment words. The fragment sentiment representations in corresponding sentences are averaged up as the SLSR (Sect. 3.1).

Then, the SLSR are attached to word semantic representations to express the words in the specific sentiment context. Finally, the concatenations are fed into LSTM for jointly learning BSSR.

3.1 Sentence-Level Sentiment Representations (SLSR) Learning

The context words within a window $[-2, 2]$ centered at each sentiment word $sent_i$ are considered as its sentiment fragment: $\{word_{i-2}, word_{i-1}, sent_i, word_{i+1}, word_{i+2}\}$. We extract these sentiment fragments, and adopt two CNN to learn sentiment representations of each sentiment fragment in English and Chinese views respectively.

CNN for SLSR Learning. The fragment sentiment representations learned with CNN in the two languages follows the same process. Let $x_i \in \mathbb{R}^d$ be the d-dimensional word representations. Then the sentiment fragment could be represented as $x = \{x_{i-2}, x_{i-1}, x_i, x_{i+1}, x_{i+2}\}$ with the word representations. A convolution operation involving a filter $w \in \mathbb{R}^{h \times d}$ is applied to the sentiment fragment to produce a new feature. We set the filter window size $h = 3$ in our work. For example, a feature c_t is generated from a window of words $x_{t:t+2}$ by $c_t = f(w \cdot x_{t:t+2} + b)$, where $x_{t:t+2}$ refers to the concatenation of x_t, x_{t+1} and x_{t+2}. Here b is a bias term and f is a non-linear function such as the hyperbolic tangent. This filter is applied to each possible window of words in the sentiment fragment to produce a feature map $c = [c_{t-2}, c_{t-1}, c_t]$, with $c \in \mathbb{R}^3$.

We apply a max pooling operation over the feature map and take the maximum value $\hat{c} = \max\{c\}$ as the feature corresponding to this filter. 200 filters are used in our work to obtain 200 features. These features are fed to a multilayer perceptron to produce the fragment sentiment representations, which are passed to a logistic regression to predict the sentiment polarities.

After obtaining the fragment sentiment representations from the CNN model for each sentence, we average them to produce the SLSR.

We also introduce LSTM to learn fragment sentiment representations to compare with CNN. The same sentiment fragment used in CNN is fed into LSTM to learn the SLSR as well.

LSTM is designed to cope with the gradients vanishing or exploding problem of RNN [18]. It introduces a gating mechanism, which comprises four components: an input gate i_t, a forget gate f_t, an output gate o_t, and a memory cell c_t. For the standard LSTM, each of the three gates receives the information from the inputs at current time step and the outputs at previous time step. In our work, we adopt an LSTM variant, which adds the "peephole connections" to the architecture [19] to let the memory cell c_{t-1} directly control the gates as follows:

$$i_t = \sigma(W^{(i)}x_t + U^{(i)}h_{t-1} + V^{(i)}c_{t-1} + b^{(i)}) \tag{1}$$

$$f_t = \sigma(W^{(f)}x_t + U^{(f)}h_{t-1} + V^{(f)}c_{t-1} + b^{(f)}) \tag{2}$$

$$o_t = \sigma(W^{(o)}x_t + U^{(o)}h_{t-1} + V^{(o)}c_t + b^{(o)}) \tag{3}$$

$$c_t = f_t \odot c_{t-1} + i_t \odot \tanh(W^{(c)}x_t + U^{(c)}h_{t-1} + b^{(c)}) \tag{4}$$

$$h_t = o_t \odot \tanh(c_t) \tag{5}$$

where W, U and V are the transition matrices for the input x_t, the hidden state vector h_{t-1}, and the memory cell c_{t-1}, respectively, b is a bias term of the hidden state vector, σ represents the sigmoid function, and $\odot$ denotes component-wise multiplication. In our work, hidden representations at every time step are fed into a mean pooling layer to obtain the final representations.

3.2 Jointly Learning Bilingual Sentiment and Semantic Representations (BSSR) of Documents

The gated recurrent architecture of LSTM is suitable for encoding sentiment and semantics of words and inherent relations of words into document representations. In this paper, we employ two LSTM for BSSR learning and the framework of jointly learning is shown in Fig. 1.

As shown in Fig. 1, one LSTM is used to produce the English representations of documents, while the other is used for the representations of their Chinese translations. These two LSTM are designed to minimize the difference between the representation pairs for capturing the cross-language semantic information. Meanwhile, the polarity labels of documents are used to guide the learning process for incorporating sentiment information into BSSR.

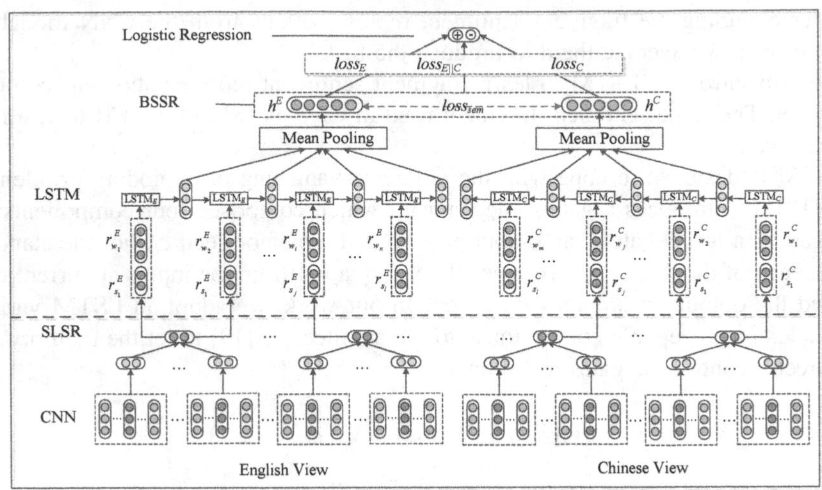

Fig. 1. The framework of jointly learning BSSR with LSTM.

For an English training document d_k^E containing p sentences and n words, it can be represented by a word sequence $\{w_1^E, w_2^E, \ldots, w_i^E, \ldots, w_n^E\}$. Once the jth SLSR $r_{s_j} \in \mathbb{R}^{d_2}$ are attached to its WSR $r_{w_i} \in \mathbb{R}^{d_1}$. The input matrix $d_k^E \in \mathbb{R}^{(d_1+d_2)\times n}$ of the English view can be represented as:

$$d_k^E = \left\{ \begin{bmatrix} r_{w_1} \\ r_{s_1} \end{bmatrix}, \begin{bmatrix} r_{w_2} \\ r_{s_1} \end{bmatrix}, \ldots, \begin{bmatrix} r_{w_i} \\ r_{s_j} \end{bmatrix}, \ldots, \begin{bmatrix} r_{w_n} \\ r_{s_l} \end{bmatrix} \right\} \tag{6}$$

If no SLSR is found during the learning process, we attached a zero vector which dimension is also d_2. The same step repeats for its Chinese translation d_k^C. Thus paired documents (d_k^E, d_k^C) are expressed by the concatenations of WSR and their SLSR, which are fed to LSTM model in English (LSTM_E) and Chinese (LSTM_C) respectively.

As for an English training document d_k^E, LSTM_E recursively computes its representations h_k^E. The same step repeats for its Chinese translation d_k^C to generate document representations h_k^C. The learned document representations h_k^E, h_k^C and the concatenation of them $[h_k^E \,||\, h_k^C]$ are used to compute the polarity probabilities $p_E(s \,|\, d_k; \xi^E)$, $p_C(s \,|\, d_k; \xi^C)$ and $p_{E||C}(s \,|\, d_k; \xi^{E||C})$ through a logistic layer, where $s = \{0, 1\}$ is the sentiment polarity of a document, and ξ^E, ξ^C, and $\xi^{E||C}$ are the hyper-parameters of the models.

The sum of three errors predicted by h_k^E ($loss_E$), h_k^C ($loss_C$), and $[h_k^E \,||\, h_k^C]$ ($loss_{E||C}$) are used as the sentiment loss function to adjust the sentiment information in BSSR learning:

$$loss_{sen} = loss_E + loss_C + loss_{E||C} \tag{7}$$

Meanwhile, we also define a semantic loss function to minimize the differences between the representation pairs to adjust the cross-language semantic information:

$$loss_{sem} = \left\|h_k^E - h_k^C\right\|^2 \tag{8}$$

To investigate the relations between sentiment and cross-language semantic information, we learn BSSR by minimizing the loss function:

$$loss = \alpha \cdot loss_{sen} + (1 - \alpha) \cdot loss_{sem} \tag{9}$$

where α is the hyper-parameter which controls the weight of sentiment information.

4 Experiment

4.1 Experimental Settings

Dataset. The proposed approach is evaluated on NLP&CC 2013 CLSC dataset[1]. The dataset consists of product reviews in three domains: Book, DVD, and Music. Each domain contains 4,000 English labelled documents as training data and 4,000 Chinese unlabeled documents as test data.

Sentiment Words. Based on the Chi-square method, 2245 English sentiment words in MPQA subjectivity lexicon [20] and 3010 Chinese sentiment words in affective lexicon ontology [21] are used for SLSR learning. Note that if there are any negative word in the sentiment segment, the polarity of sentiment word is inversed. Here, we choose the negation words as follows: no, never, none, few, little, hardly and seldom. The Chinese negation words are listed as follows: 不 (no), 没有 (none), 很少 (few), 几乎不 (seldom), 从未 (never).

Tools. In our experiments, Google Translate[2], is adopted for both English-to-Chinese and Chinese-to-English translation. All corpora are segmented by Stanford Word Segmenter[3]. CNN and LSTM are developed based on Theano [22]. The SLSR are learned for 50 epochs, and BSSR are learned for 20 epochs. To make a comparison between SLSR and WSR, we use 200-dimensional WSR to produce 200-dimensional SLSR. The details of the SLSR learning are shown in Sect. 3.1. As for the BSSR, the model has 30-dimensional SLSR and 200-dimensional WSR. Besides, publicly available Global Vectors [23] (GloVe)[4] are adopted as the pre-trained WSR in English.

---

[1] http://tcci.ccf.org.cn/conference/2013/dldoc/evsam03.zip.

[2] http://translate.google.cn/.

[3] http://nlp.stanford.edu/software/segmenter.shtml.

[4] http://nlp.stanford.edu/projects/glove/.

We pre-train the Chinese WSR on a 27 MB-sized unlabeled dataset (also included the NLP&CC 2013 CLSC training set) with GloVe toolkit.

Evaluation Metric. The performance is evaluated by the correct classification accuracy of each domain, and the average accuracy of three domains, respectively. The domain accuracy is defined as:

$$Accuracy_n = \frac{\#system\_correct}{\#system\_total} \quad (10)$$

where n is one of the three domains, $\#system\_correct$ and $\#system\_total$ stand for the number of being correctly classified reviews and the number of total reviews in the domain n, respectively.

The average accuracy is shown as:

$$Average = \frac{1}{3}\sum_n Accuracy_n \quad (11)$$

4.2 Evaluations on BSSR

We compare our model with several different representations learning methods in Table 1. The following methods are used for comparison.

Table 1. The comparison of different representations

Methods	Book	DVD	Music	Average
CHN	74.20	74.02	73.50	73.91
WSR	77.45	77.45	77.00	77.30
SLSR	74.00	75.85	66.38	72.08
LSTM-BSSR	79.88	79.38	77.58	79.40
CNN-BSSR	81.78	83.00	80.68	**81.82**
CNN-SegBSSR	78.92	78.67	77.30	78.30

CHN: The labeled English reviews are translated to Chinese. Then training and testing are performed only in Chinese view with LSTM.

WSR: Only the 200-dimensional WSR is fed into the two LSTM models to classify the document.

SLSR: Only the 200-dimensional SLSR is fed into the multilayer perceptron model to classify the document. We use the SLSR learned by the CNN model.

CNN-BSSR: SLSR learned by the CNN are used in the BSSR learning. The hyper-parameter α is set to 0.5.

LSTM-BSSR: SLSR learned by the LSTM are used in the BSSR learning. The hyper-parameter α is set to 0.5.

CNN-SegBSSR: To evaluate our model, we only concatenate the SLSR with the sentiment words and its context words in the window of $[-2, 2]$. We use the SLSR learned by the CNN model. As for the other words out of the window, we concatenate a zero vector $\mathbf{0}$ instead of the SLSR.

From Table 1, we can see that:

- The bilingual representations (**WSR**, **CNN-BSSR** and **LSTM-BSSR**) outperform the sole Chinese representations (**CHN**).
- **SLSR** gets a lower accuracy than others. The reason is that only the coarse-grained sentence-level sentiment representations could not capture polarity of the document. WSR is still needed.
- Both **CNN-BSSR** and **LSTM-BSSR** outperform **WSR** in all the three domains. The average accuracy of **CNN-BSSR** reaches 81.82%, which is 4.52% higher than that of **WSR**. These indicates that **SLSR** and **WSR** are both effective for sentiment expression and their concatenation could improve the performance further.
- **CNN-BSSR** outperforms **LSTM-BSSR**, which indicates that CNN is superior in sentence-level sentiment representation.
- **CNN-SegBSSR** cannot catch up with **CNN-BSSR**, which demonstrates that SLSR is important for each word expression in sentences.

4.3 Influences of Sentiment and Cross-Language Semantic

We investigate the influences of sentiment and cross-language semantics for BSSR learning in Fig. 2. We use the CNN-BSSR with weighting parameter α varying from 0 to 1. From Fig. 2, we can see that:

- The three curves start from the initial accuracies of the sole cross-language semantic information, and then increase to their individual highest accuracy, finally fall to the individual accuracy of the sole sentiment information. These results indicate that both the sentiment and semantic information are effective in CLSC task and the combination of the two types of information outperforms either one of them obviously.

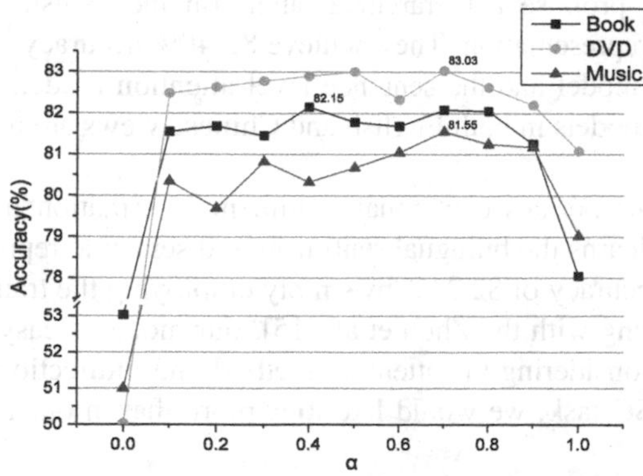

Fig. 2. Relations between sentiment and semantics with different α.

- When $\alpha = 0$, the model only has the cross-language semantic loss and produces poor results. When $\alpha = 1$, the model only has the sentiment loss and gets relatively better results than $\alpha = 0$.
- With the weighting parameter varying from 0 to 1, CLSC performances change in the Book, DVD, and Music domains. The best performances are obtained when the weighting parameter α is 0.4 (82.15% accuracy), 0.7 (83.03% accuracy), and 0.7 (81.55% accuracy) in Book, DVD and Music domains respectively. It shows that the effects of sentiment and cross-language semantics are different in various domains.

4.4 Comparison with Related Work

Table 2 shows the comparison results of our approach with some state-of-the-art systems on NLP&CC 2013 CLSC dataset.

Table 2. Cross-language sentiment classification accuracy of different methods.

	Book	DVD	Music	Average
Gui et al. [24]	80.50	82.20	79.70	80.80
Zhou et al. [11]	81.05	81.60	79.40	80.68
Zhou et al. [15]	82.10	**83.70**	81.30	82.40
Ours	**82.15**	83.03	**81.55**	82.24

Gui et al. [24] propose a transfer detection approach to learn sentiment classifiers with removing the noise from the transferred samples to avoid negative transfers (NTD) and achieved an 80.80% average accuracy. Such two-view approach is difficult to capture the common sentiment and semantics between two languages.

Zhou et al. [11] propose the bilingual sentiment word embedding algorithm based on denoising autoencoders. It learns common representations of two views by a semantic learning phase and a sentiment learning phase and achieves 80.68% accuracy. The two-phase approach is hard to adjust the relations between sentiment and semantics for different application domains.

Zhou et al. [15] propose a hierarchical attention mechanism with bidirectional LSTM for bilingual representation. They achieve 82.40% accuracy benefiting from the word-level attention model and the sentence-level attention model. However, training bidirectional LSTM models in both English and Chinese views are time consuming and high cost.

Our approach incorporates contextual sentiment information into word semantic representations, and learns the bilingual sentiment and semantic representations jointly. We achieve a high accuracy of 82.24% by simply employing the training data and their translations. Comparing with the Zhou et al. [15], our model is easy to train and has a comparable result. Considering the attention method and bidirectional architecture are effective for the CLSC task, we would like to explore that in our future work.

5 Conclusion

This paper proposes an approach to jointly learning bilingual sentiment and semantic representations in a unified framework. The learned representations are employed for calculating the sentiment polarities of documents. In the learning process, the predicted errors are used as the sentiment loss to capture the sentiment information. Meanwhile, the differences between the English and Chinese representation pairs are used as cross-language semantic loss to capture the cross-language semantic information. The experimental results demonstrate that sentiment and cross-language semantic information are both effective for CLSC, and the performance of CLSC in different domains could be optimized by adjusting sentiment and cross-language semantic loss. Attention-based methods show effectiveness in the sentiment classification task, and we leave it as future work to further boost the performance of CLSC.

Acknowledgements. This research is supported by Natural Science Foundation of China (No. 61272375).

References

1. Turney, P.D.: Thumbs up or thumbs down? Semantic orientation applied to unsupervised classification of reviews. In: Proceedings of ACL, USA, pp. 417–424 (2002)
2. Tang, D.Y., Wei, F.R., Yang, N., Zhou, M., Liu, T., Qin, B.: Learning sentiment-specific word embedding for twitter sentiment classification. In: Proceeding of ACL, USA, pp. 1555–1565 (2014)
3. Tang, D.Y., Qin, B., Feng, X.C., Liu, T.: Effective LSTMs for target-dependent sentiment classification. In: Proceeding of COLING, Japan, pp. 3298–3307 (2016)
4. Li, S., Wang, R., Liu, H., Huang, C.R.: Active learning for cross-lingual sentiment classification. In: Zhou, G., Li, J., Zhao, D., Feng, Y. (eds.) NLPCC 2013. Communications in Computer and Information Science, vol. 400, pp. 236–246. Springer, Heidelberg (2013). doi:10.1007/978-3-642-41644-6_22
5. Wan, X.J.: Co-training for cross-lingual sentiment classification. In: Proceedings of ACL, Singapore, pp. 235–243 (2009)
6. Chen, Q., Li, W.J., Lei, Y., Liu, X.L., He, Y.X.: Learning to adapt credible knowledge in cross-lingual sentiment analysis. In: Proceedings of ACL, China, pp. 419–429 (2015)
7. Chander, S., Lauly, S., Larochelle, H., et al.: An autoencoder approach to learning bilingual word representations. In: Proceeding of NIPS, Canada, pp. 1853–1861 (2014)
8. Yoshikawa, Y.Y., Iwata, T., Sawada, H., Yamada, T.: Cross-domain matching for bag-of-words data via kernel embeddings of latent distributions. In: Proceedings of NIPS, Canada, pp. 1405–1413 (2015)
9. Rajendran, J., Khapra, M., Chandar, S., Ravindran, B.: Bridge correlational neural networks for multilingual multimodal representation learning (2015). arXiv preprint arXiv:1510.03519
10. Tang, X.W., Wan, X.J.: Learning bilingual embedding model for cross-language sentiment classification. In: the 2014 IEEE/WIC/ACM International Joint Conferences on Web Intelligence and Intelligent Agent Technologies, pp. 134–141 (2014)
11. Zhou, H.W., Chen, L., Shi, F.L., Huang, D.G.: Learning bilingual sentiment word embeddings for cross-language sentiment classification. In: Proceedings of ACL, China, pp. 430–440 (2015)

12. Klementiev, A., Ivan, T., Bhattarai, B.: Inducing cross lingual distributed representations of words. In: Proceedings of COLING, Indiana, pp. 1459–1474 (2012)
13. Jain, S., Batra, S.: Cross-lingual sentiment analysis using modified BRAE. In: Proceedings of EMNLP, Portugal, pp. 159–168 (2015)
14. Zhou, X.J., Wan, X.J., Xiao, J.G.: Cross-lingual sentiment classification with bilingual document representation learning. In: Proceedings of ACL, Germany, pp. 1403–1412 (2016)
15. Zhou, X.J, Wan, X.J and Xiao, J.G.: Attention-based LSTM network for cross-lingual sentiment classification. In: Proceedings of EMNLP, USA, pp. 247–256 (2016)
16. Collobert, R., Weston, J., Bottou, L., et al.: Natural language processing (almost) from scratch. J. Mach. Learn. Res. **12**, 2493–2537 (2011)
17. Wang, X., Liu, Y.C., Sun, C.J., Wang, B.X., Wang, X.L.: Predicting polarities of tweets by composing word embeddings with long short-term memory. In: Proceedings of ACL, China, pp. 1343–1353 (2015)
18. Hochreiter, S., Schmidhuber, J.: Long short-term memory. Neural Comput. **9**(8), 1735–1780 (1997)
19. Gers, F.A., Schmidhuber, J.: Recurrent nets that time and count. In: The IEEE-Inns-Enns International Joint Conference on Neural Networks, vol. 3, pp. 189–194 (2000)
20. Wilson, T., Wiebe, J., Hoffmann, P.: Recognizing contextual polarity in phrase-level sentiment analysis. Int. J. Comput. Appl. **7**(5), 347–354 (2005)
21. Xu, L.H., Lin, H.F., Pan, Y., Ren, H., Chen, J.M.: Constructing the affective lexicon ontology. J. China Soc. Sci. Tech. Inf. **27**(2), 180–185 (2008)
22. Bastien, F., Lamblin, P., Pascanu, R., et al.: Theano: new features and speed improvements (2012). arXiv preprint arXiv:1211.5590
23. Pennington, B., Socher, R., Manning, C.D.: Glove: global vectors for word representation. In: Proceedings of EMNLP, USA, pp. 1532–1543 (2014)
24. Gui, L., Lu, Q., Xu, R., Wei, Q., Cao, Y.: Improving transfer learning in cross lingual opinion analysis through negative transfer detection. In: Zhang, S., Wirsing, M., Zhang, Z. (eds.) KSEM 2015. LNCS, vol. 9403, pp. 394–406. Springer, Cham (2015). doi:10.1007/978-3-319-25159-2_36

Stacked Learning for Implicit Discourse Relation Recognition

Yang Xu, Huibin Ruan, and Yu Hong(✉)

Natural Language Processing Lab, School of Computer Science and Technology, Soochow University, Suzhou 215006, China
andreaxu41@gmail.com, huibinnguyen@gmail.com, tianxianer@gmail.com

Abstract. The existing discourse relation recognition systems have distinctive advantages, such as superior classification models, reliable feature selection, or holding rich training data. This shows the feasibility of making the systems collaborate with each other within a uniform framework. In this paper, we propose a stacked learning based collaborative approach. By the two-level learning, it facilitates the application of the confidence of different systems for the discourse relation determination. Experiments on PDTB show that our method yields promising improvement.

1 Introduction

Discourse relation recognition aims to automatically classify the discourse relations between two adjacent arguments (abbr., Arg). In the Penn Discourse Treebank 2.0 corpus (PDTB v2.0) [16], discourse relation falls into explicit and implicit cases. See the examples as below:

(1) *Shorter maturities are considered a sign of rising rates* (Arg_1). **[Because]** *portfolio managers can capture higher rates sooner* (Arg_2).

(2) *The woman has "psychic burns" on her back from the confrontation* (Arg_1) [?] *She declines to show them* (Arg_2).

Example (1) shows an explicit *Causality* relation signaled directly by the discourse connective "*Because*". Example (2) shows an implicit *Comparison* relation. In the example, there isn't any explicit connective between the arguments, though we can imagine a possible connective as "*but*". In this paper, we focus on studying on the implicit discourse relation recognition.

Great effort has been put into the exploration of effective linguistic and structural features for relation recognition, such as polarity, verbs, inquirer tags, brown cluster pairs, word pairs, etc. [10,13–15,20,24]. Meanwhile, the validity of different classification models has been evaluated [6]. Hong et al. [4] and Wang et al. [20] mine high-quality comparable samples to enrich the reference data for estimating the relations. Li and Nenkova [8] propose a novel feature representation method to fulfill multidimensional aggregate of sparse data. Ji and Eisenstein et al. [5] propose an solution to recognize implicit discourse relation

J. Wen et al. (Eds.): CCIR 2017, LNCS 10390, pp. 161–169, 2017.
https://doi.org/10.1007/978-3-319-68699-8_13

through surface feature and distributed representations of discourse arguments and entity mentions. Recently, Zhang et al. [23] present a Shallow Convolutional Neural Network (SCNN) based approach to learn better relation classifiers and Chen et al. [3] take advantage of word embeddings to solve the problem of data sparsity and present a deep architecture to recognize discourse relation. Liu et al. [11] introduce an strategy of repeated reading by a bidirectional LSTM with multi-level attention to generalize which words in the arguments are really useful.

It has been proved that the existing implicit relation recognition systems actually have distinctive advantages. For example, some are equipped with representative features [9,10,15,24], while others employ sophisticated classification models [6].

With the aim to combine the advantages, we propose a stacked learning based collaborative approach [12,18]. The approach involves two-level learning processes, base-level and meta-level.

In the base-level, it evaluates different well-trained classification systems, so as to grasp their suitability for different types of samples (a sample refers to a pair of arguments). The suitability indicates the ability of a system to correctly recognize the relations of a specific type of samples. It is measured by the confidence of the classifiers.

In the meta-level, we regard the suitability of the base-level classifiers as novel features, and use them(optimized in the middle-level) to equip a superior classifier. This classifier, therefore, inherits the suitability of all the base-level ones when trained on the same training data. Accordingly, for every sample in the test data, the meta-level classifier can reinforce the positive effect of the most suitable base-level classifiers on the relation recognition.

2 Stacked Learning for Cooperation

We combine the advantages of the relation classifiers by two-level stacked learning. The learning process help the meta-level classifier inherit the suitability of different base-level classifiers.

2.1 Base-Level: Suitability Measurement

The main task of the base-level stacked learning is to evaluate the suitability of the basic classifiers. In our discussion, we define the suitability as the degree in which a classifier are appropriate for dealing with a type of relation samples. For example, a well-equipped classifier with the syntactic features is ideally suited for relation recognition for those adjacent arguments in long sentences, but not those in short sentences or nonadjacent arguments. It is because that such a classifier can apply rich knowledge of syntactic structure for discourse-structural relation analysis in the former case. By contrast, there is less available syntactic information in the latter case.

We use the confidence of classifiers as the measure of their suitability. Given a classifier F and a test sample s, the suitability of F for s is expressed as $\{C_p, C_n\}_{F\_s}$, where C denotes the confidence of F for its decision on the relation of s. If F accurately determines the relation, the confidence will be positive, i.e., C_p, else negative, i.e., C_n. The sum of C_p and C_n is 1.

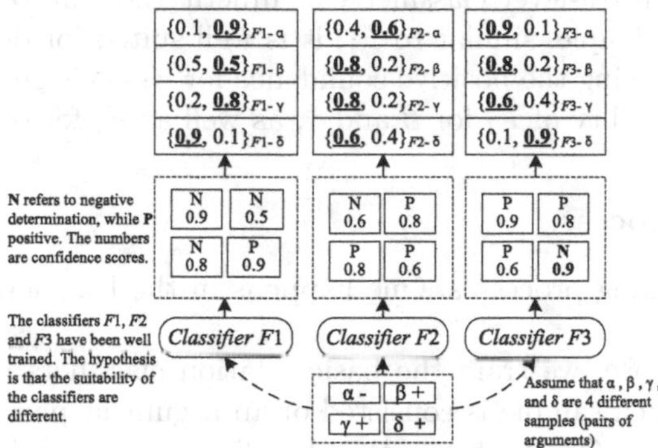

Fig. 1. Examples of suitability measurement.

For example, in Fig. 1, the confidence of the classifier F_1 for the relation determination on the sample γ is 0.9. However it is a negative determination (**N**). Thus C_n is 0.9 but C_p is 0.1, resulting in a suitability $\{0.1, 0.9\}_{F_1\_\gamma}$. This means that F_1 is quite likely to make mistakes when dealing with the samples similar to γ. In other words, F_1 is unsuitable for those samples.

It is worth noting that the confidence score is an output of the basic classifiers, attached to the determination results on discourse relation. We employ the score directly as the measure of suitability. The assertion of the positive (**P**) or negative (**N**) confidence is achieved by comparing the determination results with the ground-truth.

2.2 Meta-level: Suitability Inheritance

In Meta-level, we employ the suitability of the base-level classifiers as features to train a new classifier. We combine the suitability in a uniform confidence-based feature space. See the space built by the confidence of $F_{1\text{-}3}$ in Fig. 1 below:

$$\{R|C_p, F_1, C_n, F_1, C_p, F_2, C_n, F_2, C_p, F_3, C_n, F_3\}$$

where R denotes discourse relation type, such as the ones discussed in the paper, Expansion, Contingency, Comparison and Temporality. Accordingly, the training data represented by the suitability of F_1, F_2 and F_3 for the samples α, β, γ and δ (Fig. 1) are specified as:

$$\{R_\alpha|0.1, 0.9, 0.4, 0.6, \mathbf{0.9}, \mathbf{0.1}\}$$
$$\{R_\beta|0.5, 0.5, \mathbf{0.8}, \mathbf{0.2}, \mathbf{0.8}, \mathbf{0.2}\}$$
$$\{R_\gamma|0.2, 0.8, \mathbf{0.8}, \mathbf{0.2}, 0.6, 0.4\}$$
$$\{R_\delta|\mathbf{0.9}, \mathbf{0.1}, 0.4, 0.6, 0.1, 0.9\}$$

A well-trained meta-level classifier can inherit the suitability of the basic classifiers. As an example, similar to F_1, it is well suited for dealing with those samples like δ, achieving the positive confidence for its decision on R_δ. Similarly, it inherits the suitability of F_2 for β and γ, as well as F_3 for α and γ.

2.3 Training Process

There are two training processes. One happens in the base level, the other the meta level.

In the former case, we train the basic relation classifiers on the PDTB-v2 corpus, in which each sample is consisted of an argument pair and the ground-truth relation senses. On the basis, the classifiers conduct relation recognition on small-scale test data. The confidence of the classifiers for each test sample will be delivered to the meta-level, along with the ground-truth relation senses of the samples. They will be used as new training data to support the suitability learning of the meta-level classifier.

The scale of the training data in the meta-level, therefore, is equivalent to that of the test data in the base-level, generally much smaller than the base-level training data. In order to enrich the meta-level training data, we follow the stacked learning to conduct n-fold cross-validation based iterative training.

In the training process, we divide the base-level training data into n subsets, in which we employ $n - 1$ for training the base-level classifiers, 1 for test. By reassigning the test data for n times and n-fold cross-validation, we can obtain all

```
Input: Training set D = {(xi, yi)}, i = 1…N;
       Relation labels R1…k;
       Base-level training algorithms L1…m;
       Meta-level training algorithm LA.
Output: Meta-level classifier CD
Initialize:
   D1 … Dn ← Split training data D into n partitions.
   D~ ← ∅        //Meta-level training data
Process:
   for d ← 1…m do //Base-level training algorithms
       for t ← 1…n do //Cross validation
           Cdt ← TRAIN (Ld, D - Dt)
           for i ← 1… | Dt | do //The current test data Dt
               pti ← ∅
               for j ← 1… k do
                   ptij ← TEST (Cdt, xi | Rk)
                   pti ← pti ∪ ptij
               D~ ← D~ ∪ pti ∪ yi
   CD ← TRAIN (LA, D~).
```

Fig. 2. Training process in stacked learning.

Table 1. Settings of the relation detection systems.

Classifiers	Settings	Classifiers	Settings
Clas1 (base)	NB (f1–f6)	Clas6 (base)	SVM + (f3)
Clas2 (base)	LR (f1–f6)	Clas7 (base)	SVM + (f4)
Clas3 (base)	SVM (f1–f6)	Clas8 (base)	SVM + (f5)
Clas4 (base)	SVM (f1)	Clas9 (base)	SVM + (f6)
Clas5 (base)	SVM (f2)	Clas10 (meta)	SVM + Confidence

Features: verbs (f1), polarity (f2), inquirer tags (f3), first-last-first3 (f4), brown pairs (f5), semantic vectors (f6)

Table 2. Settings of the relation detection systems.

Systems	Training data	Base-level	Meta-level
Single-NB (baseline)	All	Clas1	N/A
Single-LR (baseline)	All	Clas2	N/A
Single-SVM (baseline)	All	Clas3	N/A
Stacking-Data	All, 1/2, 1/3	Clas3	SVM
Stacking-Algorithm	All	Clas1–3	SVM
Stacking-Feature	All	Clas4–9	SVM

the test results on the whole base-level training data, including the determination on discourse relation and the confidence scores. On the basis, we re-employ every base-level training samples and the confidence scores to build the meta-level training data. See the detailed algorithm in Fig. 2.

In the algorithm flow, p_{ti} provides the temporary storage of the confidence scores. Besides, x denotes an argument pair, while y a relation type.

3 Experiments

3.1 Experimental Settings

Data Setting: We evaluate the relation recognition systems on PDTB v2.0 corpus. We use sections 2–20 as the training data, sections 21–22 as the test, and sections 0–1 as the development for parameter optimization. We consider four discourse relation types, including Expansion (Exp.), Comparison (Comp.), Contingency (Cont.) and Temporality (Temp.).

Features: We employ 6 features to equip the relation classifiers, including verbs [7], polarity [21], inquirer tags [19], first-last-first3 [15], brown clusters pairs [1,2] and semantic vectors [17].

Classification models: We employ the Naive Bayes (NB), Logistic Regression(LR) and Support Vector Machines (SVM) and models to build the relation classifiers. We implement the models by using the toolkit WEKA [22].

In our experiments, we build 9 basic classifiers (Table 1), in which 3 were specified as the baselines (Table 2), including single NB, LR, and SVM based classifiers, i.e., Clas1–3. They are equipped with same features, f1–f6 (Table 1). By the evaluation of these 3 systems, we can confirm the optimal classification model. The model will be used to build the meta-level relation classifier in the stacked learning process. As shown in Table 2, we employ the SVM-based classifier (Clas10) in meta level.

Other 6 basic classifiers, i.e., Clas4–9 (Table 1), are built with the same classification model but equipped with different features. Each employs a distinctive linguistic feature. Clas10 in Table 1 shows the framework of the meta-level classifiers. It is built with a SVM model. And it employs the confidence as feature to learn and inherit the suitability of the base-level classifiers.

We totally built 3 meta-level relation classifiers, including Stacking-Data, Algorithm and Feature (see Table 2). Stacking-Data learns the suitability of 3 base-level classifiers. Similar to Clas3, all the 3 classifiers are built with SVM and equipped with all available features (f1–f6). The only difference is that they were trained on the training data of different sizes, 1/2, 1/3 and all of the available. Stacking-Algorithm considers the suitability only from the perspective of learning algorithm, involving Single-NB, LR and SVM in the two-level learning. Stacking-Feature only concerns the feature-oriented suitability, learning that from Clas4–9.

We evaluate the classifiers by *F-score* (F_1) and accuracy (*Acc*).

3.2 Results and Analysis

The first three rows in Table 3 indicate the performance of the individual classification model based relation recognition systems. The last three rows list the performance of the stacked learning based systems. In the individual cases, SVM is reliable when trained by small-scale training data. We use it to build meta-level classifier.

The stacked learning yields substantial performance gain for the 4 relation types. In particular, the feature-oriented stacking significantly outperforms the other two. It implies that the effect of the linguistic features on the suitability of the classifiers is more substantial than data scale or classification models. It also proves that stacked learning is conductive to the reinforcement of the effect in practice.

We compare our best system with the state of the art, including [6]'s multi-task learning based classifier and [8]'s novel feature representation based classifier. We show the performance in Table 4. It can be found that ours is comparable to [6]'s system. Nevertheless we employ smaller-scale training data, which is nearly 1/5 of that used in [6]'s system. It implies that stacked learning is more efficient than multi-task learning for task-specific linguistic knowledge understanding. Besides, [8]'s system outperforms our baselines which are equipped

Table 3. Performance comparison of stacked learning based methods with traditional learning methods.

Models	$F_1(Acc)$			
	Exp.	Comp.	Cont.	Temp.
Single-NB	.59 (.54)	.25 (.42)	.42 (.37)	.15 (.35)
Single-LR	.69 (.53)	.27 (.54)	.41 (.50)	.18 (.63)
Single-SVM	.70 (.54)	.28 (.45)	.45 (.58)	.19 (.63)
Stacking-Data	.64 (.62)	.29 (.57)	.47 (.65)	.20 (.58)
Stacking-Algorithm	.69 (.53)	.31 (.60)	.48 (.48)	.22 (.64)
Stacking-Feature	.72 (.56)	.32 (.65)	.51 (.63)	.23 (.65)

Table 4. Performance comparison of our best system with the state of the art.

Systems	$F_1(Acc)$			
	Exp.	Comp.	Cont.	Temp.
Our best	.72 (.56)	.32 (.65)	.51 (.63)	.23 (.65)
Pitler et al. (2009)	.76 (.64)	.22 (.63)	.47 (.67)	.17 (.63)
Zhou et al. (2010)	.70 (.55)	.32 (.58)	.47 (.49)	.20 (.55)
Biran et al. (2013)	.76 (.63)	.25 (.63)	.47 (.68)	.20 (.68)
Lan et al. (2013)	.70 (-)	.32 (-)	.48 (-)	.30 (-)
Li et al. (2014)	.57 (-)	.30 (-)	.48 (-)	.24 (-)

with nearly all the well-known important features. It illustrates that [8]'s multi-dimensional aggregate overcomes the problem of lack of feature learning caused by sparse data. By contrast, our best system, i.e., the feature-oriented stacking based classifier, outperforms [8]'s. Therefore it can be concluded that stacked learning helps improve the feature learning effect on sparse data.

Others include the relation classifiers of Pitler et al. [15], Zhou et al. [24] and McKeown et al. [13]. They are similar to Single-NB, LR and SVM, except employing more features. Overall, Single-NB, LR and SVM are slightly weaker than the state of the art. Tables 3 and 4 show their performance. Nevertheless, by algorithm-oriented stacking, their distinctive advantages are inherited and combined in a uniform classifier. This yields a comparable performance to the state of the art. Ording to Tables 3 and 4, our model undoubtedly gains the best result in ERC.

4 Conclusion

In the future, we will study on seed-based relation network building and relationship reasoning methods. In particular, we will develop a more advanced stacked learning method among base-level classifiers, a relation network based reasoning model (middle-level) and a meta-level classifier. By error correction in the middle, it provides reliable training data for meta-level learning, and employs the boosting method to iteratively refine the initial seed-centered relation network.

Acknowledgments. This research is supported by the National Natural Science Foundation of China, No. 61373097, No. 61672368, No. 61672367, No. 61331011. The authors would like to thank the anonymous reviewers for their insightful comments and suggestions. Yu Hong, Professor Associate in Soochow University, is the corresponding author of the paper, whose email address is tianxianer@gmail.com.

References

1. Brown, P.F., Desouza, P.V., Mercer, R.L., Pietra, V.J.D., Lai, J.C.: Class-based n-gram models of natural language. Comput. Linguist. **18**(4), 467–479 (1992)
2. Chang, C.C., Lin, C.J.: Libsvm: a library for support vector machines. ACM Trans. Intell. Syst. Technol. (TIST) **2**(3), 27 (2011)
3. Chen, J., Zhang, Q., Liu, P., Qiu, X., Huang, X.: Implicit discourse relation detection via a deep architecture with gated relevance network. In: Proceedings of the 54th Annual Meeting of the Association for Computational Linguistics (vol. 1 - Long Papers), pp. 1726–1735. Association for Computational Linguistics, Berlin, Germany (2016). http://www.aclweb.org/anthology/p16-1163
4. Hong, Y., Zhou, X., Che, T., Yao, J., Zhu, Q., Zhou, G.: Cross-argument inference for implicit discourse relation recognition. In: Proceedings of the 21st ACM International Conference on Information and Knowledge Management, pp. 295–304. ACM (2012)
5. Ji, Y., Eisenstein, J.: One vector is not enough: entity-augmented distributed semantics for discourse relations. Trans. Assoc. Comput. Linguist. **3**, 329–344 (2015). https://tacl2013.cs.columbia.edu/ojs/index.php/tacl/article/view/536
6. Lan, M., Xu, Y., Niu, Z.Y., et al.: Leveraging synthetic discourse data via multi-task learning for implicit discourse relation recognition. In: ACL, vol. 1, pp. 476–485. Citeseer (2013)
7. Levin, B.: English Verb Classes and Alternations: A Preliminary Investigation. University of Chicago Press, Chicago (1993)
8. Li, J.J., Nenkova, A.: Reducing sparsity improves the recognition of implicit discourse relations. In: Proceedings of the 15th Annual Meeting of the Special Interest Group on Discourse and Dialogue, p. 199. Citeseer (2014)
9. Li, J.J., Nenkova, A.: Reducing sparsity improves the recognition of implicit discourse relations. In: SIGDIAL Conference, pp. 199–207. The Association for Computer Linguistics (2014). http://dblp.uni-trier.de/db/conf/sigdial/sigdial2014.html#LiN14a
10. Lin, Z., Kan, M.Y., Ng, H.T.: Recognizing implicit discourse relations in the penn discourse treebank. In: Proceedings of the Conference on Empirical Methods in Natural Language Processing, pp. 343–351. Association for Computational Linguistics (2009)
11. Liu, Y., Li, S.: Recognizing implicit discourse relations via repeated reading: neural networks with multi-level attention. In: Proceedings of the 2016 Conference on Empirical Methods in Natural Language Processing, pp. 1224–1233. Association for Computational Linguistics, Austin, Texas (2016). https://aclweb.org/anthology/D16-1130

12. Martins, A.F., Das, D., Smith, N.A., Xing, E.P.: Stacking dependency parsers. In: Proceedings of the Conference on Empirical Methods in Natural Language Processing, pp. 157–166. Association for Computational Linguistics (2008)
13. McKeown, K., Biran, O.: Aggregated word pair features for implicit discourse relation disambiguation. In: Proceedings of the 51th Annual Meeting of the Association for Computational Linguistics, pp. 69–73. Association for Computational Linguistics (2013)
14. Park, J., Cardie, C.: Improving implicit discourse relation recognition through feature set optimization. In: Proceedings of the 13th Annual Meeting of the Special Interest Group on Discourse and Dialogue, pp. 108–112. Association for Computational Linguistics (2012)
15. Pitler, E., Louis, A., Nenkova, A.: Automatic sense prediction for implicit discourse relations in text. In: Proceedings of the Joint Conference of the 47th Annual Meeting of the ACL and the 4th International Joint Conference on Natural Language Processing of the AFNLP, vol. 2, pp. 683–691. Association for Computational Linguistics (2009)
16. Prasad, R., Dinesh, N., Lee, A., Miltsakaki, E., Robaldo, L., Joshi, A.K., Webber, B.L.: The penn discourse treebank 2.0. In: LREC. Citeseer (2008)
17. Socher, R., Pennington, J., Huang, E.H., Ng, A.Y., Manning, C.D.: Semi-supervised recursive autoencoders for predicting sentiment distributions. In: Proceedings of the Conference on Empirical Methods in Natural Language Processing, pp. 151–161. Association for Computational Linguistics (2011)
18. Søgaard, A.: Simple semi-supervised training of part-of-speech taggers. In: Proceedings of the 48th Annual Meeting of the Association for Computational Linguistics, pp. 205–208. Association for Computational Linguistics (2010)
19. Stone, P., Dunphy, D.C., Smith, M.S., Ogilvie, D.: The general inquirer: a computer approach to content analysis. J. Regional Sci. **8**(1), 113–116 (1968)
20. Wang, X., Li, S., Li, J., Li, W.: Implicit discourse relation recognition by selecting typical training examples. In: Proceedings of the 24th International Conference on Computational Linguistics: Posters, pp. 2757–2772. Association for Computational Linguistics (2012)
21. Wilson, T., Wiebe, J., Hoffmann, P.: Recognizing contextual polarity in phrase-level sentiment analysis. In: Proceedings of the Conference on Human Language Technology and Empirical Methods in Natural Language Processing, pp. 347–354. Association for Computational Linguistics (2005)
22. Witten, I.H., Frank, E.: Data Mining: Practical Machine Learning Tools and Techniques. Morgan Kaufmann, San Francisco (2005)
23. Zhang, B., Su, J., Xiong, D., Lu, Y., Duan, H., Yao, J.: Shallow convolutional neural network for implicit discourse relation recognition. In: Proceedings of the Conference on Empirical Methods in Natural Language Processing, pp. 2230–2235. Association for Computational Linguistics (2015)
24. Zhou, Z.M., Xu, Y., Niu, Z.Y., Lan, M., Su, J., Tan, C.L.: Predicting discourse connectives for implicit discourse relation recognition. In: Proceedings of the 23rd International Conference on Computational Linguistics: Posters, pp. 1507–1514. Association for Computational Linguistics (2010)

IR and Applications

Network Structural Balance Analysis for Sina Microblog Based on Particle Swarm Optimization Algorithm

Xia Fu(✉), Yajun Du, and Yongtao Ye

School of Computer and Software Engineering,
Xihua University, Chengdu 610039, Sichuan, China
1805256845@qq.com

Abstract. Research on structure balance of networks is of great importance for theoretical research and practical application, and received extensive attention of scholars from diverse fields in recent years. The computation and transformation of structure balance primarily aim at calculating the cost of converting an unbalanced network into a balanced network. In this paper, we proposed an efficient method to study the structure balance of the microblog network. Firstly, we model the structural balance of social network as a mathematical optimization problem. Secondly, we design an energy function incorporate with structure balance theory. Finally, considering the standard particle swarm optimization algorithm can not deal with discrete problem, we redefined the velocity and position updating rules of particles from a discrete perspective to solve the modeled optimization problem. Experiments on real data sets demonstrate our method is efficient.

Keywords: Structural balance · Signed network · Social network · Particle swarm optimization algorithm

1 Introduction

With the rapid development of internet and social medias, social network has changed the way of people live. For instance, more and more people are willing to share their life, express their viewpoints, and build their social relations with the convenience provided by social platforms such as Facebook, Renren and WeChat. The network structure is the foundation of the social network. The analysis of network structure balance will certainly give theoretical supports to other network structure issues. As a result, in this paper we will do research on network structure balance.

Perhaps the most basic theory that is applications to signed social networks but does not appear in the study of unsigned networks is social balance. Social balance theory is an old idea in sociology, which was first proposed by Heider in 1940s [1]. The theory of social balance states that relationships in plus-minus networks tend to follow patterns such as "an enemy of my friend is my enemy"

J. Wen et al. (Eds.): CCIR 2017, LNCS 10390, pp. 173–184, 2017.
https://doi.org/10.1007/978-3-319-68699-8_14

and “an enemy of my enemy is my friend”. Later, a notion called weak balance further generalizes social balance by arguing that in many cases an enemy of ones enemy can indeed act as an enemy [2], they defineded weak structural balance as follows: only triangles with exactly two positive edges are implausible in real networks, and that all other kinds of triangles should be permissible. Though, there have been more that 70 years since Heisers seminal studies, the structure balance still maintains the interest of scholars in diverse fields, such as psychology, sociology, economics, computer science, etc.

Heiders balance theory suggests that a network is balanced if it is complete and can be divided into two subsets. However, most real social networks can not be represented by complete graph. Moreover, Wasserman etc. Suggest that many signed social networks can be divided into more than two clusters [3]. Following these studies, many method [4,5] based on clustering techniques have been devised for realizing structure balance. These traditional methods based on graph parting, spectral clustering and hierarchical clustering can not deal with structure balance problem well. The reason is that these methods need to artificially specify the clusters in advance and some methods can not analysis the community topology of the network. Considering these drawbacks, some researchers focus on Evolutionary algorithms (EAs) [6,7], which inspired by principles from biology, ethology, etc., Qing et al. [8] proposed an approach involves evolutionary multiobjective optimization, the main idea is developing the multiobjective discrete particle swarm optimization framework firstly, then designed a model to select the best solution. Ma et al. [9] proposed a method based on memetic algorithm. Li et al. [10] proposed a method based on particle swarm optimization, but they do not explain the method clear. The results of these optimization methods show the outstanding techniques of optimization algorithms for solving diverse kinds of complex optimization problems in many domains.

In this paper, considering the importance of optimization and the drawbacks of traditional methods, the idea of optimization has been adopted to study the structural balance problem. The main contributions of this paper are summarized as follows:

(1) We design an energy function based on structure balance theory: A sign network can be divided into $k > 2$ clusters such that the nodes are positively linked within each other and negatively linked between the clusters.

(2) We redefined the velocity and position updating rules of particles from a discrete perspective to solve the modeled optimization problem.

The remainder of the paper is organized as follows: Sect. 2 discussed the previous work associated with this paper; in Sect. 3 the novel algorithm is proposed; the experiment results are displayed and analyzed in Sect. 4; Finally, Sect. 5 talked about the conclusion and further work.

2 Related Work

Three broad classes of related work pertinent to this study-i.e., Signed Network, Structure Balance and Particle Swarm Optimization Algorithm are briefly introduced in this section.

2.1 Signed Network

Signed network is a kind of network including edges with the property of positive or negative sign, the positive edge and negative edge represents positive and negative relationship, respectively. It is meaningful to utilize the sign property of edge to analyze, understand and predict the topology structure of complex network, and do research on signed network is important for recommendation, attitudes predicting and clustering problem. We usually use signed graph that is composed of a set of vertices and edges to model the signed network. The vertices represent the objects in the network and the edges denote the relationships between the objects.

Definition 1 (Graph). A graph commonly denoted as G = $\{V, PE, NE\}$, where V is aggregation of vertices, PE and NE is the aggregation of positive and negative edges, respectively. Using adjacency matrix to represent the relationships between objects of graph G:

$$a_{ij} = \begin{cases} w_{ij}, w_{ij} > 0 & if \;\; e_{ij} \in PE, \\ w_{ij}, w_{ij} < 0 & if \;\; e_{ij} \in NE, \\ w_{ij}, w_{ij} = 0 & if \;\; there\ is\ no\ relation\ between\ i\ and\ j. \end{cases} \tag{1}$$

where a_{ij} is the element of adjacency matrix, and w_{ij} is the weight of edge e_{ij}.

2.2 Structure Balance

The basis of Structure balance theory is theories of social psychology, which originated in the mid-20th-century, then formulated in Heiders balance model. The model divided the relationship between people or the people and things into positive and negative, and subsequently cast in graph-theoretic language by Cartwright and Harary [11], structural balance considers the possible ways in which triangles on three individuals can be signed, and assuming that triangles with three positive signs shown in Fig. 1(a) and those with one positive sign shown in Fig. 1(b) are balanced. And triangles with two positive signs shown in Fig. 1(c) and none positive signs shown in Fig. 1(d) are imbalanced. The "+" and "–' in Fig. 1 represents positive relationship and negative relationship, respectively.

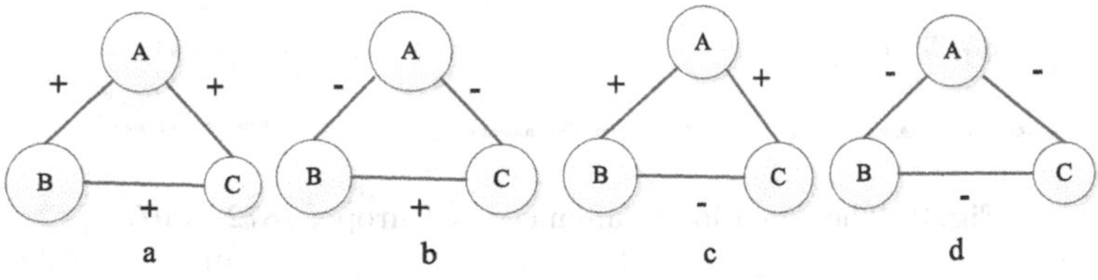

Fig. 1. Relationship patterns of triads in signed network.

Heiders balance theory provides us with two equivalent ways to view structural balance of complete graphs, i.e., the local view and global view. From the local view, a complete network is balanced if each triangles is balanced. From the global view, a complete network is balanced iff the nodes of the network can be divided into two subsets [12] X and Y, and if X and Y meet the relationship described in Fig. 2, then the network is balanced.

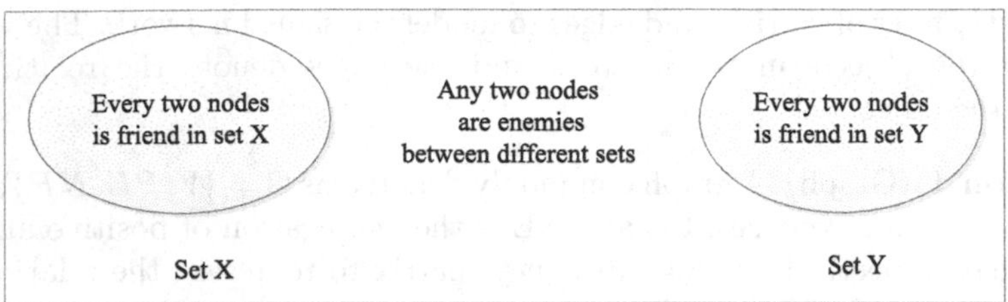

Fig. 2. A global view of balance theory.

Structure balance has become a big research field and can be used in various domains. We use the evolution of alliances in Europe in the period from 1872 to 1907 to prove the importance of structure balance, which is shown in Fig. 3, Where nodes A, B, C, D, E and F represents French, Britain, Russia, Austria-Hungary, Italy and Germany, respectively. Each node can have positive or negative opinion on another node, solid line and dotted line represents alliance and hostile, respectively.

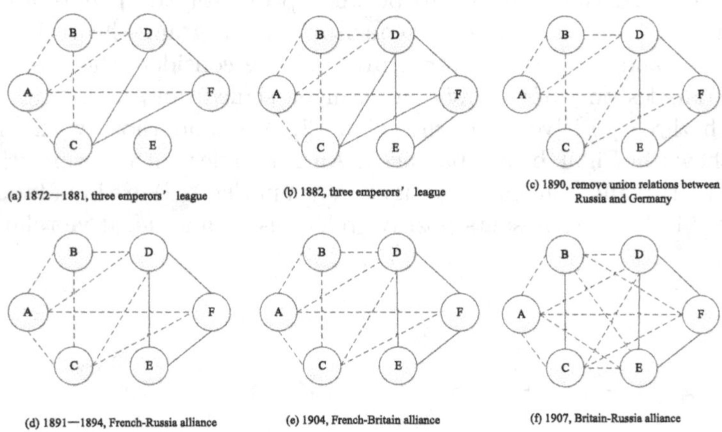

Fig. 3. The evolution of alliances in Europe, 1872–1907.

2.3 Particle Swarm Optimization Algorithm

Particle Swarm Optimization (PSO), also called Particle Swarm Optimization Algorithm, It was first proposed by Eberhart and Kennedy in 1995 [13]. The Concise and novel ideas, easy implementation and good robustness make PSO a popular optimization technique for solving complex optimization problems. The standard global PSO algorithm can be outlined as follows: suppose in a n-dimensional search space a particle swarm is comprised of M particles, the position of i-th particle is expressed as a n-dimensional vector $\boldsymbol{X_i} = (x_i^1, x_i^2, x_i^3, ..., x_i^n)$, the velocity of i-th particle is expressed as a n-dimensional vector $\boldsymbol{V_i} = (v_i^1, v_i^2, v_i^3, ..., v_i^n)$. So far the i-th particles personal best position is $\boldsymbol{Pbest_i} = (pbest_i^1, pbest_i^2, pbest_i^3, ..., pbest_i^n)$ and the global best position of the swarm is $\boldsymbol{Gbest} = (gbest^1, gbest^2, gbest^3, ..., gbest^n)$, then i-th particle adjusts its velocity and position according to the following updating rules

$$v_i^{t+1} = v_i^t + c_1 r_1 (pbest_i^t - x_i^t) + c_2 r_2 (gbest^t - x_i^t) \tag{2}$$

$$x_i^{t+1} = x_i^t + v_i^{t+1} \tag{3}$$

where c_1 and c_2 are acceleration coefficients termed as cognitive and social components, and r_1 and r_2 are two random numbers within the range from 0 up to 1.

3 Optimization Algorithm

3.1 Algorithm Framework

In this paper we propose a discrete PSO for Microblog (named MDPSO) by minimizing an energy function to locate the unbalanced edges that exist in the network, and then change their signs to obtain a structurally balanced network. A flow chart of the proposed algorithm, denoted as MDPSO is shown in Fig. 4. The proposed MDPSO algorithm is a process of constant iterative. The pseudo code of the MDPSO algorithm is given in Algorithm 1.

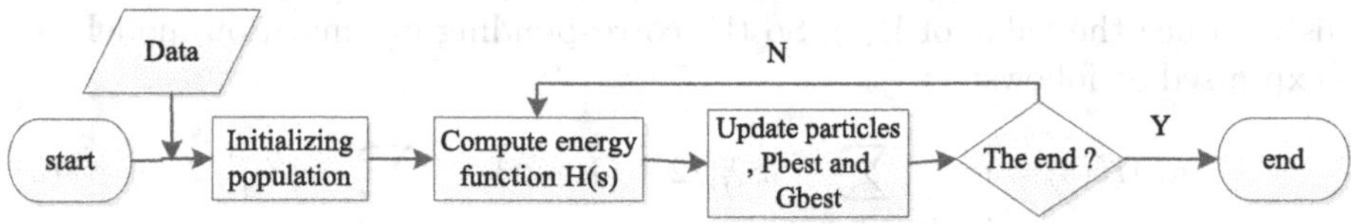

Fig. 4. The flow chart of the proposed algorithm MDPSO.

Algorithm 1. The Framework of the algorithm

Input: The adjacency matrix A of the Microblog social network G.
Output: an optimal solution S corresponding to a partition of the network G, the balanced network G.

```
Parameters : The max and min inertia weight: ωmax andωmin ; the size of population: N; the learning factor:Cp ,Cg ; max iterations: T
initialization :
Adopt the initialization described in Sect. 3.3 to initialize position of particles.
for each node do do
    for i in T do do
        Adopt the function H(s) to evaluate the degree of balance
        Utilize the updating rules described in Sect. 3.4 to update particles
    end for
    chose the solution that has minimum H
end for
```

3.2 Energy Function

Considering the weak definition of the structure balance of signed networks, we rewrite the energy function H(s) [14] previously proposed, the function is defined as follows:

$$\begin{aligned} H(s) &= -\sum_{(i,j),x_i \neq x_j} (S_{ij}^{+} x_i \cdot x_j)/2 - \sum_{(i,j),x_i = x_j} (S_{ij}^{-} x_i \cdot x_j)/2 \\ &= \sum_{(i,j),x_i \neq x_j} |S_{ij}^{+}|/2 + \sum_{(i,j),x_i = x_j} |S_{ij}^{-}|/2 \end{aligned} \tag{4}$$

where S_{ij}^{+}, S_{ij}^{-} denotes positive and negative links between node i and j, respectively; $\sum_{(i,j),x_i \neq x_j} |S_{ij}^{+}|/2$ and $\sum_{(i,j),x_i \neq x_j} |S_{ij}^{-}|/2$ represent the number of the negative links within the same cluster and the positive links across different cluster, respectively.

According to the meaning of H, we can notice that, the value of H(s) equals the number of imbalanced edges of a network. So, with the decrease of H(s), a balanced network will be achieved. Therefor, our optimization goal is to continuously reduce the value of H(s). So the corresponding optimization model can be expressed as follows:

$$minH(s) = \alpha \cdot \sum_{(i,j),x_i \neq x_j} |S_{ij}^{+}|/2 + (1-\alpha) \cdot \sum_{(i,j),x_i \neq x_j} |S_{ij}^{-}|/2 \tag{5}$$

where cost coefficient parameter α is a number between 0 and 1. If $0 \leq \alpha < 0.5$, then the cost of changing the negative link is greater than changing the positive link; If $\alpha = 0.5$, we believe that the cost of both cases is the same; Otherwise, positive links affect more.

3.3 Initialization

At the beginning of our algorithm, we need to initialization the position vectors and velocity vectors of the population. The position vector of the particle i can be defined as $\boldsymbol{X_i} = (x_i^1, x_i^2, x_i^3, ..., x_i^n)$, where $x_i^j \in [1, n]$ is an integer, and each integer represents a classification corresponding to the position of the node. If $x_i^m = x_i^n$, the nodes m and n belong to the same cluster. In our initialization step, the velocity vectors are all initialized as zero vector. Algorithm 2 describes the position initialization of network.

Algorithm 2. Initialization of position

Input: The size of population N.
Output: solutions $\boldsymbol{X} = \boldsymbol{X_1}, \boldsymbol{X_2}, \boldsymbol{X_3}, ..., \boldsymbol{X_N}$.

```
1: Parameters :1 ≤ i ≤ N, 1 ≤ j ≤ N
2: Generate N solutions X
3: for each solution X_i of X do do
4:     Generate a random sequence(e.g.r_1, r_2, ..., r_N).
5:     for each chosen node v_{r_j} of G do do
6:         x_{iu} ← x_{ir_j}, ∀u ∈ {u|Sur_j = 1}
7:     end for
8: end for
```

In line 6, u is the neighborhood of node v_{r_j}, that means the node u has a positive link with node v_{r_j}. In the end, the algorithm will generate a partition of the network. The graphical expound of initialization algorithm is shown in Fig. 5.

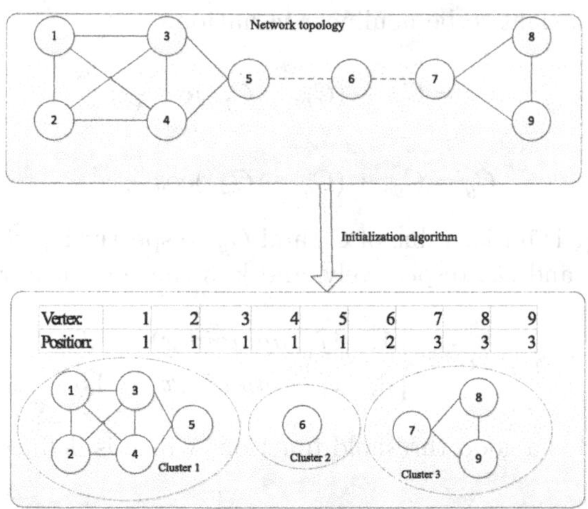

Fig. 5. The graphical expound of initialization algorithm.

3.4 Particle Status Updating Rules

According to the meaning of Algorithm 2, we know that each solution represents a partition of the network, and the position of the particle is an integer. Compared with traditional PSO algorithm, the position and velocity in our approach are both integer, so the mathematical updating rules in continuous PSO no longer suit the discrete situation. Therefore, the updating rules, given by Eqs. (2) and (3) are modified to fit discrete form:

$$v_i^{t+1} = Sign(\omega_k v_i^t + C_p R_1(P_i^t \oplus x_i^t) + C_g R_2(G^t \oplus x_i^t)) \tag{6}$$

$$x_i^{t+1} = x_i^t \otimes v_i^{t+1} \tag{7}$$

where ω_k is inertial weight, which is used to balance the global and local search ability. The calculating function of ω_k is expressed in Eq. (8). The values of C_p and C_g determine the effect of particles self cognition and social cognition on particle trajectories, and reflect the degree of information exchange between particles. The formula of C_p and C_g have been redefined in Eqs. (9) and (10). The operation $\oplus$ is defined as a XOR operator and the function Sign(x) defined in Eq. (11) returns an integer value, which indicates the positive and negative situation of parameters. The operation $\otimes$ is the core of particle updating procedure in Eq. (13), which directly affects the performance of the algorithm.

$$\omega_k = \omega_{min} + (\omega_{max} - \omega_{min})\sqrt{\frac{N-k}{N}} \tag{8}$$

where ω_{min} and ω is min and max inertia weight respectively, N and k is max and current iterations. Although the linear decreasing weight can balance the global search ability and the local search ability some extent, with the liner decreasing of iteration, the latter part of the local search ability will be worse. Therefore, we set the inertia weight to be nonlinear variation.

$$C_p = C_{ps} - (C_{ps} - C_{pe})cos\frac{k}{N} \tag{9}$$

$$C_g = C_{gs} + (C_{ge} - C_{gs})cos\frac{k}{N} \tag{10}$$

where C_{ps} and C_{gs} is initial value of C_p and C_g, respectively, while C_{pe} and C_{ge} is end value of C_p and C_g, respectively, and k is the current iterations.

$$Sign(x) = \begin{cases} 1, & if \;\; sigmoid(x) \geq 0.5, \\ 0, & if \;\; sigmoid(x) < 0.5. \end{cases} \tag{11}$$

where the sigmoid is a good threshold function, which is defined in Eq. (12)

$$sigmoid(x) = \frac{1}{1+e^{-x}} \tag{12}$$

Position $\otimes$ velocity will generate a new position which corresponds to a new solution to the optimization problem. For a network, the probability of two

Algorithm 3. Calculation of $Best(x_i^t)$

```
Input: the neighbors community label set N of node i.
Output: the community label maxlabel of node i .
Parameters :intmaxnumber = 0, intmaxlabel = 0
for i in N do do
    int number=0.
    for  k in N do do
        if i=k then
            number++
        end if
    end for
    if maxnumber < number then
        maxnumber=number
        maxlabel=i
    end if
end for
```

connected nodes belonging to a community is more than two unconnected nodes. Therefor we define the new position as follows:

$$x_i^{t+1} = x_i^t \otimes v_i^{t+1} = \begin{cases} x_i^t, & if \;\; v_i^t = 0, \\ Best(x_i^t), & if \;\; v_i^t = 1. \end{cases} \tag{13}$$

where $Best(x_i^t)$ represents the community label of the node i, which is selected from the neighbors of the node i. In reality it is more possible for one person to join the community that most of its friends join. Suppose node i has a neighborhood set $F = \{f_1, f_2, ..., f_n\}$, the definition of neighborhood is in Definition 2.

Definition 2 (Neighborhood). The neighborhood of node i is a set of nodes that have a positive link with it. So the $Best(x_i^t)$ is calculated in Algorithm 3.

4 Experiment

In this section, we test our approach both on real-word social networks and compare it with two algorithms. The one is the original Particle swarm optimization (PSO), the other is PSOSB which is proposed by Li [10].

4.1 Description of Data Sets

In our experiment, we crawled data (SMD) from sina microblog open platform via crawler that we designed and downloaded three other real-world signed networks from Stanford Network Analysis Platform (SNAP) in http://snap.stanford.edu/index.html. The statistic of these real-world social network is described in Table 1.

Table 1. Real-world signed networks.

Dataset	Description	Size	+	−
SMD	780	3280	80%	20%
Slashdot	82040	549202	77.4%	22.6%
Epinions	131828	840799	85%	15%
Wikipedia	10835	159388	76%	24%

4.2 Experiment Results and Analysis

In our method, the energy function is one of the core formula, so a suitable and effective parameter of energy function will contribute to the performance of the algorithm. The function H(s) denotes the cost for transforming an unbalanced network to a balanced network. Our aim is to minimize the H(s) in the same situation by adjusting the cost coefficient α from 0 to 1. We do experiments on our datasets based on different α. The result on several real-world datasets are shown in Table 2, according to the result, we can get the conclusion: when $\alpha = 0.4$ the cost is the smallest.

Table 2. Comparision results of α between MBPSO and the other algorithms on the four real-world networks.

Dataset	Algorithm	$\alpha = 0.2$	$\alpha = 0.4$	$\alpha = 0.6$	$\alpha = 0.8$
SMD	PSO	10,955	8706	9278	10,356
	PSOSB	9530	8019	8995	9546
	MDPSO	8423	7813	8469	8804
Slashdot	PSO	47,432	42,100	58,804	53,770
	PSOSB	35,094	28,989	40,035	45,778
	MDPSO	25,738	20,511	30,167	35,806
Epinions	PSO	97,954	90,016	93,228	96,819
	PSOSB	86,074	78,070	84,308	90,931
	MDPSO	80,026	76,533	79,039	81,003
wikipedia	PSO	30,581	22,068	35,620	41,438
	PSOSB	28,581	20,143	28,258	31,973
	MDPSO	26,374	17,559	23,760	25,920

For the proposed algorithm, the parameters, such as iteration number I_{max}, the max inertial coefficient ω_{max} and min inertial coefficient ω_{min} affect the performance of the algorithm. Through repeated experiment and comparison, we set the parameters of our algorithm as follows: the max inertial coefficient ω_{max} and min inertial coefficient ω_{min} is set to 2 and 1, respectively; the particle swarm population size and the maximum number of iterations are both set to 100; the initial value of self learning factors C_{ps} and social learning factors C_{gs} is set to 2 and 0, respectively, while the end value of self learning factors C_{pe} and social learning factors C_{ge} is set to 0 and 2.

In order to verify the convergence of the algorithm proposed in this paper, we do experiments on wikipedia and SMD datasets, the results show that the MDPSO algorithm essentially convergence after about 60 iterations, while the PSO and PSOSB essentially convergence after about 80 iterations. The detail is shown in Figs. 6 and 7.

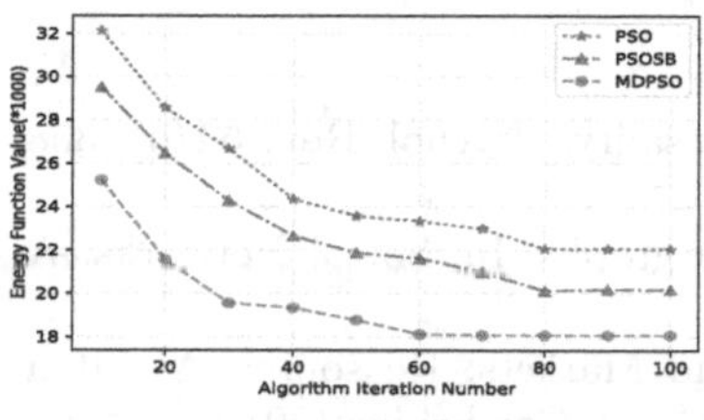

Fig. 6. wikipedia

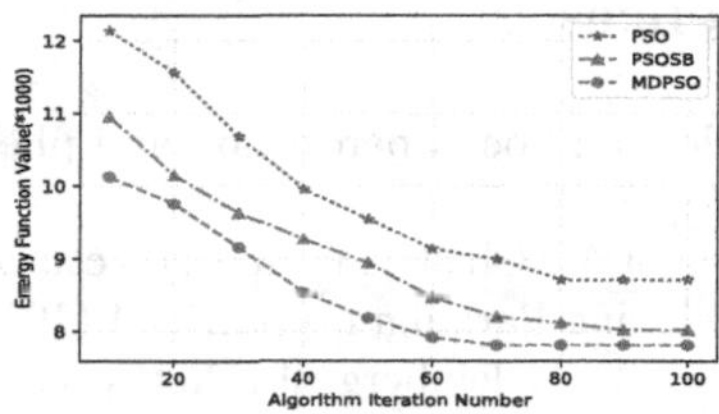

Fig. 7. SMD

In order to verify the effectiveness of the MDPSO, we do experiments on the four data sets above-mentioned, the results is shown in Fig. 8. In these experiments we give the same parameters for each algorithm. Figure 8 shows the Energy Function Value of PSOSB algorithm is smaller than PSO in every datasets, and the Energy Function Value of MDPSO algorithm is the smallest at any time. These results illustrate the effectiveness of our method.

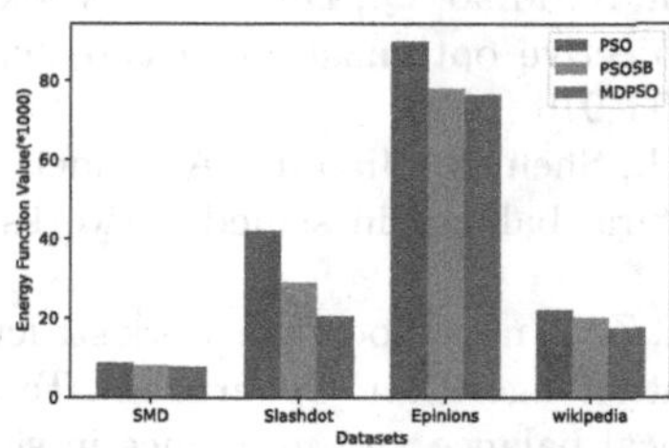

Fig. 8. Comparison the energy function value of PSO, PSOSB and MDPSO.

5 Conclusion

The study of structure balance is the foundation of complex network analysis. Because of the importance of the structure analysis, it received increasing attention of scholars from diverse fields in recent years. In this paper, we treat the network structure balance problem as an optimization problem. We extended the energy function based on the structure balance theory, and put the energy function as the fitness function of the optimization method. After that, according to the needs of discrete problem, we redefined the particle updating rules with a discrete view. We test our approach on real-world date sets, the experiment results demonstrate the effectiveness of the method. Our future work will focus on more in depth analysis of network structure, such as dynamic evolution of network structure.

Acknowledgement. This research is supported by the National Natural Science Foundation of China (Grant nos. 61472329 and 61271413) and the Innovation Fund of Postgraduate, Xihua University.

References

1. Heider, F.: Social perception and phenomenal causality. Psychol. Rev. **51**(6), 358 (1944)
2. Davis, J.A.: Clustering and structural balance in graphs. In: Social Networks: A Developing Paradigm, pp. 27–34 (1977)
3. Easley, D., Kleinberg, J.: Networks, Crowds, and Markets: Reasoning About a Highly Connected World. Cambridge University Press, Cambridge (2010)
4. Lerner, J.: Structural balance in signed networks: separating the probability to interact from the tendency to fight. Soc. Netw. **45**, 66–77 (2016)
5. Terzi, E., Winkler, M.: A spectral algorithm for computing social balance. In: Frieze, A., Horn, P., Prałat, P. (eds.) WAW 2011. LNCS, vol. 6732, pp. 1–13. Springer, Heidelberg (2011). doi:10.1007/978-3-642-21286-4_1
6. BoussaïD, I., Lepagnot, J., Siarry, P.: A survey on optimization metaheuristics. Inf. Sci. **237**, 82–117 (2013)
7. Črepinšek, M., Liu, S.-H., Mernik, M.: Exploration and exploitation in evolutionary algorithms: a survey. ACM Comput. Surv. (CSUR) **45**(3), 35 (2013)
8. Cai, Q., Gong, M., Ruan, S., Miao, Q., Du, H.: Network structural balance based on evolutionary multiobjective optimization: a two-step approach. IEEE Trans. Evol. Comput. **19**(6), 903–916 (2015)
9. Ma, L., Gong, M., Du, H., Shen, B., Jiao, L.: A memetic algorithm for computing and transforming structural balance in signed networks. Knowl. Based Syst. **85**, 196–209 (2015)
10. Xing, L.Z., Le, H.L., Hui, Z.: A novel social network structural balance based on the particle swarm optimization algorithm. Cybern. Inf. Technol. **15**(2), 23–35 (2015)
11. Harary, F., et al.: On local balance and n-balance in signed graphs. Mich. Math. J. **3**(1), 37–41 (1955)
12. Cartwright, D., Harary, F.: Structural balance: a generalization of heider's theory. Psychol. Rev. **63**(5), 277 (1956)
13. Kennedy, J., Eberhart, R.: Particle swarm optimization. In: IEEE International Conference on Neural Networks, Proceedings, vol. 4, pp. 1942–1948 (1995)
14. Facchetti, G., Iacono, G., Altafini, C.: Computing global structural balance in large-scale signed social networks. Proc. Natl. Acad. Sci. **108**(52), 20953–20958 (2011)

Vision Saliency Feature Extraction Based on Multi-scale Tensor Region Covariance

Shimin Wang(✉), Mingwen Wang, Jihua Ye, and Anquan Jie

College of Computer Information and Engineering,
Jiangxi Normal University, Nanchang 330022, China
wsmyangxi@126.com

Abstract. In the process of extracting image saliency features by using regional covariance, the low-level higher-order data are dealt with by vectorization, however, the structure of the data (color, intensity, direction) may be lost in the process, leading to a poorer representation and overall performance degradation. In this paper we introduce an approach for sparse representation of region covariance that will preserve the inherent structure of the image. This approach firstly calculates the image low-level data (color, intensity, direction), and then uses multi-scale transform to extract the multi-scale features for constructing tensor space, at last by using tensor sparse coding the image bottom features are extracted from region covariance. In the paper, it compares the experimental results with the commonly used feature extraction algorithms' results. The experimental results show that the proposed algorithm is closer to the actual boundary of the object and achieving better results.

Keywords: Saliency feature · Region covariance · Tensor space · Multi-scale transform

1 Introduction

When people observe natural scenes, predictions of objects are called saliency predictions or detections, and the saliency studying has attracted a lot of computer vision scientists and neuroscience scientists. Highly predictive accuracy is important for many applications, such as object detection and segmentation, content-based compression, scene understanding, robot navigation, and image/video quality assessment.

In the existing literatures the saliency models include: (1) pure bottom-up image saliency features; (2) top-down semantic saliency features; (3) combination models. These models use a variety of visual features, including low, medium and advanced features. However, each single model has its own assumptions and methods, respectively, to solve the different aspects of human visual attention to the problem. In addition,

This work was supported in part by the National Natural Science Foundation of China under Grant 61650105, Grant 61462042, and Grant 61462045, in part by the Jiangxi Education Science and Technology Research Project under Grant GJJ160324.

J. Wen et al. (Eds.): CCIR 2017, LNCS 10390, pp. 185–197, 2017.
https://doi.org/10.1007/978-3-319-68699-8_15

some researchers selected fewer data sets to validate their models. Thus, there is a data set deviation in their model [1]. Because of these assumptions and constraints, each model has its own image set [2] and the corresponding weaknesses. The bottom-up saliency model of the use of biological low-level features is mainly based on the principle proposed by Itti and others [3–5]. Assuming that the saliency region is significant in terms of color, intensity, or orientation, Itti et al. used the visual saliency of the center-surround difference across multi-scale image features. Harel and others [6] based on the local image block different from the surrounding image block to establish the image of the visual saliency (model GBVS), which uses the Markov chain to measure the dissimilarity between local blocks. Similarly, Liu and others [7] proposed a set of novel features including multi-scale contrast, center-surround histogram, and color spatial distribution to describe a salient object locally, regionally, and globally. Bruce and Tsotsos [8] presented a saliency model (AIM) is based on Shannon's self-information measure and is achieved in a neural circuit, which is demonstrated as having close ties with the circuitry existent in the primate visual cortex. Hou and Zhang [9] introduced the incremental coding length (ICL) to measure the perspective entropy gain of each feature. Zhang and others [10] proposed a definition of saliency by considering what the visual system is trying to optimize when directing attention. Kong and others [11] proposed an integration approach to detect salient region based on three principles from psychological evidence and observations of images, including colour contrast in a global context, spatially compact colour distribution, and multi-scale image abstraction. Fang and others [12] proposed a novel video saliency detection model based on feature contrast in compressed domain. Four types of features including luminance, color, texture, and motion are extracted from the discrete cosine transform coefficients and motion vectors in video bitstream. Jiang and others [13] proposed a novel automatic salient object segmentation algorithm which integrates both bottom-up salient stimuli and object-level shape prior. The above models perform well, and each model has its own most suitable image set due to different assumptions. For example, the Itti [3] model focuses primarily on local dissimilarity and is insufficient in detecting global visual saliency, whereas the AIM [8] model shows insufficient performance in detecting local dissimilarity.

In general, these bottom-up models cannot detect saliency areas with semantic information. Itti model cannot distinguish between the human body and its surroundings. A large number of researchers recently solved the saliency detection by combining low-level and advanced features. Cerf and others [14] improved the performance by adding face detection to the Itti model. Judd and others [15] proposed a new saliency detection model by using the support vector machine (SVM) to learn the optimal weight of all the combined features. Similarly, Chang and others [16, 17] proposed object-based saliency models using object metrics. While the top-down model emphasizes the importance of advanced and semantic features, the goal of detecting failure is usually a saliency object that has not yet been trained. In addition, the performance of the combined model in detecting non-semantic bottom-up saliency regions may not be as good as pure bottom-up model. For example, the Judd [15] model does not well detect local dissimilarity in some cases.

Before the study most of algorithms dealt with the low-level higher-order data by vectorization, the structure of the data may be lost in the process, which leads to a poorer representation and overall performance degradation. In this paper we use regional covariance to extract image saliency features, and we introduce an approach for sparse representation of region covariance which will preserve the inherent structure of the image. In the paper, it compares the experimental results with the commonly used feature extraction algorithms. The experimental results show that the proposed algorithm is closer to the actual boundary of the object and achieving better results.

2 Algorithm

In the process of saliency image feature extraction, it is necessary to save the structural features of the saliency region texture as much as possible. At the same time, the local low-level features of the image should be preserved. From the statistical point of view, the features of visual saliency detections are very similar to the statistical components with the Supper Gaussian distribution, because the "super-Gaussian" in statistics and "sparse, structured" in image processing basically have the same meaning. This approach uses multi-scale transform to extract the multi-scale features for constructing tensor space, at last by tensor sparse coding the image bottom features are extracted from region covariance.

2.1 Region Covariance

2006 Erkut and Aykut [18] first proposed covariance as a compact regional descriptor, and since then it has been effectively using and solving various advanced computer vision problems. Description of region covariance:

$$\boldsymbol{Q}(x,y) = \boldsymbol{F}(\boldsymbol{I},x,y) \tag{1}$$

The $\boldsymbol{I}$ represents an image, the target area in the image is R, its size is $M \times N$, each pixel to generate d-dimension eigenvector $\boldsymbol{Q}(x,y)$. The M and N is the width and height of the target area, $\boldsymbol{Q}(x,y)$ is a description of the pixel in the target area, and its element may be the spatial position information, the intensity, the color, the direction, and may be the response value of the filter of the pixel, etc. The choice of eigenvectors can be determined according to the needs of the actual application. A pixel $\boldsymbol{Q}_i$ can be expressed as follows:

$$\boldsymbol{Q}_i = [x, y, R(x,y), G(x,y), B(x,y), T(x,y), \ldots] \tag{2}$$

The (x,y) is the coordinates of the pixel; $R(x,y)$, $G(x,y)$, $B(x,y)$ are the corresponding pixel (x,y) RGB color values; $T(x,y)$ is the corresponding pixel brightness feature. The covariance matrix of the target region can be expressed as:

$$C_R = \frac{1}{MN}\sum_{i=1}^{MN}(\boldsymbol{Q}_i - \boldsymbol{u}_R)^T(\boldsymbol{Q}_i - \boldsymbol{u}_R) \tag{3}$$

Formula: $\boldsymbol{u}_R = \frac{1}{MN}\sum_{i=1}^{MN}\boldsymbol{Q}_i$, the covariance matrix $\boldsymbol{C}_R$ is a real symmetric positive definite matrix, the values on the diagonal in the covariance matrix represent each individual feature, and the value on the non-diagonal indicates the correlation between the features, The size of $\mathbf{C}_R$ depends only on the dimension of the selected feature vector d, regardless of the size of the target area.

2.2 Multi-scale of Image Low-Level Features

The input image is assumed to be $\boldsymbol{I}(x, y)$, the red, green, and blue channel images are $(\boldsymbol{R}_0, \boldsymbol{G}_0, \boldsymbol{B}_0)$.

Color features: set up $\boldsymbol{R}$ (red), $\boldsymbol{G}$ (green), $\boldsymbol{B}$ (blue), $\boldsymbol{Y}$ (yellow) 4 color channels, such as:

$$\boldsymbol{R} = \boldsymbol{R}_0 - (\boldsymbol{G}_0 + \boldsymbol{B}_0)/2 \tag{4}$$

$$\boldsymbol{G} = \boldsymbol{G}_0 - (\boldsymbol{R}_0 + \boldsymbol{B}_0)/2 \tag{5}$$

$$\boldsymbol{B} = \boldsymbol{B}_0 - (\boldsymbol{G}_0 + \boldsymbol{R}_0)/2 \tag{6}$$

$$\boldsymbol{Y} = (\boldsymbol{G}_0 + \boldsymbol{B}_0)/2 - (\boldsymbol{G}_0 - \boldsymbol{B}_0)/2 - \boldsymbol{B}_0 \tag{7}$$

Intensity features: use $\boldsymbol{L}$ to describe the brightness of the original image:

$$\boldsymbol{L} = (\boldsymbol{R}_0 + \boldsymbol{G}_0 + \boldsymbol{B}_0)/3 \tag{8}$$

Direction features: the paper uses a two-dimensional Gabor filter [19] for image processing, the two-dimensional image data can be processed directly, and operational efficiency can be improved. The calculation process is as follows:

$$\boldsymbol{G}_{u,v}(z) = \boldsymbol{I}(z) * \psi_{u,v}(z) \tag{9}$$

$*$ is convolution operation, $z = (x, y)$ is grey value, $\psi_{u,v}(z)$ Defines the Gabor kernel of orientation u and scale v.

The definition of multi-scale features $\boldsymbol{GF}(i), i = 1, 2, \ldots, 5p, p = u * v$. In the paper, we use $u = 8, v = 5, p = 40$. Using image low-level features set $(\boldsymbol{R}, \boldsymbol{G}, \boldsymbol{B}, \boldsymbol{Y}, \boldsymbol{L})$ and $\psi_{u,v}(z)$ convolution operation get $\boldsymbol{GF}(i), (i = 1, 2, \ldots, 200)$.

According to $\boldsymbol{GF}(i)$ and region covariance (Sect. 2.1), each pixel can generate 200-dimensions eigenvector $\boldsymbol{Q}(x, y)$ and region covariance $\boldsymbol{C}$.

2.3 Multi-scale Tensor Region Covariance

We begin with a known dictionary consisting of k $n * n$ positive definite matrices $\{A_i\}_{i=1}^{k}$, where each A_i is referred to as a dictionary atom. Given a positive definite matrix (region covariance) C, our goal is to represent the new matrix as a linear combination of the dictionary atoms, as follow:

$$C = \sum_{i=1}^{k} y_i A_i \tag{10}$$

$y_i (i = 1, 2, \ldots, k)$ are the vector of coefficients.

According to the literature [20, 21], it proposed tensor sparse coding to get sparse representation of C. We can get the corresponding value y_i of each pixel. The eigenvector representation is $\varphi(C) = \{y_i\}_{i=1}^{k}$.

In order to calculate the distance between R_i and R_j, the definitions of x_i and x_j as the center of the area R_i and R_j. The distance $G(R_i, R_j)$ can be calculated as follow:

$$G(R_i, R_j) = ||\varphi(C_i) - \varphi(C_j)||/(1 + ||x_i - x_j||) \tag{11}$$

Then by calculating the distance with the nearest m neighboring area and getting the saliency of the region, the formula is:

$$S(R_i) = (1/m) \sum_{j=1}^{m} G(R_i, R_j) \tag{12}$$

3 Experiment

3.1 Experimental Data Set

This paper uses the CAT2000 dataset with 20 categories of scenes [22], including: (1) action, (2) emotion, (3) art, (4) black and white, (5) cartoon, (6) irregular shape, (7) indoors, (8) drawings, (9) low resolution, (10) noisy, (11) object, (12) artificial outdoors, (13) natural outdoors, (14) pattern, (15) free, (16) satellite, (17) sketch, (18) social. Where the resolution of the image is 1920 * 1080 pixels and different categories of scenes are suitable for studying the different aspects of saliency behavior. Part of the CAT2000 data set shown in Fig. 1.

Fig. 1. Part of the CAT2000 data set

3.2 Measure Principles

Theoretically, any model can be measured by any standard, and the saliency model produces a saliency graph S, which is usually quantified by comparison with the eye movement data G. In the literature, Judd et al. have the following measures of the saliency model [23]: (1) AUC-Judd; (2) SIM; (3) EMD; (4) AUC-Borji; (5) sAUC; (6) CC; (7) NSS; (8) KL.

3.3 Visualize Measurement Data

In order to visualize the saliency of the statistical data, the visualization of the measurement criteria data is based on the Benchmark et al. method [24]. The experimental visualization data are as follows (Fig. 2):

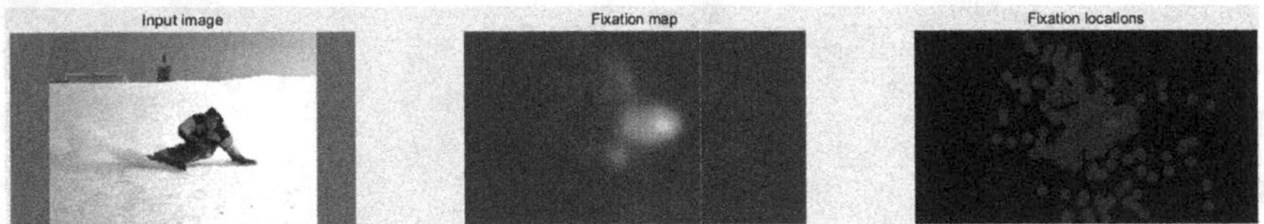

Fig. 2. Input image, fixation map, fixation locations

Visualize area under ROC curve (AUC) computation: A set of AUC visualizations that plot the level sets of the saliency map at different thresholds, the corresponding true positives and false negatives and the resulting ROC curve, the results are shown in Fig. 3.

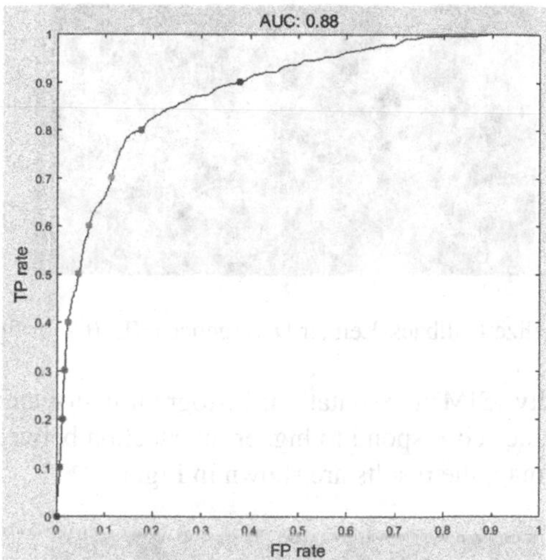

Fig. 3. Visualize area under ROC curve

Visualize normalized scan path saliency (NSS) computation: NSS plotted in parula color scale; higher values mean higher normalized saliency at fixated locations, the results are shown in Fig. 4.

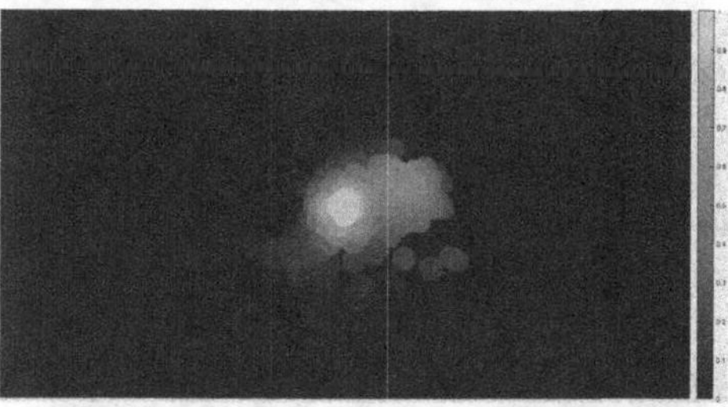

Fig. 4. Visualize normalized scan path saliency (NSS) (Color figure online)

Visualize Kullback-Leibler Divergence (KL) computation: KL heat map, with brighter red values corresponding to higher KL divergence (salMap fails to sufficiently approximate fixMap at those pixels), the results are shown in Fig. 5.

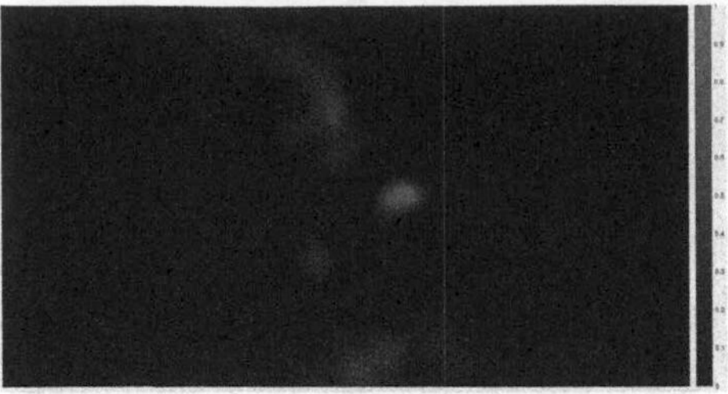

Fig. 5. Visualize Kullback-Leibler Divergence (KL) (Color figure online)

Visualize Similarity (SIM) computation: histogram intersection, plotted in parula color scale; higher values correspond to higher intersection between saliency map and ground truth fixation map, the results are shown in Fig. 6.

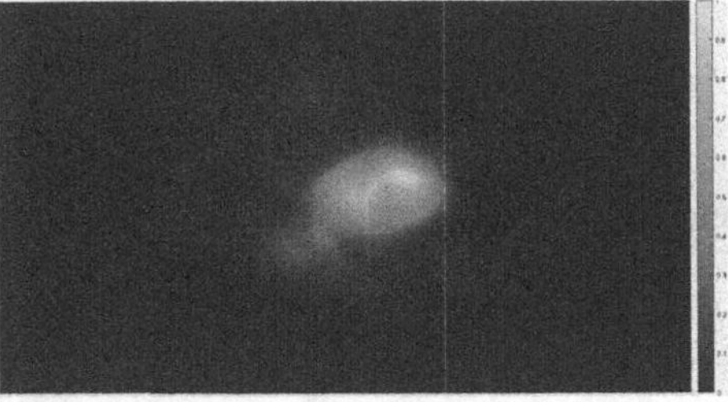

Fig. 6. Visualize similarity (SIM) (Color figure online)

Visualize Pearson's Correlation Coefficient (CC) computation: correlation coefficient, plotted in parula color scale; higher values correspond to higher correlation between saliency map and ground truth fixation map, the results are shown in Fig. 7.

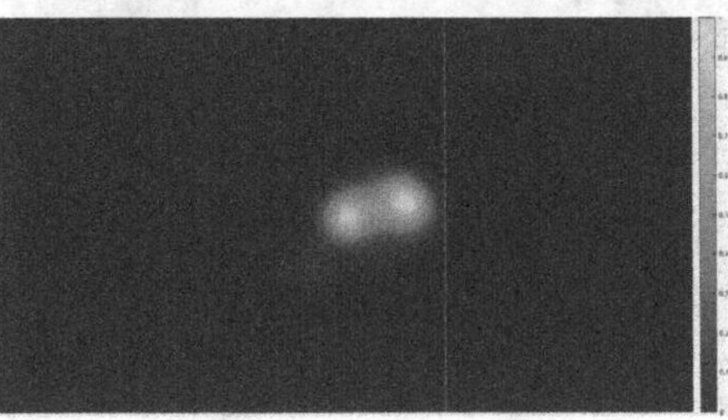

Fig. 7. Visualize Pearson's correlation coefficient (CC) (Color figure online)

Visualize Earth Mover's Distance (EMD) computation: in green are regions of the saliency map from which density needs to be moved to the regions in red to transform salMap to match fixMap, The results are shown in Fig. 8.

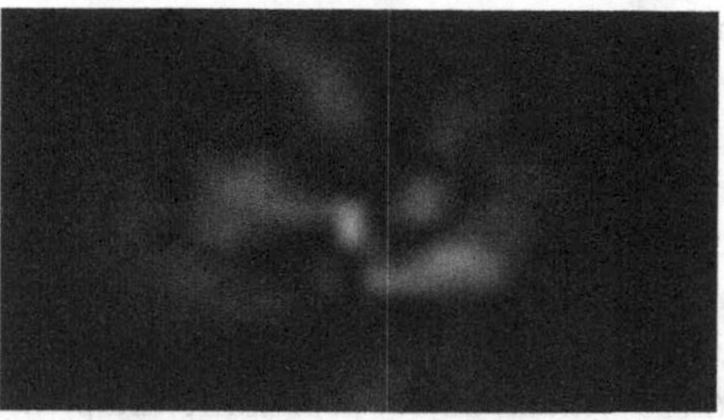

Fig. 8. Visualize earth mover's distance (EMD) (Color figure online)

Visualize Information Gain (IG) computation: visualize information gain over baseline map cyan color scheme corresponds to gain relative to baseline (salMap better predicts fixMap at these pixels) red color scheme corresponds to loss relative to baseline (baseMap better predicts fixMap at these pixels), the results are shown in Fig. 9.

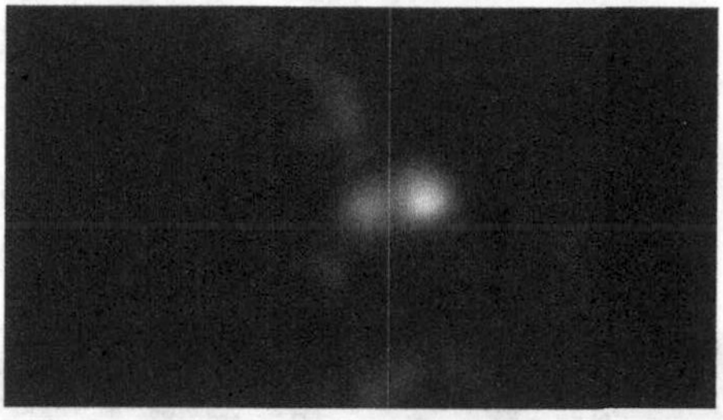

Fig. 9. Visualize information gain (IG) (Color figure online)

3.4 Model Comparison

By comparing the saliency of the CAT2000 data set, the saliency maps are obtained and compared with the other algorithms. As shown in Fig. 10, an image is selected from each type of image set. Color feature as an important feature of the image, the algorithm focuses on the use of color features, the test images are color images, excluding the gray images of CAT2000 data set.

Table 1 and Fig. 10 show the results of the comparison with other up-to-date methods on the CAT2000 dataset. In this paper, we compare with the following algorithms: Goferman and others [25] proposed CAS (PAMI), which presented a detection algorithm which is based on four principles observed in the psychological literature, the benefits of the proposed approach are evaluated in two applications where the context of the dominant objects is just as essential as the objects themselves; Erdem and others [26] proposed CovSal, they proposed to use covariance matrices of simple image features (known as region covariance descriptors in the computer vision community; Hage and others [6] based on the local image block different from the surrounding image block to establish the image visual saliency (model GBVS), which uses the Markov chain for measuring the heterogeneity between local blocks; Fang and others [27] proposed to estimate image saliency (model LDS)by learning a set of discriminative subspaces that perform the best in popping out targets and suppressing distractors; Riche and others [28] proposed a novel saliency prediction model, called RARE2012, which selects information worthy of attention based on multi-scale spatial rarity; Murray and others [29] proposed an efficient model of color appearance in human vision, which contains a principled selection of parameters as well as an innate spatial pooling mechanism, can be generalized to obtain a saliency model that outperforms state-of-the-art models; Ran and others [30] proposed an approach for saliency detection that combines previously suggested patch distinctness with an object probability map(model CGI2012).

Table 1. Correlation results by using the CAT2000 data set.

Model name	AUC-Judd	SIM	EMD	AUC-Borji	CC	NSS	KL
Judd Model	0.84	0.46	3.60	0.84	0.54	1.30	0.94
GBVS	0.80	0.51	2.99	0.79	0.50	1.23	0.80
CAS	0.77	0.50	3.09	0.76	0.42	1.07	1.04
IttiKoch	0.56	0.34	4.66	0.53	0.09	0.25	6.71
Murray model	0.70	0.43	3.79	0.70	0.30	0.77	1.14
CGI2012	0.77	0.46	4.17	0.76	0.43	1.14	1.53
RARE2012	0.82	0.54	2.72	0.81	0.57	1.44	0.76
LDS	0.83	0.58	2.09	0.79	0.62	1.54	0.79
The paper	0.87	0.46	1.78	0.79	0.36	1.89	2.62

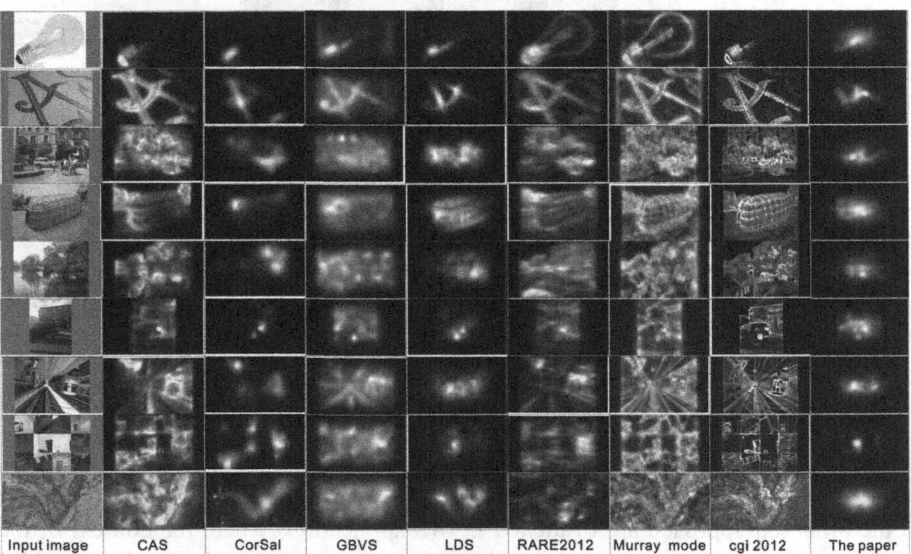

Fig. 10. The saliency map comparison with other up-to-date algorithms.

As shown in Fig. 10 and Table 1, the interior of the original image region is relatively uniform, and the extraction of the saliency feature is relatively difficult. CAS (PAMI) is based on four principles observed in the psychological literature, but the emphasis of saliency feature is not obvious enough, GBVS, Murray mode and CGI2012 also encounter this problem; RARE2012 algorithm selects information worthy of attention based on multi-scale spatial rarity, due to the influence of noise, the saliency extraction is ineffective; CovSal, LDS and the algorithm proposed can be very good to obtain the clear boundary and clear features, and they are good to distinguish features and disturbances, especially the proposed algorithm has been almost targeted to the specific saliency of the object, other disturbances have been excluded, the saliency effect is very obvious.

4 Conclusion

Now the most of algorithms dealt with the low-level higher-order data by vectorization, the structure of the data may be lost in the process, which leads to a poorer representation and overall performance degradation. In this paper we use regional covariance to extract image saliency features, and we introduce an approach for sparse representation of region covariance which will preserve the inherent structure of the image. Firstly it obtains the image low-level data (color, intensity, direction, etc.), and then it uses the multi-scale filtering to process all of the low-level data, on the one hand it filters out the noises of the low-level features, on the other hand it obtains image features for tensor sparse coding from the multi-band perspective, it is advantageous to preserve the inherent structure of the multi-band from the image. At last, it compares the experimental results with the commonly used feature extraction algorithms. The experimental results show that the proposed algorithm is closer to the actual boundary of the object and achieving better results.

References

1. Torralba, A., Efros, A.A.: Unbiased look at dataset bias. In: Proceedings of the IEEE Conference on Computer Vision Pattern Recognition (CVPR), pp. 1521–1528, June 2011
2. Borji, A., Sihite, D.N., Itti, L.: Salient object detection: a benchmark. In: Proceedings of the European Conference on Computer Vision (ECCV), pp. 414–429 (2012)
3. Itti, L., Koch, C., Niebur, E.: A model of saliency-based visual attention for rapid scene analysis. IEEE Trans. Pattern Anal. Mach. Intell. **20**(11), 1254–1259 (1998)
4. Koch, C., Ullman, S.: Shifts in selective visual attention: towards the underlying neural circuitry. Hum. Neurobiol. **4**(4), 219–227 (1985)
5. Treisman, A.M., Gelade, G.: A feature-integration theory of attention. Cognit. Psychol. **12**(1), 97–136 (1980)
6. Harel, J., Koch, C., Perona, P.: Graph-based visual saliency. In: Proceedings of the Advances in Neural Information Processing Systems (NIPS), pp. 542–552 (2006)
7. Liu, T., et al.: Learning to detect a salient object. IEEE Trans. Pattern Anal. Mach. Intell. **33**(2), 353–367 (2011)

8. Bruce, N.D.B., Tsotsos, J.K.: Saliency based on information maximization. In: Proceedings of the Advances in Neural Information Processing Systems (NIPS), pp. 155–162 (2005)
9. Hou, X., Zhang, L.: Dynamic attention: searching for coding length increments. In: Proceedings of the Advances in Neural Information Processing Systems (NIPS), pp. 681–688 (2008)
10. Zhang, L., Tong, M.H., Marks, T.K., Shan, H., Cottrell, G.W.: SUN: a Bayesian framework for saliency using natural statistics. J. Vis. **8**(7), 1–20 (2008)
11. Kong, L., Duan, L., Yang, W., Dou, Y.: Salient region detection: an integration approach based on image pyramid and region property. IET Comput. Vis. **9**(1), 85–97 (2015)
12. Fang, Y., Lin, W., Chen, Z., Tsai, C.-M., Lin, C.-W.: A video saliency detection model in compressed domain. IEEE Trans. Circuits Syst. Video Technol. **24**(1), 27–38 (2014)
13. Jiang, H., Wang, J., Yuan, Z., Liu, T., Zheng, N., Li, S.: Automatic salient object segmentation based on context and shape prior. In: Proceedings of the British Machine Vision Conference (BMVC), pp. 1–12 (2011)
14. Cerf, M., Frady, E.P., Koch, C.: Faces and text attract gaze independent of the task: experimental data and computer model. J. Vis. **9**(12), 1–15 (2009)
15. Judd, T., Ehinger, K., Durand, F., Torralba, A.: Learning to predict where humans look. In: Proceedings of the IEEE 12th International Conference on Computer Vision (ICCV), pp. 795–825, September/October 2009
16. Chang, K.-Y., Liu, T.-L., Chen, H.-T., Lai, S.-H.: Fusing generic abjectness and visual saliency for salient object detection. In: Proceedings of the International Conference on Computer Vision (ICCV), pp. 914–921, November 2011
17. Alexe, B., Deselaers, T., Ferrari, V.: What is an object? In: Proceedings of the IEEE Conference on Computer Vision and Pattern Recognition (CVPR), pp. 73–80, June 2010
18. Tuzel, O., Porikli, F., Meer, P.: Region covariance: a fast descriptor for detection and classification. In: Leonardis, A., Bischof, H., Pinz, A. (eds.) ECCV 2006. LNCS, vol. 3952, pp. 589–600. Springer, Heidelberg (2006). doi:10.1007/11744047_45
19. Daugman, J.G.: Uncertainty relation for resolution in space, spatial frequency, and orientation optimized by two-dimensional visual cortical filters. J. Opt. Soc. Am. A **2**(7), 1160–1169 (1985)
20. Wang, S., Cheng, B., Ye, J., Wang, M.: Method of face image feature extract based on weighted multi-scale tensor subspace. J. Data Acquis. Process. **31**(04), 791–7 98 (2016)
21. Sivalingam, R., Boley, D., Morellas, V., Papanikolopoulos, N.: Tensor sparse coding for region covariances. In: Daniilidis, K., Maragos, P., Paragios, N. (eds.) ECCV 2010. LNCS, vol. 6314, pp. 722–735. Springer, Heidelberg (2010). doi:10.1007/978-3-642-15561-1_52
22. Borji, A., Itti, L.: CAT2000: a large scale fixation dataset for boosting saliency research. In: CVPR 2015 Workshop on "Future of Datasets" (2015). arXiv preprint arXiv:1505.03581
23. Bylinskii, Z., Judd, T., Oliva, A., Torralba, A., Durand, F.: What do different evaluation metrics tell us about saliency models? arXiv preprint arXiv:1604.03605 (2016)
24. Judd, T., Durand, F., Torralba, A.: A benchmark of computational models of saliency to predict Hum. Fixat. (2012)
25. Goferman, S., Zelnik-Manor, L., Tal, A.: Context-aware saliency detection. IEEE Trans. Pattern Anal. Mach. Intell. **34**(10), 1915–1926 (2012)
26. Erdem, E., Erdem, A.: Visual saliency estimation by nonlinearly integrating features using region covariances. J. Vis. **13**(4), 103–104 (2013)
27. Fang, S., Li, J., Tian, Y., et al.: Learning discriminative subspaces on random contrasts for image saliency analysis. IEEE Trans. Neural Netw. Learn. Syst. (2016)

28. Riche, N., Mancas, M., Duvinage, M., et al.: RARE2012: a multi-scale rarity-based saliency detection with its comparative statistical analysis. Sig. Process. Image Commun. **28**(6), 642–658 (2013)
29. Murray, N., Vanrell, M., Otazu, X., et al.: Saliency estimation using a non-parametric low-level vision model. In: The IEEE Conference on Computer Vision and Pattern Recognition (CVPR 2011), Colorado Springs, 20–25 June, DBLP, pp. 433–440 (2011)
30. Ran, M., Zelnik-Manor, L., Tal, A.: Saliency for image manipulation. Vis. Comput. **29**(5), 381–392 (2013)

Combining Large-Scale Unlabeled Corpus and Lexicon for Chinese Polysemous Word Similarity Computation

Huiwei Zhou(✉), Chen Jia, Yunlong Yang, Shixian Ning, Yingyu Lin, and Degen Huang

School of Computer Science and Technology, Dalian University of Technology, Dalian 116024, Liaoning, China
{zhouhuiwei,huangdg}@dlut.edu.cn, {jiachen,SDyyl_1949,ningshixian}@mail.dlut.edu.cn, lyydut@sina.com

Abstract. Word embeddings have achieved an outstanding performance in word similarity measurement. However, most prior works focus on building models with one embedding per word, neglect the fact that a word can have multiple senses. This paper proposes two sense embedding learning methods based on large-scale unlabeled corpus and Lexicon respectively for Chinese polysemous words. The corpus-based method labels the senses of polysemous words by clustering the contexts with *tf-idf* weight, and using the HowNet to initialize the number of senses instead of simply inducing a fixed number for each polysemous word. The lexicon-based method extends the *AutoExtend* to Tongyici Cilin with some related lexicon constraints for sense embedding learning. Furthermore, these two methods are combined for Chinese polysemous word similarity computation. The experiments on the Chinese Polysemous Word Similarity Dataset show the effectiveness and complementarity of our two sense embedding learning methods. The final Spearman rank correlation coefficient achieves 0.582, which outperforms the state-of-the-art performance on the evaluation dataset.

Keywords: Sense embeddings · Chinese word similarity evaluation · Chinese polysemous words · Large-scale unlabeled corpus · Lexicon

1 Introduction

Many Nature Language Processing tasks benefit from word embeddings [1] because they help capture syntactic and semantic characteristics of words by projecting words into a low-dimensional vector space. However, most of the previous researches in word embeddings associate each word with a single embedding. Such single embedding representation method is forced to express polysemous word with an uneasy central vector between its various meanings [2]. This will lead to a text understanding problem, since polysemous words may have different meanings in different contexts.

In recent years, there has been an increasing interest in learning sense embeddings for polysemous words from large-scale unlabeled corpus [3, 4]. These researches learn

J. Wen et al. (Eds.): CCIR 2017, LNCS 10390, pp. 198–210, 2017.
https://doi.org/10.1007/978-3-319-68699-8_16

sense embeddings based on two fundamental steps: (1) Label the sense of the polysemous words by clustering the contexts. (2) Learn sense embeddings for polysemous words. Such methods simply include uniform *K* senses for all polysemous words, ignoring the real sense number of each polysemous word. Besides, the simple context representations without regard to the word frequency limit the performance of clusters, and therefore prevent improving sense embeddings. Meanwhile, sense embeddings trained on large-scale unlabeled corpus suffers from the unsatisfied quality of embeddings of low frequency senses.

This paper proposes a sense embedding learning method based on large-scale unlabeled corpus for Chinese polysemous words. The proposed corpus-based method clusters the contexts of polysemous words based with *tf-idf* (Term Frequency-Inverse Document Frequency) weight [5], and using the HowNet [6] to initialize the number of senses for each polysemous word. Then, to address low frequency senses, we extend the *AutoExtend* [7] to the Tongyici Cilin [8] with some related lexicon constraints and use it as an annotated knowledge base to learn lexicon-based sense embeddings. Finally, these two methods are combined for Chinese polysemous word similarity computation [9]. The experimental results show that the corpus-based and lexicon-based methods are both effective for capturing the sense-level word similarities, and their combination could further improve similarity computation performance.

2 Related Work

Bengio et al. [10] first propose to learn word embeddings by using neural networks. Continuous Word2vec model [11, 12] and GloVe model [13] are the two efficient models for learning word embeddings. However, representing each word with a single embedding fails to capture polysemy. Reisinger et al. [3] and Huang et al. [4] introduce a kind of sense embedding learning method by clustering the contexts of polysemous words and then learning embeddings of the sense-labeled words. Neelakantan et al. [14] improve sense embeddings by performing word sense clustering and embedding learning jointly. But these methods usually assume that contexts are directly represented as sum (or average) of the surrounding words' vectors, neglecting that the context words do not contribute equally to the sense of the polysemous word. Zheng et al. [15] tackle this issue by learning the context representations with a neural network architecture before learning sense embeddings. Guo et al. [9] propose a novel approach for Chinese sense embedding leaning by exploiting bilingual parallel data. However, these corpus-based models are generally unable to learn some low frequency senses which rarely occurrence in regular corpus and bilingual parallel data.

Lexicon-based models drawing this issue by using sense-specific knowledge. Chen et al. [16] use WordNet glosses to learn sentence-level embeddings as the initializations of sense embeddings. Rothe et al. [7] take advantage of the architecture of WordNet to extend word embeddings to embeddings of synsets and lexemes. However, these models generally suffer for inaccurate semantic representations for sense items in the lexicon. Pei et al. [17] combine semantic lexicon and word embeddings to compute Chinese word similarity, but they neglect polysemy.

As discussed above, there are only a few researches on learning sense embeddings, and almost all of them focus on English words. We address these issues and introduce a corpus-based and a lexicon-based sense embedding learning methods, which are demonstrated effective and complementary for Chinese polysemous word similarity computation.

3 Methods

Figure 1 briefly illustrates the general architecture of our Chinese polysemous word similarity computation system. It consists of three components: (1) the corpus-based sense embedding learning, (2) the lexicon-based sense embedding learning and (3) the similarity computation and combination.

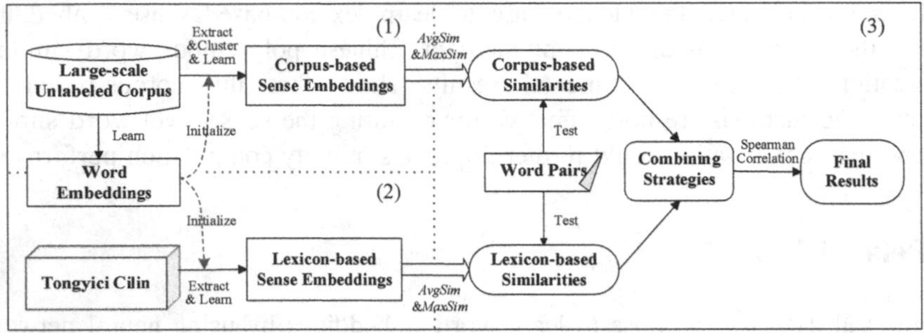

Fig. 1. Architecture of our Chinese polysemous word similarity computation system.

3.1 Corpus-Based Sense Embedding Learning

Based on an enlightening idea that word sense is associated with its context [3, 4], the corpus-based sense embedding learning uses the contexts of the polysemous words to distinguish senses. The detail of our method can be described into three sections:

- Context Representation Extraction: Context key words of the current polysemous word are extracted based on *tf-idf* weight for context representations.
- Sense Clustering: The context representations are clustered according to the sense number of each polysemous word to generate clusters, which are then used to label the sense of polysemous words.
- Sense Embedding Learning: Regarding each sense of a word as a new word, the sense embeddings of polysemous words and the single embeddings of the other monosemous words are learned and unified in one vector space with the GloVe model [13].

Context Representation Extraction. Consider a word w_t and its contexts $c_t = \{w_{t-R_t}, \ldots, w_{t-1}, w_{t+1}, \ldots, w_{t+R_t}\}$, where R_t represents the semi-width of the context window. The context representation of word w_t is the sum of its context word embeddings:

$$v_{context}(w_t) = \sum_{w \in c_t} v_g(w) \tag{1}$$

where $v_g(w)$ is the pre-trained word embedding of $w \in c_t$.

Although the large context window contains the comprehensive meaning of the contexts, too many context words also obfuscate the representation of the meaning. For example, some non-essential words such as "是 (is or are)" and "有 (has or have)" definitely have a negative effect on the context representations.

This paper defines a selectivity factor R_t^* to choose the key words in the contexts. To be specific, we use *tf-idf* to compute the weights of context words [5]. And then choose $R_t^*(R_t^* \leq R_t)$ words $c_t^* = \{w_{t-R_t^*}^*, \ldots, w_{t-1}^*, w_{t+1}^*, \ldots, w_{t+R_t^*}^*\}$ with the largest R_t^*-th weights as the context representations:

$$v_{context}(w_t) = \sum_{w \in c_t^*} v_g(w) \tag{2}$$

Sense Clustering. For each polysemous word, the extracted context representations are then used for sense clustering. This paper uses *k*-means clustering algorithm [3].

Here, assume that S_j $(j = 1, 2, \ldots, K)$ is the jth cluster whose center is μ_j. K is the number of senses. Instead of fixing the number of all clusters to a single parameter K, we use the HowNet [6] to initialize the number of senses for each polysemous word. The clustering rule can be represented as:

$$J = \sum_{j=1}^{K} \sum_{v_{context} \in S_j} sim(\mu_j, v_{context}) \tag{3}$$

where $sim(\cdot, \cdot)$ is cosine similarity. The object of clustering is to maximize the clustering rule J. And then each polysemous word in the unlabeled corpus is labeled a sense tag according to the sense clusters.

Sense Embedding Learning. Once a sense labeled corpus are generated, the GloVe model [13] is used to learn the sense embeddings of the polysemous words and the single embeddings of the other monosemous words in one vector space. The GloVe model is directly based on the statistics of word (or sense) occurrences in the sense labeled corpus. Let the matrix of word-sense co-occurrence counts be denoted by X, whose entries X_{ij} tabulate the number of times that word (or sense) j occurs in the context of word (or sense) i. And $X_i = \sum_k X_{ik}$ is the number of times that any word (or sense) appears in the context of word (or sense) i. Finally, $P_{ij} = P(j|i) = X_{ij}/X_i$ is the probability that word (or sense) j appears in the context of word (or sense) i.

We use a simple example to show how certain senses of a polysemous word can be extracted directly from co-occurrence probabilities with various probe words k. Consider two senses i = 打 (*fetch*) and j = 打 (*hit*) of the word "打". For words k related to i = 打 (*fetch*) but not j = 打 (*hit*), say k = 水 (water), we expect the ratio P_{ik}/P_{jk} will be large. However, for words k related to j = 打 (*hit*) but not i = 打 (*fetch*), say k = 人 (man), the ratio should be small.

3.2 Lexicon-Based Sense Embedding Learning

We utilize the HIT-CIR Tongyici Cilin (Extended) [8] as the lexical resource to learn sense embeddings of Chinese polysemous words. Figure 2 briefly illustrates the process of lexicon-based sense embedding learning.

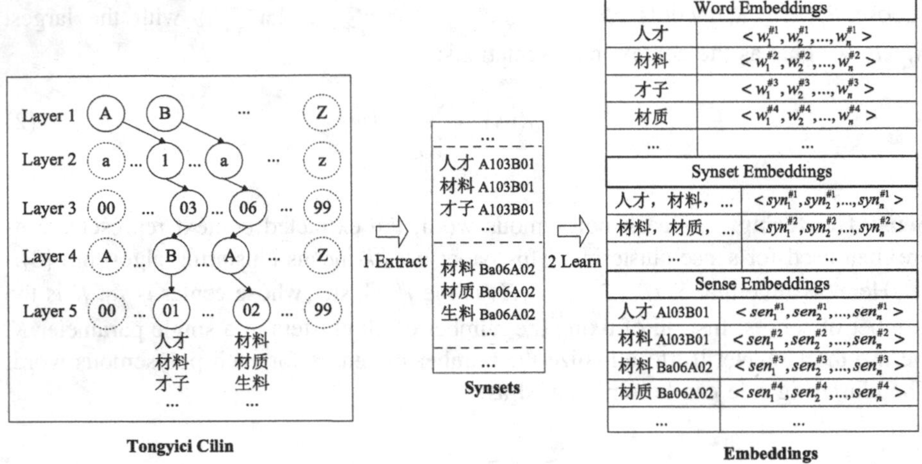

Fig. 2. Lexicon-based sense embedding learning.

The architecture of the Tongyici Cilin is shown in the left part of Fig. 2. Cilin encodes a group of words which have the relationships like "synonym" or "relevance" in a 5-layer hierarchical tree structure. The nodes of the upper 4 layers in the Cilin represent the abstract categories, while only the bottom leaf nodes are the words. A set of synonyms are organized as a synset and coded as the same leaf node in the hierarchical tree. While the polysemous words are coded as the different leaf nodes according to their different senses, that is to say, the same word may have different codes in Cilin. As shown in Fig. 2, the polysemous word "材料" is coded in different synsets "Al03B01 = 人才 (talents), 材料 (stuff) …" and "Ba06A02 = 材料 (material), 材质 (material) …" etc.

We take advantage of the architecture of Cilin by assuming that words with the same code form the synsets, and polysemous words with different synset codes could provide the senses for sense embedding learning. Inspired by *AutoExtend* [7], our model learn sense embedding $sen^{(i,j)} \in \mathbb{R}^n$ of a word together with its word embedding $w^{(i)} \in \mathbb{R}^n$ and the relevant synset embedding $syn^{(j)} \in \mathbb{R}^n$ by premising that $w^{(i)} = \sum_j sen^{(i,j)}$ and $syn^{(j)} = \sum_i sen^{(i,j)}$.

Let $W \in \mathbb{R}^{|W| \times n}$ be a matrix of word embeddings with the size of $|W|$ and $S \in \mathbb{R}^{|S| \times n}$ be a matrix of synset embeddings with the size of $|S|$. We can use a rank 4 tensor $E \in \mathbb{R}^{|S| \times n \times |W| \times n}$ to encode the matrix W into the matrix S and similarly use a rank 4 tensor $D \in \mathbb{R}^{|W| \times n \times |S| \times n}$ to decode the matrix S into the matrix $\overline{W}$. We can state the learning objective under the synset constraints as follows:

$$loss_{syn} = \arg\min_{E,D} \|D \otimes E \otimes W - W\| \tag{4}$$

where $\otimes$ is tensor product.

The sense embeddings are defined when transitioning from W to S or transitioning from S to $\overline{W}$. Aligning these two representations, we could get the sense constraints as follows:

$$loss_{sen} = \arg\min_{E,D} \|E \otimes W - D \otimes S\| \tag{5}$$

The above two constraints are very similar to *AutoExtend* architecture [7]. However, the architecture of Cilin leads to a problem that there is a large number of synsets which have only one word (sense). This makes it hard to learn high quality embeddings for single-word synsets. To remedy this problem, we use the relation constraints that the synsets in Cilin which are with the similar codes (either the upper 3 layer codes or 4 layer codes are the same) should have the similar senses and the distribution in the vector space should be close as well. Let $R \in \mathbb{R}^{r \times |S|}$ be the relation matrix, where r is the number of relation tuples. For each row, the dimension corresponding to the original and related synsets are set to 1 and -1, respectively. The relation constraints could be written as follows:

$$loss_{rel} = \arg\min_{E} \|RE \otimes W\| \tag{6}$$

The final training objective is minimizing the sum of synset constraints, sense constraints and relation constraints and more explicitly by:

$$loss = \alpha \cdot loss_{syn} + \beta \cdot loss_{sen} + (1 - \alpha - \beta) \cdot loss_{rel} \tag{7}$$

where α is the weight of synset constraints and β is the weight of sense constraints. By giving the three constraints different weights, we can easily learn the sense embeddings and synset embeddings.

3.3 Similarity Computation and Combination

Similarity Computation. Word similarity is calculated using the *AvgSim* and *MaxSim* metrics [3] respectively:

$$AvgSim(u, v) = \frac{1}{K_u \times K_v} \sum_{i=1}^{K_u} \sum_{j=1}^{K_v} \cos(u^i, v^j) \tag{8}$$

$$MaxSim(u, v) = \max_{1 \leq i \leq K_u, 1 \leq j \leq K_v} \cos(u^i, v^j) \tag{9}$$

where K_u and K_v are the number of senses for words u and v, $\cos(u^i, v^j)$ is the cosine similarity between the ith sense embedding of word u and the jth sense embedding of word v.

Similarity Combination. The best similarity results of corpus-based and lexicon-based methods are combined using 5 strategies based on fundamental math operations. For each similarity score pair Sim_c and Sim_l of the corpus-based and lexicon-based model, we calculate a combination score Sim according to the combining strategy. The 5 strategies are defined as follows:

Max

$$Sim = max(Sim_c, Sim_l) \tag{10}$$

Average

$$Sim = \frac{Sim_c + Sim_l}{2} \tag{11}$$

Max and Average

$$Sim = \begin{cases} max(Sim_c, Sim_l) & Sim_c \neq 0, Sim_l \neq 0 \\ \frac{Sim_c + Sim_l}{2} & Sim_c = 0 \ or \ Sim_l = 0 \end{cases} \tag{12}$$

Replace 0

$$Sim = \begin{cases} \frac{Sim_c + Sim_l}{2} & Sim_c \neq 0, Sim_l \neq 0 \\ Sim_l & Sim_c = 0 \\ Sim_c & Sim_l = 0 \end{cases} \tag{13}$$

Improved Geometric Mean

$$Sim = \begin{cases} \sqrt{Sim_c * Sim_l} & Sim_c \neq 0, Sim_l \neq 0 \\ Sim_l & Sim_c = 0 \\ Sim_c & Sim_l = 0 \end{cases} \tag{14}$$

4 Experiments

4.1 Experimental Setup

We pre-train word embeddings and learn sense embeddings both on two corpora: the Sogou Chinese news corpus[1] and Chinese Wikipedia[2]. All corpora are word segmented by Stanford Word Segmenter[3]. In corpus preprocessing, we remove the common punctuations and some rare symbols. All the numbers in the corpus are replaced by "XXXX". The embedding training datasets is relatively large, with a total of 400 million tokens. With GloVe tool[4], we build a vocabulary of the 150,000 most frequent words whose word frequency are no less than 50. The initial embedding dimension is set to 300, and the sampling window size is set to 10 for the co-occurrence statistics.

In sense embedding learning, the semi-width R_t of the context window is set to 5 and the initialization weights of synset constraints α, sense constraints β and relation constraints $(1 - \alpha - \beta)$ are set to 0.2, 0.5 and 0.3, respectively.

Following Guo et al. [9] we use Spearman correlation ρ for evaluation. The proposed methods are evaluated on Chinese Polysemous Word Similarity Dataset [9]. Chinese polysemous words in the dataset are extracted according to their sense definitions in HowNet [6]. Then several other polysemous words (from related to unrelated) are selected to form word pairs with each polysemous word. Finally, 401 word pairs are manually annotated with their similarity scores from 0.0 to 10.0 by human judgments. E.g., word "制服 (control)" paired with "征服 (conquer)" gets the similarity score of 8.60. But if paired with "重点 (focus)", the similarity score will only be 0.12.

4.2 Evaluation Results

We first evaluate similarity performance by computing the nearest neighbors across corpus-based and lexicon-based sense embeddings respectively. Table 1 lists the nearest word (W/), sense (Sen/) and synset (Syn/) of some senses of polysemous words.

As can be seen from Table 1, for word "开阔", both of the two models could find concrete and abstract senses well. As for word "制服", which has different Part-of-Speech senses, the corpus-based model could distinguish between none and verb senses but the lexicon-based model could not. As for word "材料", which has different aspects of senses, the lexicon-based model could find the sense refers to stuff but the corpus-based could not. This inspires us to combine the two models using some combination strategies.

The Results of Corpus-Based Sense Embedding Learning. Figure 3 shows the similarity results, in which the selectivity factor R_t^* is varied from 1 to 5 with an interval of 1. From the Fig. 3 we can see that:

---

[1] http://www.sogou.com/labs/dl/c.html.

[2] https://dumps.wikimedia.org/zhwiki/latest/zhwiki-latest-pages-articles.xml.bz2.

[3] http://nlp.stanford.edu/software/segmenter.shtml.

[4] https://nlp.stanford.edu/projects/glove/.

Table 1. Nearest neighbors of the senses of some polysemous words.

Methods	Center word	Nearest neighbors
Corpus-based	制服1 (uniform)	W/队服 (jersey), W/礼服 (full dress), W/军装 (military uniform)
	制服2 (control)	W/歹徒 (gangster), W/擒获 (catch), W/抓捕 (arrest)
	材料1 (material)	W/超硬 (super-hard), W/核级 (nuclear), W/金属 (metal)
	材料2 (data)	W/文件 (file), W/资料 (data), W/提交 (submit)
	开阔1 (wide)	W/绿树 (green trees), W/山峦 (mountains), W/环绕 (surround)
	开阔2 (broad)	W/视野 (view), W/实践 (practice), W/学习 (study)
Lexicon-based	制服1 (control)	Sen/歹徒 (gangster), W/歹徒 (gangster), Syn/穿,穿着 (wear)
	材料1 (material)	W/材料 (material), Syn/材料,材质 (material), Sen/材质 (texture)
	材料2 (data)	Syn/资料,材料 (material), W/材料 (material), Sen/资料 (data)
	材料3 (stuff)	W/材料 (material), Sen/人才 (talent), Syn/人才,材料 (talent)
	开阔1 (wide)	W/开阔 (open), Syn/广阔,宽阔 (vast), Sen/广阔 (vast)
	开阔2 (broad)	W/开阔 (wide), Sen/视野 (view), W/视野 (view)

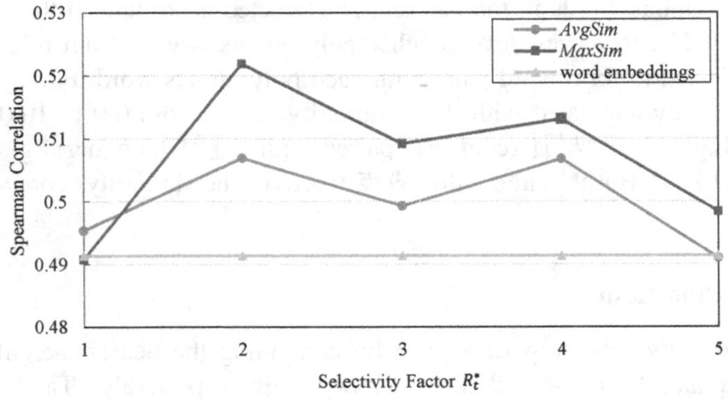

Fig. 3. Spearman correlation ρ based on large-scale corpora.

- The sense embeddings learned on the large-scale corpus could improve the performance of the single word embeddings, which only achieves the Spearman correlation ρ of 0.491.
- The best performances are under the conditions $R_t^* = 2$. It indicates that context represented by the words with the largest R_t^**-th tf-idf* weights could improve the cluster performance and therefore obtain the superior performance of sense embeddings than represented by the total words ($R_t^* = 5$). The maximum similarity scores under the conditions $R_t^* = 2$ are used as the corpus-based similarity scores for the combination of the two methods in the following experiments.

The results of lexicon-based sense embedding learning. In this section, we use the initialization weights of synset constraints, sense constraints and relation constraints ($\alpha = 0.2$ and $\beta = 0.5$). If the test pairs contain *out of vocabulary* (OOV) words, we simple assign the similarity with the minimum score of 0. *AvgSim* and *MaxSim* similarity measurement of sense embeddings and synset embeddings are listed in Table 2. For word embeddings, each word pair has one word similarity since each word has one word embedding. The Spearman correlation ρ of word embeddings is 0.471, which is not listed in Table 2.

Table 2. Spearman correlation ρ based on Tongyici Cilin.

Embeddings	*AvgSim*	*MaxSim*
Sense embeddings	**0.479**	**0.517**
Synset embeddings	0.161	0.161

From Table 2, we can see that:

- Sense embeddings significantly outperform word embeddings and synset embeddings using either *MaxSim* or *AvgSim* measurement. Sense embeddings could capture the differences senses of polysemous words and therefore get the best performance.
- Synset embeddings get the worst results. The reason is that synset embeddings of the word pair appears in the same synset are the same, which fails to distinguish synonyms.

The similarity scores of sense embeddings using *MaxSim* measurement are used as the lexicon-based similarity scores for the combination of the two methods in the following experiments.

The Effects of the Combination Strategies. Table 3 shows the similarity results of the 5 combination strategies. From Table 3 we can see that all of the combination strategies could improve the performance significantly. Average strategy achieves the best ρ values of 0.578 among all of the combination strategies. This is because it could balance between the similarity results of the two methods which belong to numerical data. This indicates that the lexicon-based and corpus-based methods are both effective and complementary for capturing the senses of polysemous words.

Table 3. Combination results based on 5 strategies.

No.	Strategy	ρ
1	Max	0.561
2	Average	**0.578**
3	Max and average	0.556
4	Replace 0	0.576
5	Improved geometric mean	0.565

We further investigate the impact of the parameters α and β that control the weighting of synset constraints vs. sense constraints vs. word relation constraints. The Average and the Replace 0 strategies are employed as these two strategies show better results than other strategies.

As shown in Fig. 4, the best performance weighting area of the 2 combination strategies are both in the center and therefore all the three constraints are effective for sense embedding learning. It should be noted that average strategy achieves the best ρ values of 0.582 with the parameters $\alpha = 0.5$ and $\beta = 0.4$. This indicates that the lexicon-based and corpus-based methods are effective and complementary for capturing the multiple senses of polysemous words.

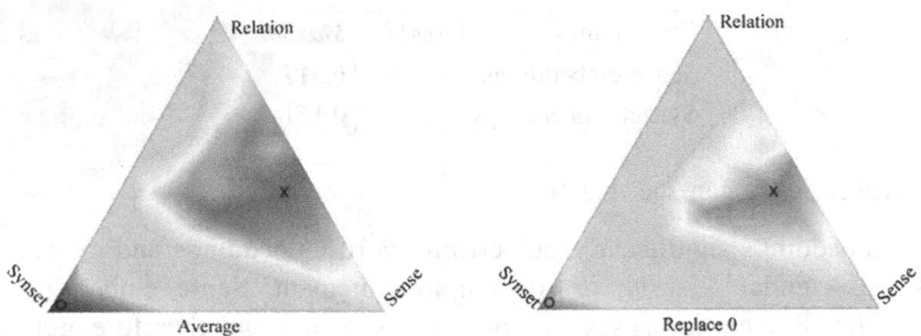

Fig. 4. Performance of different combination strategies with adjusting weightings of three constraints. "x" indicates the maximum and "o" indicates the minimum.

4.3 Compare with Related Work

As can be seen in Table 4, compared with word embedding methods, our method achieves a significant improvement on the baseline system of GloVe [13] and other methods such as Skip-Gram model of Word2vec [11, 12]. This indicates that sense embeddings is more beneficial to polysemous words similarity computation than single word embeddings. For other sense embedding methods, Huang et al. [4] cluster the contexts of polysemous word with a uniform parameter K (the number of clusters), and then learn embeddings of the sense-labeled words. Comparing with Huang et al. [4], our corpus-based method clusters the effective context representations with the real number of polysemous senses, and achieves a ρ value of 0.522. Guo et al. [9] propose a novel approach for Chinese sense embeddings leaning by exploiting bilingual parallel data. Their method heavily depends on high quality bilingual parallel data. Benefiting from the complementary relationship of corpus-based and lexicon-based methods, we improve the ρ value to 0.582 with Average combination strategy.

Table 4. Spearman correlation ρ on the dataset compared with the previous methods.

Methods	ρ
GloVe (*baseline*)	0.492
Word2vec	0.497
Huang et al. [4]	0.407
Guo et al. [9]	0.554
Corpus-based	0.522
Lexicon-based	0.517
Combination	**0.582**

5 Conclusions

This paper proposes a framework for Chinese polysemous words similarity measurement, including a corpus-based method, a lexicon-based method. Evaluation on the Chinese Polysemous Word Similarity Dataset shows both the two methods could capture the senses of polysemous words well. Furthermore, we investigate the complement relation of the two methods and improve the ρ value up to 0.582 by combining the two methods, which outperforms the state-of-the-art performance on the dataset. In the future, we would like to exploit more rich knowledge resources for sense embedding learning and investigate more effective combining strategies for Chinese polysemous words similarity computation.

Acknowledgements. This research is supported by Natural Science Foundation of China (No. 61272375).

References

1. Turian, J., Ratinov, L., Bengio, Y.: Word representations: a simple and general method for semi-supervised learning. In: Proceedings of ACL, pp. 384–394 (2010)
2. Li, J., Jurafsky, D.: Do multi-sense embeddings improve natural language understanding. In: Proceedings of EMNLP, pp. 1722–1732 (2015)
3. Reisinger, J., Mooney, R.J.: Multi-prototype vector-space models of word meaning. In: Proceedings of NAACL-HLT, pp. 109–117 (2010)
4. Huang, E.H., Socher, R., Manning, C.D., Ng, A.Y.: Improving word representations via global context and multiple word prototypes. In: Proceedings of ACL, pp. 873–882 (2012)
5. Salton, G., Buckley, C.: Term-weighting approaches in automatic text retrieval. Inf. Process. Manage. **24**(5), 513–523 (1988)
6. Dong, Z.D., Dong, Q.: HowNet and the computation of meaning. In: World Scientific, pp. 85–95 (2006)
7. Rothe, S., Schütze, H.: Autoextend: extending word embeddings to embeddings for synsets and lexemes. In: Proceedings of ACL, pp. 1793–1803 (2015)
8. Che, W.X., Li, Z.H., Liu, T.: LTP: a Chinese language technology platform. In: Proceedings of COLING, pp. 13–16 (2010)
9. Guo, J., Che, W.X., Wang, H.F., Liu, T.: Learning sense-specific word embeddings by exploiting bilingual resources. In: Proceedings of COLING, pp. 497–507 (2014)
10. Bengio, Y., Ducharme, R., Vincent, P., Jauvin, C.: A neural probabilistic language model. J. Mach. Learn. Res. **3**, 1137–1155 (2003)
11. Mikolov, T., Chen, K., Corrado, G., Dean, J.: Efficient estimation of word representations in vector space. In: Workshop at ICLR (2013)
12. Mikolov, T., Sutskever, I., Chen, K., Corrado, G., Dean, J.: Distributed representations of words and phrases and their compositionality. In: NIPS, pp. 3111–3119 (2013)
13. Pennington, J., Socher, R., Manning, C.D.: Glove: global vectors for word representation. In: Proceedings of EMNLP, pp. 1532–1543 (2014)
14. Neelakantan, A., Shankar, J., Passos, A., McCallum, A.: Efficient non-parametric estimation of multiple embeddings per word in vector space. In: Proceedings of EMNLP, pp. 1059–1069 (2014)

15. Zheng, X.Q., Feng, J.T., Chen, Y., Peng, H.Y., Zhang, W.Q.: Learning context-specific word/character embeddings. In: Proceedings of the AAAI 2017, pp. 3393–3399 (2017)
16. Chen, T., Xu, R.F., He, Y.L., Wang, X.: Improving distributed representation of word sense via WordNet gloss composition and context clustering. In: Proceedings of ACL, pp. 15–20 (2015)
17. Pei, J.H., Zhang, C., Huang, D.G., Ma, J.J.: Combining word embedding and semantic lexicon for Chinese word similarity computation. In: Proceedings of NLPCC, pp. 766–777 (2016)

Latent Dirichlet Allocation Based Image Retrieval

Jing Hao and Hongxi Wei(✉)

School of Computer Science, Inner Mongolia University, Hohhot 010021, China
cswhx@imu.edu.cn

Abstract. In recent years, Bag-of-Visual-Word (BoVW) model has been widely used in computer vision. However, BoVW ignores not only spatial information but also semantic information between visual words. In this study, a latent Dirichlet allocation (LDA) based model has been proposed to obtain the semantic relations of visual words. Because the LDA-based topic model used alone usually degrade performance. Thus, a visual language model (VLM) is combined with LDA-based topic model linearly to represent each image. On our dataset, the proposed approach has been compared with state-of-the-art approaches (such as BoVW, LLC, SPM and VLM). Experimental results indicate that the proposed approach outperforms the original BoVW, LLC, SPM and VLM.

Keywords: Image retrieval · Latent dirichlet allocation · Visual language model · Query likelihood model · Smoothing

1 Introduction

In recent years, image retrieval has attracted more and more attention. But, how to improve the performance is still one of the hot and difficult points in the computer vision.

Since the Bag-of-Word model in the field of text processing has obtained a huge success, researchers transplanted it to the field of computer vision that per image expressed into a visual histogram. And it has shown impressive levels of performance. Due to the bag-of-word model disregard of spatial information relative to its visual words [1, 2], researchers have proposed approaches for improving performance. In [3], the authors generated a higher-level lexicon to reduce the ambiguity in representation. Cao Y et al. demonstrated a novel technique that developing a new class of bag-of-features to encode geometric information of objects within an image [4]. Lazebnik et al. proposed an algorithm named Spatial Pyramid Matching (SPM) that utilizes spatial pyramid [7] for image matching, identification and classification [5]. Wang et al. presented an approach named Locality-constrained Linear Coding (LLC) that similar basis of local descriptor is selected by locality constraint to reconstruct each descriptor [6]. Ren et al. presented an approach named bag-of-bags of words (BBoW) that consider spatial constraints in the image space rather than feature space, color and limited spatial information are incorporated into image representation for image retrieval [8]. Jégou et al. presented a graph-structured quantize to ameliorate the performance of assign SIFT descriptors to visual words [9].

J. Wen et al. (Eds.): CCIR 2017, LNCS 10390, pp. 211–221, 2017.
https://doi.org/10.1007/978-3-319-68699-8_17

In this paper, a method of linear combination of LDA-based topic model and visual language model to represent an image was taken. First of all, visual words of each image are built into a visual language model, which draws on the language model applied in text information retrieval. Subsequently, the LDA-based topic model is leveraged to extract the semantic relationship of visual words. Finally, the corresponding topic model and the visual language model are linearly united for representing an image. The detailed process is shown in Sect. 2.

The rest of the paper is organized as follows. Section 2 describes our method that combine Latent Dirichlet Allocation (LDA) with a visual language model linearly for image representation. Section 3 shows our experimental results. Finally, Sect. 4 concludes the paper.

2 Method

This section introduces our method of image retrieval. We extracted descriptors from training set by using Scale-Invariant Feature Transform (SIFT) [11], and then k-means is performed on those SIFT descriptors to build the codebook. Afterwards, each image is modeled by visual language model and LDA-based topic model respectively. Finally, the images are represented by combining the two models linearly. The aforementioned details are presented in the following sub-sections.

2.1 Visual Language Model of Image

The visual language model is used to establish the probability distribution of an image by the following equation.

$$P(v_i|d) = \frac{\text{count}(v_i)}{\sum_{v \in d} \text{count}(v)} \tag{1}$$

Where $\text{count}(v_i)$ means the occurrence frequency of the $i^{th}(1 \leq i \leq n)$ visual word in an image d, $\sum_{v \in d} \text{count}(v)$ means the total of visual words in the image d.

At the retrieval stage, if a visual word of the query does not appear in an image, and then it will be assigned a zero probability. Generally, a smoothing scheme can deal with the visual language models of images before being retrieved to avoid zero probabilities. In our experiment, we use the Dirichlet smoothing method [12] to avoid zero probabilities. The formula of the Dirichlet smoothing method is defined as follows:

$$P(v_i|d) = \frac{\text{count}(v_i) + \mu P(v_i|C)}{\sum_{v \in d} \text{count}(v) + \mu} \tag{2}$$

Where μ is a parameter which is ranged from 1000 to 2000. $P(v_i|C)$ means the occurrence frequency of the visual word in the collection of all images.

By using the visual language model, each image can be represented as a distribution of probabilities. In Sect. 3, the value of the parameter μ is set to 1000 by manually selecting the optimal value.

2.2 LDA-Based Topic Model of Image

Latent Dirichlet allocation (LDA) is a topic model that be proposed by reference [13]. LDA was first applied to the text fields originally, it can represent per document's topic as a probability distribution. We calculate the probability of a word in a document by the following formula.

$$P\left(w\middle|d, \hat{\theta}, \hat{\varphi}\right) = \sum_{z=1}^{Z} P(w|z, \hat{\varphi}) \cdot P\left(z\middle|\hat{\theta}, d\right) \tag{3}$$

Where z is a topic selected from the Dirichlet distribution θ_d and Z is the total number of topics. $\hat{\theta}$ and $\hat{\varphi}$ are the posterior estimates of θ which is a multinomial distribution over document and φ which is a multinomial distribution over topic, respectively. The LDA model is difficult to calculate exactly due to its complexity. So, it is an approximate method which is used to calculate, we obtain $\hat{\theta}$ and $\hat{\varphi}$ directly by using Gibbs sampling [14]. Their formulas are defined as follows.

$$\hat{\varphi} = \frac{n_j^{(w)} + \beta}{\sum_{v=1}^{V} n_j^{(v)} + V \cdot \beta}, \hat{\theta} = \frac{n_j^{(d)} + \alpha}{\sum_{t=1}^{T} n_t^{(d)} + T \cdot \alpha} \tag{4}$$

Where V and T are the number of words and topics, respectively. $n_j^{(w)}$ is the number of word w assigned to topic j and $n_j^{(d)}$ is the number of words in document d assigned to topic j. And α and β are hyper-parameters. Therefore, $\sum_{v=1}^{V} n_j^{(v)}$ is the total number of words assigned to topic j and $\sum_{t=1}^{T} n_t^{(d)}$ is the total number of words in document d.

When (3) is applied to the image, w means a visual word and d means an image. The formula of combine visual language model with LDA-based topic model linearly is defined as follows.

$$P(v|d) = \lambda \cdot \hat{P}(v|d) + (1 - \lambda) \cdot \left(\sum_{z=1}^{Z} P(v|z, \hat{\varphi}) \cdot P\left(z\middle|\hat{\theta}, d\right)\right) \tag{5}$$

Where λ (0 < λ < 1) is the parameter controlling proportion of linear combination. In the Eq. (4), we can see that the larger λ is the proportion of representation of visual language model is much greater.

2.3 Retrieval Scheme

After the aforementioned processing, each image is represented as a probability histogram. Retrieval scheme is used the query likelihood model (QLM). The formula of rank images for a given query is defined as follows.

$$P(q|d) = \prod_{v \in q} P(v|d) \tag{6}$$

Where q is an image as a query, and v is a visual word occurred in the query q. According to (4) and (5), the Eq. (6) can be rewritten as follows.

$$P(q|d) = \prod_{v \in q} \left(\lambda \cdot \hat{P}(v|d) + (1-\lambda) \cdot \left(\sum_{z=1}^{Z} P(v|z,\hat{\varphi}) \cdot P\left(z\middle|\hat{\theta}, d\right)\right)\right) \tag{7}$$

At the retrieval stage, a ranking list of images can be formed in descending order of (7).

3 Experimental Results

In this section, we considered the original BoVW (Bag-of-Visual-Word) model as a baseline. Other methods, such as LLC (Locality-constrained Linear Coding), SPM (Spatial Pyramid Matching), Visual Language Model and LDA-based representation model, have been compared with the baseline.

3.1 Dataset

Our experiment was performed on the Caltech-101 database [10]. This database is composed of 102 categories and each category contains about 32 to 800 images. Size of each image is roughly 300 * 200 pixels. Due to the dataset contains variable categories, such as faces, cannon, etc., and the number of images reasonable, this database was selected as our dataset. The BACKGROUND_Google was removed, and rest in each category of the 80% for training and 20% as the query images to be tested.

In our experiment, the number of the clusters, namely a size of codebook, rises from 500 to 10000 and increases 500 each time. We evaluated performance on the training set, the full set, and the testing set respectively. And evaluation metric is mean Average Precision (mAP) [16].

3.2 Performance of the BoVW

In the original BoVW model, each image was transformed into a histogram of visual words using the tf-idf (term frequency–inverse document frequency) [16] scheme for weighting.

In Fig. 1, the performance of the original BoVW on the training set is ranged from 10.66% to 12.85%, on the full set is ranged from 11.06% to 13.56%, and on the testing set, the MAP is ranged from 11.37% to 14.56%.

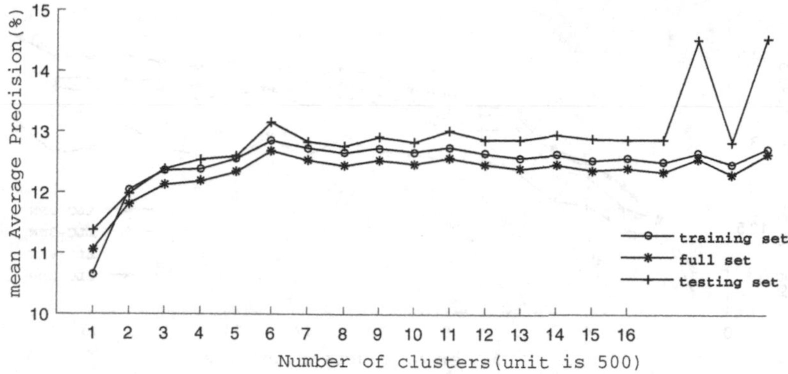

Fig. 1. The performance of BoVW on three sets

We can know that on the three datasets, performance curves are increased to the cluster equals to 3000, and then leveled off. Fluctuation appears in performance curve when cluster equals to 9000 and 10000. MAP on the testing set betters than other sets.

3.3 Performance of the LLC

In Fig. 2, the best performance on the training set is attained to 14.87% when the K (K nearest neighbor visual words of a feature) is set to 3. In Fig. 3, the best performance on the full set is attained to 14.65% when the K is set to 3. In Fig. 4, the best performance on the training set is attained to 15.10% when the K is set to 3. As we can see, LLC get the best performance when K equals to 3.

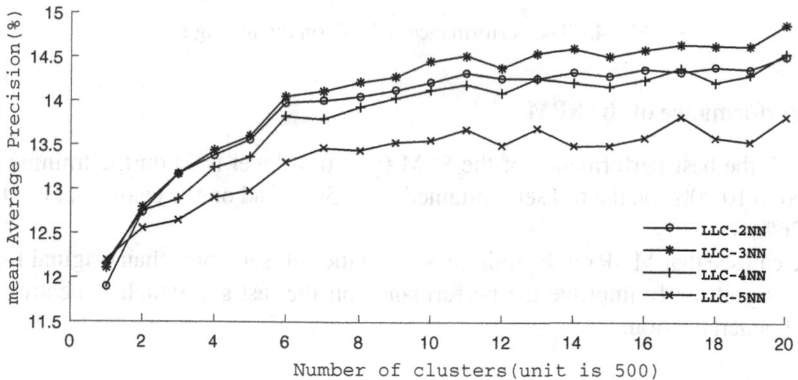

Fig. 2. The performance of LLC on the training set

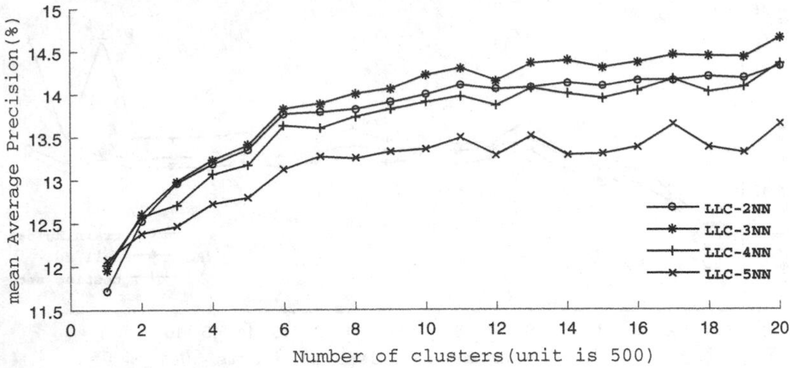

Fig. 3. The performance of LLC on the full set

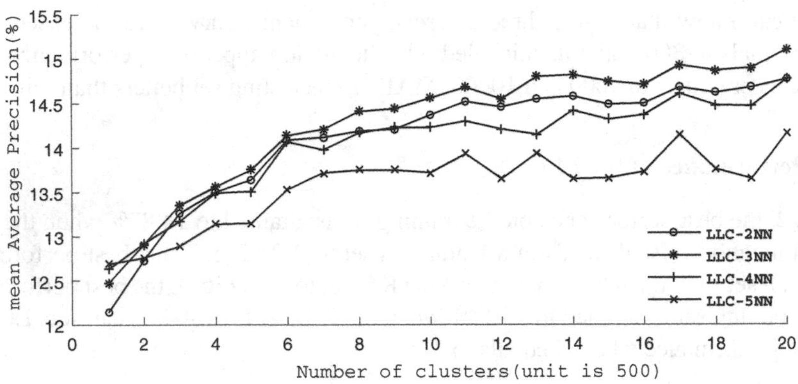

Fig. 4. The performance of LLC on the testing set

3.4 Performance of the SPM

In Fig. 5, the best performance of the SPM (partition level is 2) on the training set is attained to 10.90%, on the full set is attained to 11.50%, and on the testing set is attained to 16.26%.

We can see that MAP on the training set and the full set worse than original BoVW, but can significantly improve the performance on the test set, which indicating SPM good at generalization.

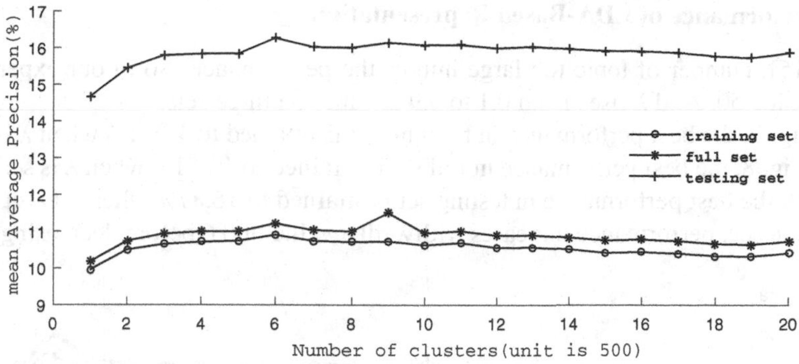

Fig. 5. The performance of SPM on three sets

3.5 Performance of the Visual Language Model

In Fig. 6, the performance of the VLM on the training set is ranged from 9.02% to 11.94%, on the full set is ranged from 8.56% to 11.55%, and on the testing set is ranged from 8.61% to 12.25%.

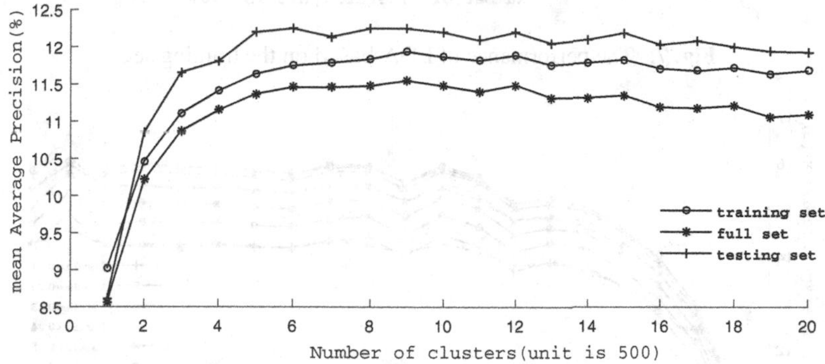

Fig. 6. The performance of VLM on three sets

From Fig. 6, we can conclude that if visual language model is used alone, the performance will worse than original BoVW.

3.6 Performance of LDA-Based Representation

From [15], number of topic too large injures the performance. So in our experiment, topics value 50. And λ rises from 0.1 to 0.9 as same on three sets.

In Fig. 7, the best performance in training set is attained to 17.11% when λ is set to 0.6. In Fig. 8, the best performance in full set is attained to 16.51% when λ is set to 0.6. In Fig. 9, the best performance in testing set is attained to 16.87% when λ is set to 0.6. As we can see, performance increases with λ till a value of 0.6 before decreasing again.

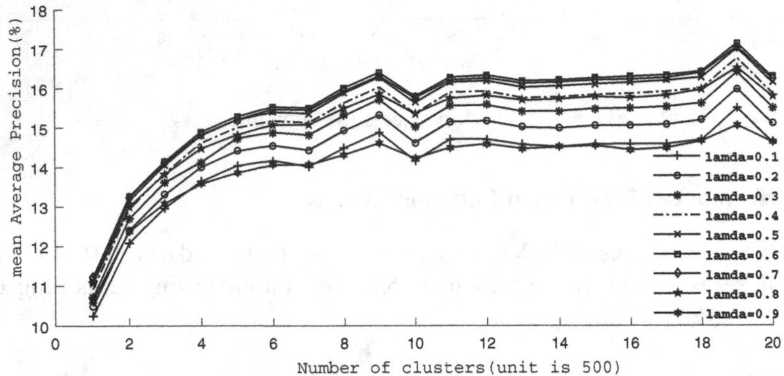

Fig. 7. The performance of LDA-based on the training set

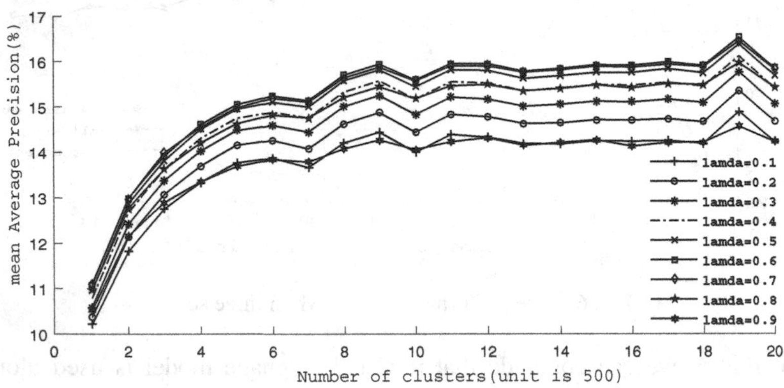

Fig. 8. The performance of LDA-based on the full set

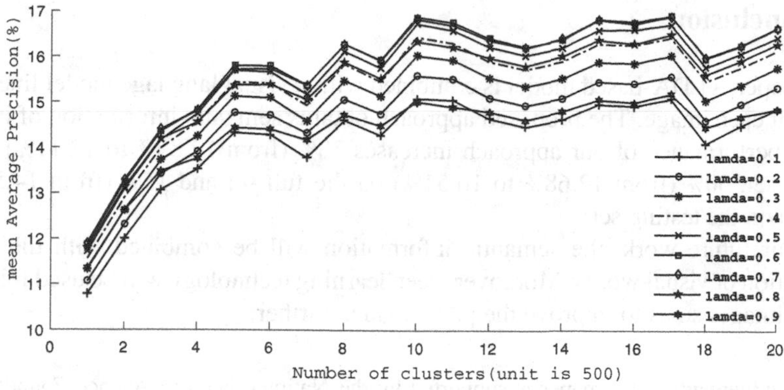

Fig. 9. The performance of LDA-based on the testing set

From the Tables 1, 2, 3, the best performance and the corresponding number of cluster centers of five models are shown, respectively.

Table 1. Best performance and the corresponding number of cluster centers on the training set

Model	MAP	Number of clusters
BoVW	12.85%	3000
LLC	14.87%	10000
SPM	10.90%	3000
VLM	11.94%	4500
LDA-Based	17.11%	9500

Table 2. Best performance and the corresponding number of cluster centers on the full set

Model	MAP	Number of clusters
BoVW	12.68%	3000
LLC	14.65%	10000
SPM	11.21%	3000
VLM	11.55%	4500
LDA-Based	16.51%	9500

Table 3. Best performance and the corresponding number of cluster centers on the testing set

Model	MAP	Number of clusters
BoVW	14.56%	10000
LLC	15.10%	10000
SPM	16.26%	3000
VLM	12.25%	4500
LDA-Based	16.87%	5000

4 Conclusion

In this paper, a LDA-based model is combined with a visual language model linearly to represent each image. The proposed approach obtains semantic information of images.

The performance of our approach increases 33% (from 12.85% to 17.11%) on the training set, 30% (from 12.68% to 16.51%) on the full set and 15% (from 14.56% to 16.87%) on the testing set.

In our future work, the semantic information will be combined with the spatial information of visual words. Moreover, deep learning technology will be used for representing images so as to improve the performance further.

Acknowledgements. The paper is supported by the National Natural Science Foundation of China under Grant 61463038.

References

1. Chen, X., Hu X., Shen, X.: Spatial weighting for bag-of-visual-words and its application in content-based image retrieval. In: Proceedings of PAKDD 2009, pp. 867–874. ACM Press, New York (2009)
2. Willamowski, J., Arregui, D., Csurka, G., et al.: Categorizing nine visual classes using local appearance descriptors. In: Proceedings of ICPR Workshop on Learning for Adaptable Visual Systems. IEEE Press, New York (2004)
3. Yuan, J., Wu, Y., Yang, M.: Discovery of collocation patterns: from visual words to visual phrases. In: Proceedings of CVPR 2007, pp. 1–8. IEEE Press, New York (2007)
4. Cao, Y., Wang, C., Li, Z., et al.: Spatial-bag-of-features. In: Proceedings of CVPR 2010, pp. 3352–3359. IEEE Press, New York (2010)
5. Lazebnik, S., Schmid, C., Ponce, J.: Beyond bags of features: spatial pyramid matching for recognizing natural scene categories. In: Proceedings of CVPR 2006, pp. 2169–2178. IEEE Press, New York (2006)
6. Wang, J., Yang, J., Yu, K., et al.: Locality-constrained linear coding for image classification. In: Proceedings of CVPR 2010, pp. 3360–3367. IEEE Press, New York (2010)
7. Harada, T., Ushiku, Y., Yamashita, Y., et al.: Discriminative spatial pyramid. In: Proceedings of CVPR 2011, pp. 1617–1624. IEEE Press, New York (2011)
8. Ren, Y., Bugeau, A., Benois-Pineau, J.: Bag-of-bags of words irregular graph pyramids vs spatial pyramid matching for image retrieval. In: Proceedings of IPTA 2014, pp. 1–6. IEEE Press, New York (2014)
9. Jégou, H., Douze, M., Schmid, C.: Improving bag-of-features for large scale image search. Int. J. Comput. Vis. **87**(3), 316–336 (2010)
10. Li, F.F., Fergus, R., Perona, P.: Learning generative visual models from few training examples: an incremental bayesian approach tested on 101 object categories. J. Comput. Vis. Image Underst. **106**(1), 59–70 (2007)
11. Lowe, D.G.: Object recognition from local scale-invariant features. In: Proceedings of ICCV 1999, pp. 1150–1157. IEEE Press, New York (1999)
12. Zhai, C., Lafferty, J.: A study of smoothing methods for language models applied to ad hoc information retrieval. In: Proceedings of SIGIR 2001, pp. 334–342. ACM Press, New York (2001)

13. Blei, D.M., Ng, A.Y., Jordan, M.I.: Latent dirichlet allocation. J. Mach. Learn. Res. **3**, 993–1022 (2003)
14. Wei, X., Croft, W.B.: LDA-based document models for ad-hoc retrieval. In: Proceedings of SIGIR 2006, pp. 178–185. ACM Press, New York (2006)
15. Wei, H., Gao, G., Su, X.: LDA-based word image representation for keyword spotting on historical mongolian documents. In: Hirose, A., Ozawa, S., Doya, K., Ikeda, K., Lee, M., Liu, D. (eds.) ICONIP 2016. LNCS, vol. 9950, pp. 432–441. Springer, Cham (2016). doi: 10.1007/978-3-319-46681-1_52
16. Manning, C.D., Raghavan, P., Schütze, H.: An Introduction to Information Retrieval. Cambridge University Press, Cambridge (2009)

Query Processing and Analysis

Leveraging External Knowledge to Enhance Query Model for Event Query

Wang Pengming[1,2(✉)], Li Peng[1,2], Li Rui[1,2], and Wang Bin[1,2]

[1] Institute of Information Engineering,
Chinese Academy of Sciences, Beijing 100093, China
{wangpengming,lipeng,lirui,wangbin}@iie.ac.cn
[2] School of Cyber Security,
University of Chinese Academy of Sciences, Beijing 100029, China

Abstract. Retrieval based on event query has recently become one of the most popular applications in information retrieval domain, whose goal is to retrieve event-related documents according to the given query about some specific event. However, using conventional retrieval method for this kind of task would usually be demonstrated with poor performance. To enhance query model and improve retrieval effectiveness for event query, an adaptive learning approach of PLSA model is presented in this paper. Through leveraging the knowledge of known coarse-grained events from external resource, the new approach can adaptively adjust the topic generative process of PLSA model on pseudo-relevance feedback documents, and learn the accurate language model for a particular topic, i.e., target event, which can be used to update the representation of users intention and finally improve the retrieval results. Experimental results on standard TREC collections show the proposed approach consistently outperform the state-of-the-art methods.

Keywords: Language model · Event-oriented retrieval · Pseudo relevance feedback · Topic model

1 Introduction

Nowadays, once a public event (e.g., natural disaster, protests, accident, etc.) happens, an explosive growth of online related reports would cause information overload for individual users. Therefore, there emerge a series of event-driven research subjects, such as event summarization, sub-event detecting, event monitoring. The rapid rise of the practical scenarios concentrating upon events has also generated much interest in retrieval based on event query, which is the background of this paper.

The purpose of an retrieval system is to retrieve the documents relevant to user's information need expressed in the form of a query, as shown in Fig. 1(a). Retrieval based on event query can be regarded as a specific type of retrieval task, i.e., event-oriented retrieval. Specifically, event-oriented information need often involves special properties of event object, such as the variability of the time

J. Wen et al. (Eds.): CCIR 2017, LNCS 10390, pp. 225–236, 2017.
https://doi.org/10.1007/978-3-319-68699-8_18

and space, which cannot easily be expressed with keyword-based event query. For instance, Given an event query "Queensland floods" by a user who takes an interest in the developments of flood event in Queensland, it's clear that we could not only think about the query keyword "Queensland" and "floods" during the retrieval process, the time and place where the event is taking place, including each stage of the evolution on subordinate states of Queensland, should also be considered. Empirically, using conventional retrieval method for event-oriented retrieval task would usually be demonstrated with poor performance.

This can be explained by the fact that, the result of event-oriented retrieval should be the documents with high $sim(e, d)$ instead of documents with high $sim(q, d)$, that is, the retrieved documents should describe relevant content of target event. Intuitively, there is a target event hidden behind the initial event query, and there might be a discrepancy between term distribution in target event and empirical query model. Motivated by the effectiveness of pseudo-relevance feedback (Fig. 1(b)), the initial event query could also be updated to improve the retrieval result, what differ is using accurate term distribution in target event instead of the term distribution in feedback documents, as shown in Fig. 1(c).

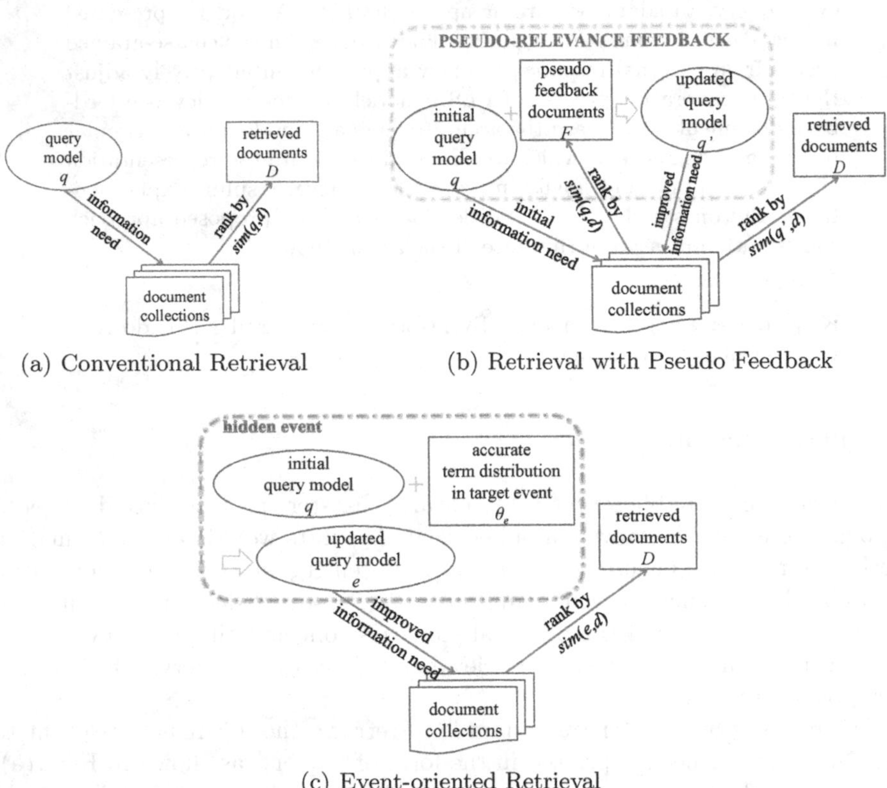

(a) Conventional Retrieval

(b) Retrieval with Pseudo Feedback

(c) Event-oriented Retrieval

Fig. 1. Some retrieval scenarios

Then, the crux of the matter is how to accurately obtain the term distribution in target event hidden behind the initial event query. It is naturally think that the probability distribution of a group of terms could be estimated through topic model, in which target event can be treated as a particular topic. PLSA is a widely-used topic model which can be applied to an arbitrary set of documents to learn a set of latent semantic topic models. However, PLSA is essentially an unsupervised clustering algorithm, in other words the topics would be learned without any guidance. The motivation of our work comes from an interesting idea that the learning process of PLSA might be guided by the information of known coarse-grained events, to get the topic which is really wanted. Thus, ***for the first time***, we propose an adaptive approach of PLSA model in this paper, which makes use of the prior knowledge acquired from external resources, to adaptively learn the accurate term distribution for target event.

The remainder of the paper is organized as follows. In the next section we review the related work and discuss the merits and drawbacks of existing methods for dealing with event-oriented retrieval task. We introduce our approach in detail in Sect. 3 and its implementation in Sect. 4. An experimental evaluation and its results are presented in Sect. 5. A summary and outlook concludes the paper in the last section.

2 Related Work

2.1 Event Identification

Recently, many researchers focus attention on event identification task included in Topic Discovery and Track (TDT) [1]. That task is essentially a cluster task, which is enthusiastic in clustering online-media content, such as newswire and radio, into events based on abundant background information which combined text feature and non-text feature, w.r.t. appropriate document similarity metrics [2].

However, the event-oriented retrieval task studied in this paper essentially remains a retrieval task which needs to return the most relevant documents according to the given query, so the correlativity between them is not significant.

2.2 Methods for Event-Oriented Retrieval

In addition to this paper, there are some researches focused on event-oriented retrieval task recently.

In [3,4], authors proposed a graph-based model for event-centered information retrieval. [3] structures both queries and documents as graphs of event mentions and employ graph kernels to measure the query-document similarity, and [4] novel proposes a bipartite graph to exclusively describe an event. However, these two methods are both sophisticated and time-consuming.

Aiming at the demand for event information by users, [5] proposes a method: local analysis-based event-oriented (LA-EO) query expansion, which divides query terms into two categories: event terms and qualifying term. [6] proposes an event extraction method which combines user reliability and timeline analysis. [7] proposes a simple model to represent real-world news events using two sources of information: the locations that are mentioned in the event (where the event occurs), and the locations of users that discuss or comment on it.

Although the above methods indeed made some achievements by the use of some characteristics of event object, there exists a common problem in these methods that most of them are heuristics, which cannot be explained in principled framework.

3 The New Approach

As aforementioned, from the perspective of pseudo-relevance feedback, there exists significant relevance between feedback documents and target event. Nevertheless, the term distribution in feedback documents might be biased to target event. Along the way, we propose an adaptive learning approach (ALA for short) of PLSA model to enhance query model and improve retrieval effectiveness for event query, whose whole process can be divided into three steps, as depicted in Fig. 2.

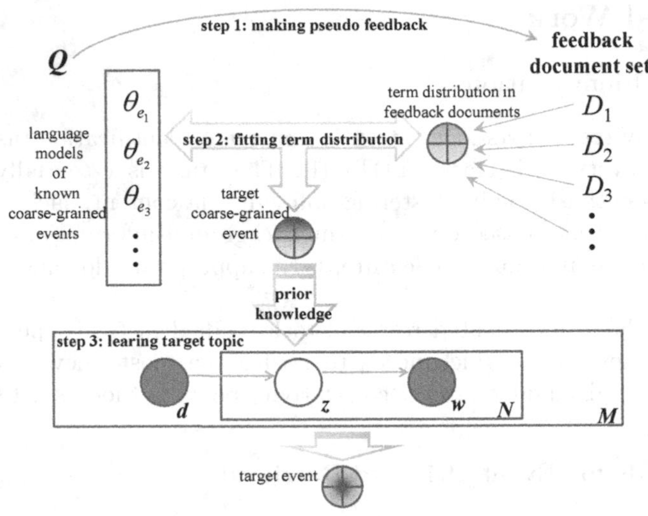

Fig. 2. The framework of ALA

Note that the notion of "coarse-grained events" (marked as $\theta_{e_1}, \theta_{e_2}, \theta_{e_3}, \ldots$ in figure, and referred to hereafter as "coarse event") is critical for the effectiveness of ALA. The argument rests on the assumption that the target event might usually be a poorly understood event, or even an event has yet to happen, so the query model of target event would be quite difficult to get directly. However, the target event tends to be a member of known coarse event. For instance, the event "Flight MH370 Losing Contact" belongs to coarse event "Traffic Accident", while event "Queensland flood" mentioned above should be assigned to coarse event "Flood". The knowledge of such coarse events can always be acquired from external resource, such as Wikipedia. As a consequence, we should pre-build, and real-time update the language models of known coarse events before the execution of ALA.

3.1 Pre-building Event Language Model

An event language model refers to a multinomial distribution of all terms in a event, in other words, if we have to describe a event, the terms we used should follow the term distribution in corresponding event language model. Formally, given the vocabulary $V = \{w_1, w_2, \cdots\}$ and an event e, the language model of that event can be written as: $\theta_e = \{p(w_i)\}_{i=1}^{|V|}$, where $\sum_{i=1}^{|V|} p(w_i) = 1$.

There are many options to build language models for known events. After many tries, we find that event language model isn't very sensitive to building approaches because high frequency terms in different resulting models are almost the same. In this case, we use that approach as follow in this paper: Pulling out a fixed number of documents (for instance, 100) for every known event from event-annotated corpus, and establishing the language model of an event through counting terms' frequency in corresponding documents.

It is important to note that new event would emerge continuously, so we cannot build language model for all events. Consequently, we assume that there also exists an unknown event $e_{unknown}$ and corresponding language model $\theta_{unknown}$. In this paper, the term distribution in $\theta_{unknown}$ is supposed to be consistent with the term distribution in the whole given document collection[1].

3.2 Identifying Target Coarse Event

With $e_{unknown}$, the coarse event set E_{coarse} can be regarded as a complete set, we argue that there exists a probability distribution in event set E_{coarse}, so the probability of generating a term w is:

$$p(w|\theta_{combined}) = \sum_{i=1}^{k} \lambda_i p(w|\theta_{e_i}). \tag{1}$$

where θ_{e_i} is the language model of ith event.

Thus the log-likelihood function for the entire set of feedback documents is:

$$\log p(F|\theta_{combined}) = \sum_{w \in V} c(w, F) \log(\sum_{i=1}^{k} \lambda_i p(w|\theta_{e_i})). \tag{2}$$

where $c(w, F)$ is the count of term w in the set of feedback documents F.

[1] This is reasonable as, in general, we only need to confirm the term distribution in unknown event not be consistent with any known event language model.

Intuitively, λ_i indicates how much faith should we put in $e_{target} = e_i$, so we can finally determine what event e_{target} is by that weight. Quite evidently, there exist a group of latent variables between events and terms which can represent the probability of a term generated by individual coarse event, so it is naturally think that we can realize the log-likelihood function's maximization through expectation-maximization (EM) algorithm. Specifically, the EM algorithm can start with a random initialization of $\{\lambda_i\}_{i=1}^{k}$[2], and then improve $\theta_{combined}$ by iteratively alternating between an E-step and an M-step.

In E-step, we would use the following equation to compute the posterior probability of a term w being generated using $p(\cdot|\{\theta_{e_i}\}_{i=1}^{k})$ based on the current estimate of $\theta_{combined}$:

E-step: $p(z_{w,e_i} = 1) = \dfrac{\lambda_i^{(n)} p(w|\theta_{e_i})}{\sum_{i=1}^{k} \lambda_i^{(n)} p(w|\theta_{e_i})}$

where $z_{w,e_i} \in [0,1]$ is a hidden variable indicating whether term w is generated using ith event language model θ_{e_i} ($z_{w,e_i} = 1$).

Naturally, we try to "guess" which model has been used to generate term w. However, $p(z_{w,e_i} = 1)$ does not tell us for sure whether term w is generated using θ_{e_i} , it's only a expectation probability. Thus, in M-step, we would use a discounted term count (i.e., $c(w, F)p(z_{w,e_i} = 1)$) for estimating λ_{e_i}.

M-step: $\lambda_i = \dfrac{\sum_{w \in V} p(z_{w,e_i} = 1) c(w, F)}{\sum_{w \in V} c(w, F)}$

The EM algorithm is guaranteed to converge to a local maximum of the likelihood function [8]. In our case, the likelihood function has just one local maximum, so we are guaranteed to find the global maximum and get the most optimal weight set $\{\lambda_i\}_{i=1}^{k}$, and the target coarse event can be identified as the event whose weight is maximal in weight set, except when the two biggest event weights are close to each other.

3.3 Adaptive Topic Learning

Along with the basic idea of PLSA, we assume that there are k latent topics in a set of documents, and each topic would be represented using a unigram language model, $\theta_i (i = 1, \cdots, k)$. In turn, we can fit such a mixture model to a collection of documents to discover the exact term distributions of the k latent topics. Formally, let $C = D_1, \cdots, D_{|C|}$ be a collection of documents. Clearly, the log-likelihood of the collection is

$$\log p(C|\Lambda) = \sum_{D \in C} \sum_{w \in V} c(w, D) \log \sum_{i=1}^{k} p(i|D) p(w|\theta_i). \quad (3)$$

where Λ denotes the set of all parameters, i.e., $\Lambda = \{p(i|D)\}_{i \in [1,k], D \in C} \cup \{\theta_i\}_{i \in [1,k]}$.

[2] In this paper, we don't have any prior knowledge in this step, so we can only use the same initial value for all λ.

To perform latent semantic analysis with PLSA, we would solve the optimization problem efficiently with the EM algorithm:

$$p(z_{D,w} = i) = \frac{p^{(n)}(i|D)p^{(n)}(w|\theta_i)}{\sum_{i'=1}^{k} p^{(n)}(i'|D)p^{(n)}(w|\theta_{i'})} \quad (4)$$

$$p^{(n+1)}(w|\theta_i) = \frac{\sum_{D\in C} c(w,D)p(z_{D,w} = i)}{\sum_{w'\in V}\sum_{D\in C} c(w',D)p(z_{D,w'} = i)} \quad (5)$$

$$p^{(n+1)}(i|D) = \frac{\sum_{D\in C} c(w,D)p(z_{D,w} = i)}{\sum_{i'=1}^{k}\sum_{w\in V} c(w,D)p(z_{D,w} = i')} \quad (6)$$

where $z_{D,w} \in [1,k]$ is a hidden variable indicating which topic model has been used to generate term w in document D, and the superscripts n and $n+1$ indicate the iterations.

In the scenario of event-oriented retrieval task, however, we are only concerned with a particular topic, that is, target event. For this purpose, a solution falls into the following strategy: we can naturally incorporate the information of coarse event through defining priors on parameters and adaptive learning a particular topic.

Specifically, we can use $p(\Lambda)$ to specify our preferences on what kinds of topics we would like to discover and add our knowledge about which document covers which topic. Since Dirichlet distribution is a conjugate prior for multinomial distributions, we only need to slightly modify the M-step for re-estimating θ_i in the EM algorithm to incorporate this prior [9]:

$$p^{(n+1)}(w|\theta_i) = \frac{\sum_{D\in C} c(w,D)p(z_{D,w} = i) + \alpha_w}{\sum_{w'\in V}(\sum_{D\in C} c(w',D)p(z_{D,w'} = i) + \alpha_w)}. \quad (7)$$

Clearly, if we parameterize α_w based on a prior distribution $p(w|\varphi_i)$ such that $\alpha_w = \mu p(w|\varphi_i)$, we would essentially favor an estimated θ_i close to φ_i. In ALA, we treat the language model of target coarse event identified in previous step as the prior distribution, μ indicates the strength on the prior and in effect, would balance the pseudo counts from the prior ($\mu p(w|\varphi_i)$) and the collected counts from the EM algorithm aforementioned (i.e., $\sum_{D\in C} c(w,D)p(z_{D,w} = i)$).

4 ALA for Event-Oriented Retrieval

At present, with the learned language model of target event, ALA adopts an interpolation method, which is widely adopted in pseudo-relevance feedback, to enhance initial query model, that is, interpolating an existing query model with the learned topic model [10].

Specifically, let θ_Q be the current query model and θ_e be the language model of target event learned in the previous step. The updated new query model θ'_Q is given by

$$p(w|\theta'_Q) = (1-\alpha)p(w|\theta_Q) + \alpha p(w|\theta_e). \quad (8)$$

where $\alpha \in [0,1]$ is an interpolation parameter to control the amount of updating. For efficiency, we would only keep the highest probability terms according to $p(w|\theta'_Q)$, so the setting of interpolation parameter α and truncation number would has significant impact on the retrieval performance, we will set multiple groups of these super-parameters to verify the validity of ALA in experimental section.

5 Experiment and Analysis

5.1 Experimental Setting

Because of lacking the available event-annotated corpus, we use three datasets for TREC Temporal Summarization task[3], which called TREC-TS 2013, TREC-TS 2014 and TREC-TS 2015 respectively. According to the website, the filtering methods used to create the three datasets are not the same, in other words, these datasets can be seen as mutually independent entities. As a consequence, the experimental results could be reliable if ALA can perform consistently in all those datasets. On the whole, these datasets consist of 939,263 event-annotated documents, which respectively belong to 47 events and 10 coarse events.

We establish the language model for all coarse events with the method described in Sect. 3.1. Since the imbalance between term numbers in different coarse events, we retain top 1,000 terms for each coarse event and re-normalize their probabilities. To avoid over-fitting and obtain accurate results, we randomly generate 10 subsets from the whole dataset, each of which contains the same number of documents individually. The experimental results given below are all based on an average of 10-fold cross-validation method on the 10 subsets.

We implemented our document pre-processing with *Lucene* toolkit[4], and use the event names as initial event queries. Following the TREC standard, we retrieve 1,000 documents for each query, and use mean average precision (MAP) as the primary performance measure. In all experiments, we first use the basic KL-divergence method without feedback to retrieve a ranked list of documents for each query as pseudo feedback document set in which document number is set to 100 in this paper. Dirichlet Prior smoothing is used and the smoothing parameter is set to 2000 as recommended in [11].

5.2 The Accuracy of ALA to Identify Target Coarse Event

As stated above, in the second step of ALA, EM algorithm would be used to identify target event. The iteration numbers of all event queries are showed in Fig. 3.

From this figure we can see that all events can reach the convergence situation in only a small quantity of iterative numbers, so this step of ALA is efficiency feasible.

---

[3] See http://www.trec-ts.org/ for details.

[4] Available at http://lucene.apache.org/.

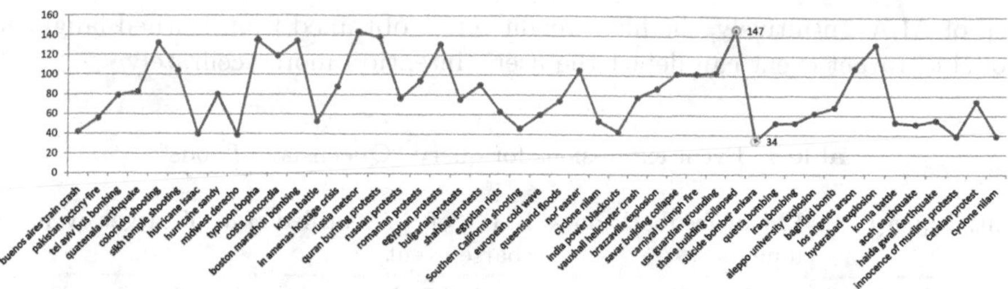

Fig. 3. Iteration Num of All Queries. EM algorithm is regarded as converged when the difference of likelihood values between two iterations is less than 10^{-13} continuous for ten times. The maximal iteration number is 147 in query "thane building collapsed", while the minimum value is 34 in query "suicide bomber ankara", the mean number is 81.15.

Of course, equally important as practicability is the accuracy of ALA to identify target coarse event. During the identification process, we record the identification result for each query, which labels as "identify correctly", "identify incorrectly" or "identified as unknown". The identification situations in different datasets are depicted in the three subgraphs of Fig. 4.

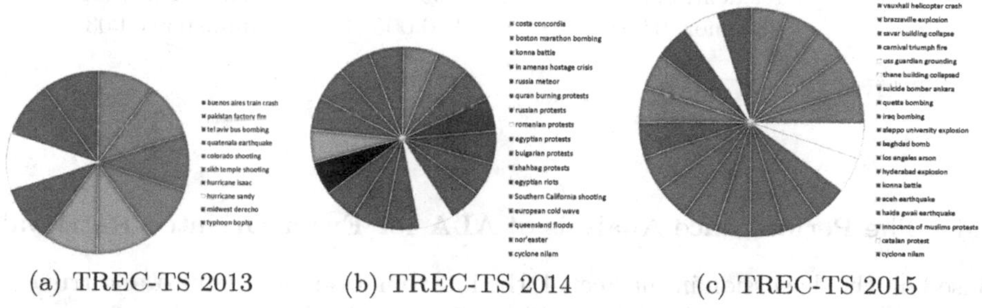

(a) TREC-TS 2013 (b) TREC-TS 2014 (c) TREC-TS 2015

Fig. 4. Identification situation for queries. Colored sector means correct-identified query (each coarse event symbolized by a unique color), white sector means unrecognized query, and incorrect-identified query is plotted with gradient color. (Color figure line)

To our surprises, although a small number of queries are identified as unknown event, there is no query being identified incorrectly. The correct-identified rate in all datasets as a whole is 89.36%, which show ALA is effective.

5.3 The Effectiveness of ALA to Learn Target Event

In order to evaluate the quality of generated language model of target event, we take the event query "Queensland floods" from TREC-TS 2014 dataset as example. Table 1 lists some of the most representative terms for that query at each step of ALA. Intuitively, the high-weight terms obtained from learned language model of target event can depict the user's intention more accurately.

Table 1. Event expressions for query "Queensland floods"

Initial query	LM of feedback documents	LM of identified coarse event	LM of target event
Queensland, 0.5 flood, 0.5	flood, 0.447 queensland, 0.427 Australia, 0.022 aid, 0.013 disaster, 0.011 pollution, 0.011 helicopter, 0.008 water, 0.008 rescue, 0.006 food, 0.006 loss, 0.005 Brisbane, 0.005 storm, 0.004 hospital,0.004 reef, 0.003 resident, 0.002 governor, 0.002 ...	flood, 0.762 disaster, 0.033 aid, 0.021 Danger, 0.018 evacuate, 0.017 rescue, 0.015 waterlevel, 0.013 natural, 0.013 environment, 0.012 waterflow, 0.011 flood-control, 0.011 loss, 0.011 climate, 0.008 peak, 0.008 transfer, 0.007 rise, 0.005 threat, 0.005 ...	flood, 0.427 queensland, 0.415 Australia, 0.024 aid, 0.015 rescue, 0.012 disaster, 0.011 helicopter, 0.01 pollution, 0.008 loss, 0.007 Brisbane, 0.006 Girard, 0.005 hospital, 0.005 resident, 0.005 governor, 0.004 evacuate, 0.004 Bundaberg, 0.003 minister, 0.003 ...

5.4 The Performance Analysis of ALA for Event-Oriented Retrieval

Based on the discussion in the Sect. 4, the interpolation parameter α and truncation term number n would have significant impact on the retrieval performance. Therefore, we use 6 different setting of these two parameters to evaluate the effectiveness of ALA used for event-oriented retrieval task. In this paper we use general Kullback-Leibler divergence as ranking metric, and naturally, we compare ALA with (1) ad-hoc retrieval with initial query (KL), (2) retrieval directly used the language model of feedback set (FM), and (3) two-component mixture feedback model (SMM), which is considered the-state-of-art pseudo feedback method by far [12]. To be fair, the improved query model learned from these comparative methods are all incorporating with knowledge of identified coarse event (CE) by interpolation. The MAP results of all event queries are shown in Table 2.

Overall, the retrievals with improved query model have obvious advantage in performance over ad-hoc retrieval in all parameter settings, in the meanwhile, we notice that ALA is always better than other comparative methods, and the improvement is statistically significant in almost all cases, so it is obviously that

Table 2. The performance comparation of ALA and other methods (MAP)

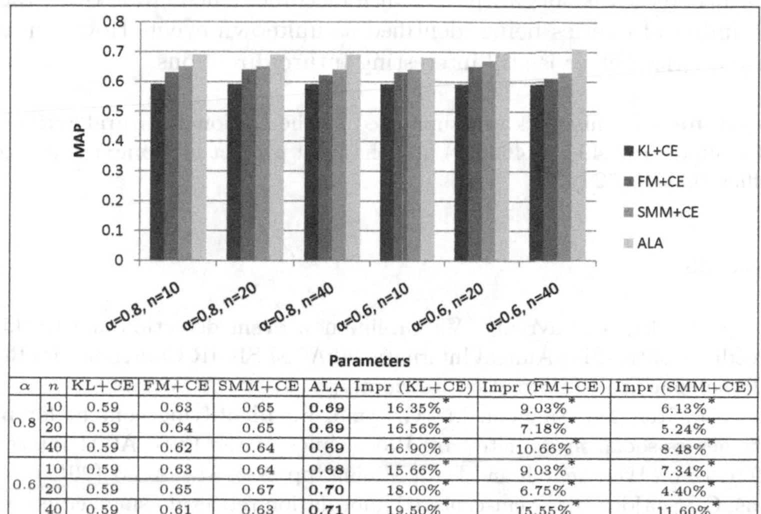

α	n	KL+CE	FM+CE	SMM+CE	ALA	Impr (KL+CE)	Impr (FM+CE)	Impr (SMM+CE)
0.8	10	0.59	0.63	0.65	**0.69**	16.35%*	9.03%*	6.13%*
	20	0.59	0.64	0.65	**0.69**	16.56%*	7.18%*	5.24%*
	40	0.59	0.62	0.64	**0.69**	16.90%*	10.66%*	8.48%*
0.6	10	0.59	0.63	0.64	**0.68**	15.66%*	9.03%*	7.34%*
	20	0.59	0.65	0.67	**0.70**	18.00%*	6.75%*	4.40%*
	40	0.59	0.61	0.63	**0.71**	19.50%*	15.55%*	11.60%*

*The improvement is statistically significant at the level of 0.05 according to the Wilcoxon signed rank test.

based on event query, the learned language model of target event through ATA is effective and robust w.r.t. different setting of parameters for feedback, which would be able to provide considerable benefits for event-oriented retrieval task.

In addition to this, we should also take note that in contrast to other methods, ALA has better stability. Specifically, from Table 2, when truncation number increases from 10 to 20, the MAP values of FM+CE and SMM+CE have a marked increase, while truncation number increases from 20 to 40, their MAP values reduce considerably. One possible explanation is that as we use more terms truncated from pseudo feedback set (FM) or fitted topic model (SMM), the more noise integrated into the initial query, so that the affect becomes more harmful for retrieval performance. In the meanwhile, we can see that when truncation number increases from 10 to 40, MAP value of ALA is monotonically increased, and the growth rates remain stable. The possible explanation is that the learned language model through ALA is more consistent with user's intention, so there is hardly any noise in truncated terms.

6 Conclusions

In this paper, we propose an adaptive learning approach of PLSA model, to deal with the invalidation problem of conventional retrieval method for event-oriented retrieval task. The experimental results in several available large-scale datasets show that compared with other methods, retrieval using learned language model of target event can constantly acquire better retrieval performance. Nevertheless, we should also note that though in this paper there is no event query being identified incorrectly, is this all just chance or coincidence? Moreover, there are a small number of queries being identified as unknown event. How can we deal with these queries better is still interesting future directions.

Acknowledgment. This work was supported by the National Natural Science Foundation of China (61572494, 61462027) and the fund project of Jiangxi Province Education Office (GJJ160529).

References

1. Allan, J., Papka, R., Lavrenko, V.: On-line new event detection and tracking. In: Proceedings of the 21st Annual International ACM SIGIR Conference on Research and Development in Information Retrieval, pp. 37–45. ACM (1998)
2. Becker, H., Iter, D., Naaman, M., Gravano, L.: Identifying content for planned events across social media sites. In: Proceedings of the Fifth ACM International Conference on Web Search and Data Mining, pp. 533–542. ACM (2012)
3. Glavaš, G., Šnajder, J.: Event-centered information retrieval using kernels on event graphs. In: TextGraphs-8 at Empirical Methods in Natural Language Processing (EMNLP 2013) (2013)
4. Yang, W., Li, R., Li, P., Zhou, M., Wang, B.: Event related document retrieval based on bipartite graph. In: Cui, B., Zhang, N., Xu, J., Lian, X., Liu, D. (eds.) WAIM 2016. LNCS, vol. 9658, pp. 467–478. Springer, Cham (2016). doi:10.1007/978-3-319-39937-9_36
5. Zhong, Z., Zhu, P., Li, C., Guan, Y., Liu, Z.: Research on event-oriented query expansion based on local analysis. J. China Soc. Sci. Tech. Inf. **31**(2), 151–159 (2012)
6. Tsolmon, B., Lee, K.S.: An event extraction model based on timeline and user analysis in latent Dirichlet allocation. In: International ACM SIGIR Conference on Research and Development in Information Retrieval, pp. 1187–1190 (2014)
7. Quezada, M., Poblete, B.: Location-aware model for news events in social media. In: The International ACM SIGIR Conference, pp. 935–938 (2015)
8. Dempster, A.P., Laird, N.M., Rubin, D.B.: Maximum likelihood from incomplete data via the EM algorithm. J. R. Stat. Soc. Ser. B (Methodological). 1–38 (1977)
9. Mclachlan, G.J., Krishnan, T.: The EM algorithm and extensions. Biometrics **382**(1), 154–156 (2008)
10. Zhai, C., Lafferty, J.: Model-based feedback in the language modeling approach to information retrieval. In: Proceedings of the Tenth International Conference on Information and Knowledge Management, pp. 403–410. ACM (2001)
11. Zhai, C., Lafferty, J.: A study of smoothing methods for language models applied to ad hoc information retrieval. In: Proceedings of the 24th Annual International ACM SIGIR Conference on Research and Development in Information Retrieval, pp. 334–342. ACM (2001)
12. Lv, Y., Zhai, C.: Negative query generation: bridging the gap between query likelihood retrieval models and relevance. Inf. Retr. J. **18**(4), 359–378 (2015)

Musical Query-by-Semantic-Description Based on Convolutional Neural Network

Jing Qin[1,2], Hongfei Lin[1(✉)], Dongyu Zhang[1], Shaowu Zhang[1], and Xiaocong Wei[3]

[1] School of Computer Science and Technology, Dalian University of Technology, Dalian, China
qinjing@dlu.edu.cn, hflin@dlut.edu.cn

[2] College of Information Engineering, Dalian University, Dalian 116622, China

[3] School of Software Engineering, Dalian University of Foreign Language, Dalian 116044, China

Abstract. We present a new music retrieval system based on query by semantic description (QBSD) system, by which a novel song can be used as query and transformed into semantic vector by a convolutional neural network. This method based on Supervised Multi-class labeling (SML), which a song can be annotated by some semantically meaningful tags and retrieved relevant song in semantically annotated database. CAL500 data set is used in experiment, we can learn a deep learning model for each tag in semantic space. To improve the annotation effect, loss function adjustment algorithm and SMOTE algorithm are employed. The experiment results show that this model can get songs with high semantically similarity, and provide a more nature way to music retrieval.

Keywords: Query by semantic description · Convolutional neural network · Supervised multi-class labeling · Semantically retrieval

1 Introduction

As the size of audio and music collection dramatically increase in internet, millions of songs are available to consumers online and it is difficult to find and discover new songs. In music industry, the music retrieval systems mainly used manual retrieval [1], which based on mete-data such as artist name, song name etc. A few methods used semantic feature like music style. Tags and other text information are used in music retrieval. Content-based Music Information Retrieval (CBMR) is gaining widespread attention and could be helpful, since it forsakes the need of keyword.

However, there is few research on retrieval model on query-by-semantic description or other CBMR fusion method. Pitch or tempo, or other low-level feature is used in CBMR, these features cannot be transformed to high level semantic features directly. Relative research [2] shows that there may have a "semantic gap", which means the lack of association between low level physical features and semantic concept.

To solve the semantic gap problem, many recent researches on MIR is focus on music content and semantic expressions. Jun [3] proposed an ontology where the low-level and high-level descriptors collaborate to support semantics-based MIR. Buccoli [4] propose a Dimensional Contextual Semantic Model for defining semantic relations among descriptors in a context-aware fashion. This model is used for developing a

J. Wen et al. (Eds.): CCIR 2017, LNCS 10390, pp. 237–248, 2017.
https://doi.org/10.1007/978-3-319-68699-8_19

semantic music search engine [5]. Miotto and Lanckriet [6] proposed a Dirichlet mixture model (DMM) to improve automatic music annotation. In [7], they proposed a new approach that retrieves music using fuzzy music-sense features and audio features, which is a new method on semantic and CBMR. Foster [8] proposed string compressibility as a descriptor of temporal structure in audio, for determining musical similarity. The descriptors are based on computing track wise compression rates of quantized audio features, using multiple temporal resolutions and quantization granularities. In those related work, Turnbull [9–11] presented a concept of Query-by-Semantic-Description (QBSD), it is a natural way to retrieval in large music database, to overcome the lack of a cleanly-labeled, publicly-available, heterogeneous data set of songs and associated annotations, they collected CAL500 data set by having humans listen to and annotate songs using a survey designed to capture semantic associations between music and words. The methods adapted the supervised multi-class labeling (SML) model, used the CAL500 data to learn a model, annotated a novel song with meaningful words and retrieve relevant songs given a multi-word, text-based query.

The study shows that, music retrieval focusses on narrow the gap between music content and music semantic meaning. We present a query by example, a semantic vector is extracted by example song and used as query, searched in auto-annotated music database, the output of the retrieval system is a song list with the most similar songs in semantic vector space. Unlike other QBSD system, our system has two distinct advantages: firstly, instead of query by text, query by example song is a more convenient way for MIR. On the other hand, the similarity in semantic space which means more semantic same tags, is more exact than several words. The core problem lies on the auto-annotation efficiency by SML. If the model could not annotate the songs exactly, the retrieval result cannot satisfy the needs. Thus, the motivation of this paper is to improve effectiveness and the system feasibility. In consideration of the current study in deep learning methods on audio, we proposed a SML based on Convolutional Neural Network and improve the annotation algorithm by SMOTE. Experiments show the performance of this model is well, which provide a novel method for the QBSD system.

2 Related Work

2.1 The Architecture of the QBSD System

The architecture of the QBSD system is shown in Fig. 1. First, low-level feature, such as MFCCs, is used as the input of a well-studied deep learning network, then CNN map the audio signal into semantic space, by annotated the songs with semantic tags. Meanwell, music in database is also auto-annotated by the same mode. At last, query example and the songs in database are matched in similarity of semantic space, the most similar songs are the feedback as the retrieval or recommend results. By this system, users could receive better experience in a natural way, and get a result they really want.

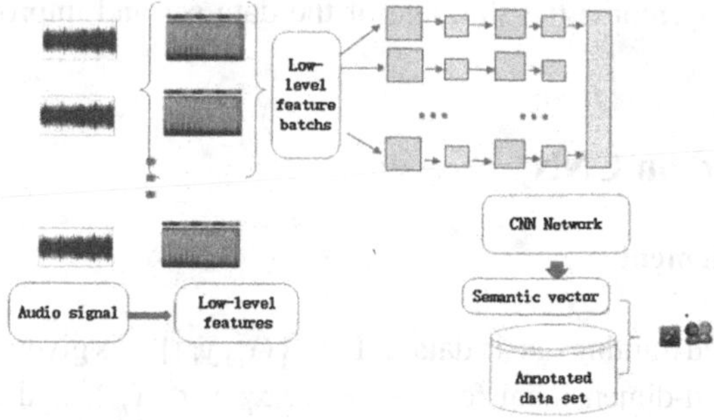

Fig. 1. The architecture of query by semantic description system

2.2 The Deep Learning Algorithms in Audio Signal Process

In recent years, deep learning method is widely employed in research image classification, image semantic mapping, audio classification and recognition, and get excellent efficiency. Deep learning method on audio process always use convolutional deep belief network to extract features for classification and retrieval. In [12], Deep Belief Network (DBN) was used to feature extraction, the learned features perform significantly better than MFCCs and obtained a classification accuracy of 84.3% on the Tzanetakis dataset. Dieleman [13] built a convolutional network that is then trained to perform artist recognition, genre recognition and key detection, improved accuracy for the genre recognition and artist recognition tasks. Hu Zhen [14] presented a hybrid model based on deep belief network (DBN) and stacked denoising autoencoder (SDA) to identify the composer from audio signal, the model got an accuracy of 76.26% in testing data set. Humphrey [15] learned a robust Tonnetz-space transform for automatic chord recognition. Hamel [16] proposed a feature extraction system consists of a Deep Belief Network (DBN) on Discrete Fourier Transforms (DFTs) of the audio, and the learned features perform significantly better than MFCCs. In a word, hidden Markov models (HMMs) to deal with the temporal variability of speech and Gaussian mixture models (GMMs) traditionally, however, as Hinton [17] said, Deep neural networks (DNNs) that have many hidden layers and are trained using new methods have been shown to outperform GMMs on a variety of speech recognition benchmarks, sometimes by a large margin. But deep learning algorithms should be learned by enough data and the structure of network needed finetuning. Therefore, a suitable network should be designed for the specific problem requirement.

To compare with GMM, we experiment on the CAL500 data set [11], which is composed by 500 songs of 500 unique artists, each annotated by a minimum of 3 individuals using a 174-tag vocabulary representing genres, instruments, emotions and other musically relevant concepts. The number of songs is minor. On the other hand, the number of frames in each music piece is huge (10000 frames in a song). Consider by the data characteristics, we employ the synthetic minority oversampling technique(SMOTE) [18], which is an over-sample method and the core idea was to construct the synthetic minority samples through the minority training data and its k nearest neighborhoods, to overcome the shortage of the data set and improve the annotation efficiency.

3 QBSD Based on CNN

3.1 Problem Statement

Assume that a labeled training music dataset $D \equiv \left\{(x_i, y_j)\right\}_{j=1}^{c}$ is given, where each music is represented by a d-dimensional feature vector x_i, $x_i \in X$, X is the data set, i is the number of music pieces, y_j is distinct semantic labels available for training, j is the number of labels. The semantic vectors $S \equiv \langle y_0, y_1, \ldots, y_c \rangle$ is learned, c is the number of labels in semantic vectors. QBSD by example is addressed by learning a mapping from input features x_{input} to semantic label vectors S_x using a CNN model on the human annotation data set D, and the output of the system is a song list X_{list}, according to similarity between semantic label vectors S_x in data set and the input.

In the supervised multi-class labeling (SML) model [9], the probability distribution of each label in the semantic space was calculated by Gauss mixture model (GMM). The drawback of the nonparametric estimation technique is that the number of mixture components in the word-level distribution grows with the size of the training database. In practice, it may have to evaluate thousands of multivariate Gaussian distributions for each feature vectors of a novel query track. The training data should be subsampled or used mixture hierarchies' estimation, but they are not efficient in annotation or time cost.

3.2 Model Description

Audio signal has short-time stationarity and periodic features in long-time. The audio signals are segmented by a window, and lower-level features, include zero-crossings, centroid, rolloff and MFCC etc., are extracted traditionally. While, self-similarity is a common property in music pieces, melody may be repetitive in a song. CNN could be used to learn features in local receptive fields, thus music piece is the process unit in our algorithm and try to find new features for SML model.

39-dimensional MFCCs feature is extracted from the segmented audio signal frame, and five frames are cascaded as a long-frame that has 195-dimensional MFCCs. Then fifty long-frames is treated as a music piece, which is a two-dimension vector with a size of 195 × 10. The music piece is the input of a CNN, and the output of the net is:

$$h_{ij}^{k} = \theta((W^{k} * x)_{ij} + b_k) \quad (1)$$

Where k is the k-th filter, x is the input music piece feature batch, W^k denote the parameter (or weight) associated with filter k. b_k is the bias associated with filter k. (i, j) is the location on feature batch. Convolutional features are calculated after the activation function θ, which usually uses sigmoid or tanh function. We choose relu, because it results in the neural network training several times faster [19], without making a significant difference to generalization accuracy. The architecture based on CNN is shown in Fig. 2.

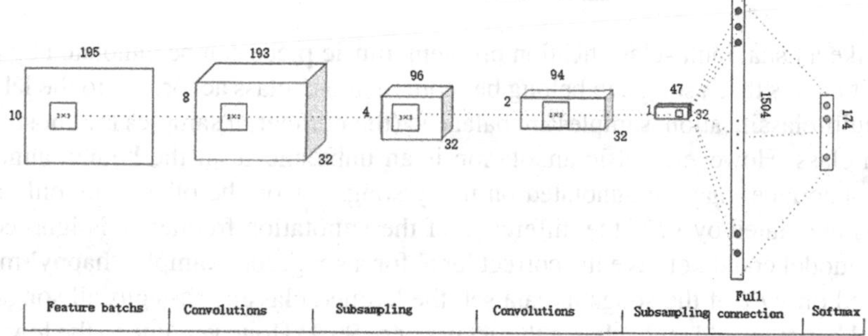

Fig. 2. The architecture of CNN

There are two convolutional layers, two pooling layers, one fully connected layer and a loss layer in this network, the size of each layer is shown in Fig. 2. We choose 32 convolutional filters in one layer, the shape of the filter is 3×3 and the max pooling shape is 2×2. $x_{flatten}$ donate the output of the fully connected layer, then the output of the last lay y_{out} is shown as follow:

$$y_{out} = softmax(x_{flatten}) = \frac{1}{1 + e^{-x_{flatten}}} \tag{2}$$

The value of y_{out} is in (0, 1), the network training penalizes the deviation between y_{out} and the human annotated semantic label y_j, c the number of labels:

$$E^N = \frac{1}{2}\sum\nolimits_{n=1}^{N}\sum\nolimits_{k=1}^{c}\left(y_j^{nk} - y_{out}^{nk}\right)^2 \tag{3}$$

E^N is the total loss of the N training songs, y_j^{nk} denote human annotated label of the n th sample on the kth label, y_{out}^{nk} is the output of the n th sample on the kth label, then the network weights are fine-tuned according to the loss.

Loss function adjustment. Consider imbalanced distribution as prior information. For each song S in data set, S is annotated by N labels, according to CNN structure, the output from last level softmax might be $\{\ldots, 0\ldots, \frac{1}{N}, \ldots, \frac{1}{N}, \ldots, 0\ldots\}$. We adjust the output by adding a weight w_i, which in inverse proportion of annotation frequency. Let f_i be the number of annotation samples for label i, the smaller f_i is, and the bigger w_i is. The output of the net turns into $\{\ldots, 0\ldots, w_i\frac{1}{N}, \ldots, w_j\frac{1}{N}, \ldots, 0\ldots\}$, $\sum_{i=0}^{N} w_i\frac{1}{N} = 1$, then:

$$w_i = 1 - \frac{f_i}{\sum_{i=0}^{N} f_i} + \frac{1}{N} \tag{4}$$

The loss function is adjusted as:

$$E^N = \frac{1}{2} \sum_{n=1}^{N} \sum_{k=1}^{c} w_i \left(y_i^{nk} - y_{out}^{nk}\right)^2 \tag{5}$$

Unlike a usual multi-classification problem, music piece can be annotated by many labels, it means that a song can belong be more than one class according to the label. In most multi-classification, samples are balanced (the number of samples is almost equal) in each class. However, music annotation is an unbalanced, in the human annotated dataset, one label may be annotated on many songs, or on the other side, only a few songs is annotated by it. If the difference of the annotation frequency is ignored, the learned model could not give the correct label for a song, for example, 'happy' may be annotated on 99% of the songs in data set, the learned classifier assigns all songs with the label happy would still achieve the accuracy of 99%. However, due to the low recall ratio for the minority, such extreme result is not what we have desired.

To solve imbalance learning problem, we employ two strategies for the learning model. First, the adjustment of the weights of errors in the loss function, which is a direct way to reduce the impact of imbalance. Another method is synthetic minority oversampling technique(SMOTE) proposed by Chawla etc. [18]. CAL500 is a small data set, human annotation is costly work, we conduct the synthetic minority samples for training data, thus the model would learn more information from minority labels.

SMOTE algorithm. SMOTE is a typical oversampling method with universal applications and the concrete process for generating the synthetic samples can be described as shown in Table 1:

Table 1. SMOTE algorithm

Algorithm 1: SMOTE ALGORITHM
Input: Each sample in data set x_i;
Output: New synthetic sample for the minority x_{new};
Description:
1.Calculate the K-nearest neighbors for each sample x_i,
2.Select a random number δ, generated from a uniform distribution U[0,1];
3.Output a new synthetic sample for minority as:
$x_{new} = x_i + \delta(x_i' - x_i)$

The result of SMOTE, implemented by [20], is shown in Fig. 3. The number of samples increased and the data from each class becomes balanced.

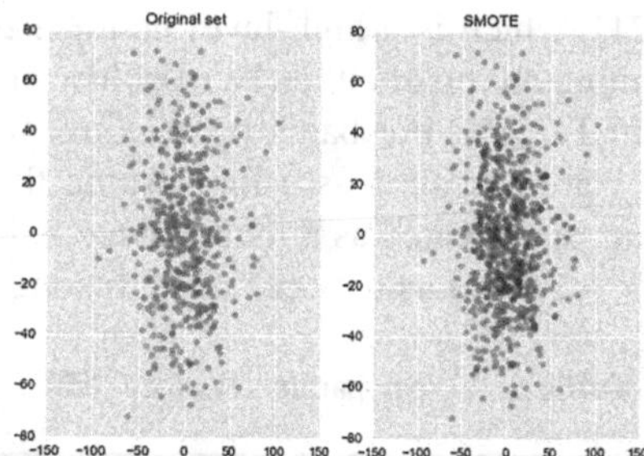

Fig. 3. Result of SMOTE algorithm

3.3 Retrieval Algorithm

Query by example algorithm is shown in Table 2:

Table 2. Retrieval algorithm

Algorithm 2: Retrieval algorithm with example

Input: A labeled training music dataset $D \equiv \{(x_i, y_j)\}_{j=1}^{c}$, each music is represented by a d-dimensional feature vector x_i, $x_i \in X$, X is the data set, i is the number of music pieces, j is the number of labels; A new query song q;
Output: The output of the system is a song list X_{list};
Description:

1. Training a CNN network shown in Fig.2 ,according to annotation relation between x_i,y_j,get the net parameters W^k, b_k , and the input y_{out}^{p};
2. Calculate semantic vectors of each song x_i:

$$S_{sementic}^{X} = \frac{1}{p}\sum_{j=1}^{p} y_{out}^{p}$$

3. Calculate semantic vector of the query q, $S_{sementic}^{q}$;
4. Compute the Cosine similarity R_s between $S_{sementic}^{X}$ and $S_{sementic}^{q}$;
5. Let X_{list} be a list of top x candidate songs with the highest R_s;
6. Return X_{list}.

4 Experiments and Analysis

4.1 Data Set and Features

To evaluate the performance of the proposed approach, we use CAL500 data set [9], which has 500 songs by 500 unique artists each annotated by minimum of 3 individuals using a 174-tag vocabulary. A song is annotated with a tag if 80% of the human annotators agree that the tag is relevant, the value is 1 in semantic vector, otherwise the value is 0. In our experiments, 39-dimensional MFCCs are used as CNN input, 174-dimensional semantic vector as retrieval model output. CNN model is implemented by Keras [21], which is a high-level neural networks API, written in Python and capable of running on top of either TensorFlow or Theano. Hyper-parameters for CNN is shown in Table 3. For batch processing, we align each song with 10000 frames, five frames are cascaded as a long-frame that has 195-dimensional MFCCs. Then 50 long-frames is treated as a music piece, which is a two-dimension vector with a size of 195 × 10, 200 music pieces in each song.

Table 3. Hyper-parameters for CNN

Hyper-parameter	Value
size of batch	4
number of classes	174
number of training epoch	30
number of convolutional filters to use	32
size of pooling area for max pooling	2
convolution kernel size	3×3
optimizer	SGD
learning rate	0.1
decay	1e-6
momentum	0.9
nesterov	True

4.2 Evaluation of Annotation and Retrieval

Annotation performance is measured following the procedure described by Coviello etc. [22]. Annotation accuracy is reported by computing precision, recall and F-score for each tag and then averaging over all tags. Per-tag precision is the probability that the model correctly uses the tag when annotating a song. Per-tag recall is the probability that the model annotates a song that should have been annotated with the tag. Precision, recall and F-score measure for a tag are defined as:

$$P = |W_C| / |W_A|, R = |W_C| / |W_H|,$$

$$F = 2((P)^{-1} + (R)^{-1})^{-1} \quad (6)$$

Where $|W_H|$ is the number of tracks that have W in the ground truth, $|W_A|$ is the number of times our annotation system uses when W automatically tagging a song, and $|W_C|$ is the number of times is W correctly used.

To evaluate retrieval performance, we report mean average precision (MAP), area under the receiver operating characteristic curve (AROC) and averaged over all the query tags. The ROC curve is a plot of true positive rate versus false positive rate as we move down the ranked list. The AROC is obtained by integrating the ROC curve, and it is upper bounded by 1. Random guessing would result in an AROC of 0.5.

4.3 Result Analysis

Annotation and retrieval on tags. First, we search each tag in the data set, list is ranked by decision value, and results are averaged on tags. The results are shown in Table 4, CNN-AD (loss function adjustment on CNN), CMM-SMOTE (SMOTE on train data) are better than SVM and HEM-GMM on annotation precision and AROC.

Table 4. Annotation and retrieval results for various algorithms on the CAL500 data set

	Annotation			Retrieval	
Model	P	R	F-score	AROC	MAP
HEM-GMM	**0.49**	0.33	0.26	0.66	0.45
SVM	0.46	0.46	0.44	**0.72**	0.50
CNN-AD	**0.49**	0.46	**0.45**	0.71	**0.52**
CNN-SMOTE	0.49	0.47	0.46	**0.72**	**0.52**

Retrieval by tag 'ACOUSTIC GUITAR' results is shown in Table 4, CNN-AD and CNN-SMOTE are better than CNN-SVM.

Retrieval by query example. Query based on example is used the whole song as a query, the example is annotated by the learned model and transformed into a semantic vector, then we calculate the similarity between the example semantic vector and the human-annotated data set, a song list is ranked by Cosine similarity. In our experiment, if the annotation accuracy is well, the same human-annotated song would be the output. Thus, we define Hit@k as the percentage that the same song output from retrieval, could be used to compare different models. The results are shown in Fig. 4. CNN-SMOTE and CNN-AD are better than CNN-SVM on Hit@10, and Hit@10 values are more than 92%, which means query by example is better than several text descriptions and nearby human-annotation (Table 5).

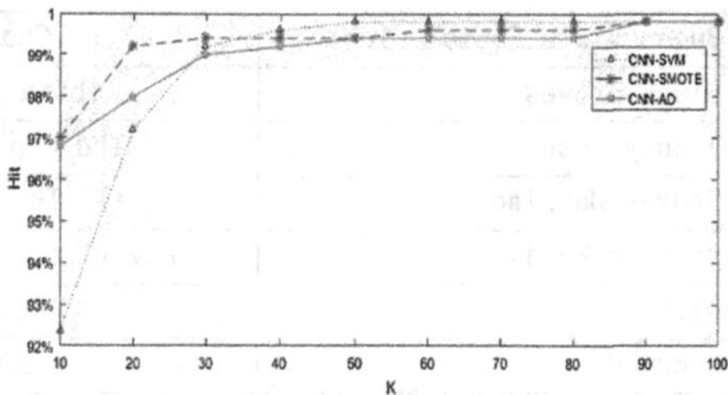

Fig. 4. Hit@k performance of our model on CAL500

Table 5. Top-10 retrieved songs for "ACOUSTIC GUITAR." Songs with acoustic guitar are marked in bold

Rank	CNN-SVM	
	Artist	Song Name
1	**Myles cochran**	**Getting stronger**
2	**Van morrison**	**And It Stoned Me**
3	**Buena Vista Social Club**	**El Cuarto de Tula**
4	**The Black Crowes**	**Thorn in My Pride e**
5	**Johnny Cash**	**The Man Comes Around**
6	**George Harrison**	**All Things Must Pass**
7	R. E. M	Camera
8	The Monkees	A Little Bit Me a Little Bit You
9	LOVE	You Set the Scene
10	Wicked Allstars	Happy
Rank	**CNN-AD**	
	Artist	**Song Name**
1	**Black Crowes**	**Thorn in My Pride**
2	**Gram parsons**	**$ 1000 wedding**
3	**Neil Young**	**Razor Love**
4	**Bob Dylan**	**I'll Be Your Baby To-night**
5	Mr Gelatine	Knysnamushrooms
6	**New Order**	**Blue Monday**
7	**Myles Cochran**	**Getting stronger**
8	The Rolling Stones	Little by Little
9	They Might Be Giants	I Should Be Allowed to Think
10	Ultravox	Dancing with Tears in My Eyes
Rank	**CNN-SMOTE**	
	Artist	**Song Name**
1	**Myles Cochran**	**Getting stronger**
2	**Van Morrison**	**And It Stoned Me**
3	**Buena Vista Social Club**	**El Cuarto De Tula**
4	**Black Crowes**	**Thorn in My Pride**
5	**Johnny Cash**	**The Man Comes Around**
6	**George Harrison**	**All Things Must Pass**
7	Brenton Wood	Lovey Dovey Kind of Love
8	Dirt	THE STOOGES
9	Buddy Holly	Peggy Sue
10	**Air**	**Sexy Boy**

5 Conclusions

In this paper, we proposed a music retrieval model based on query example semantic description, CNN is used to learn a Supervised Multi-class labeling system, by which example query is transformed to a semantic vector, and searched in the data set. To improve the annotation accuracy, loss function adjustment and SMOTE algorithm are employed, the results show that an example song instead of only a few texts, could get a result more semantically similarity and it is a more natural way to find what we want. Loss function adjustment method based on CNN, the model should adjust the weight by labeling frequency, low frequency tags would not be ignored in learning process. SMOTE algorithm produces more samples for low frequency tags and get better annotation result, but it should be used on each tag in vocabulary, which means we should learn different models for each tag and costs much more time. In future work, we will design and test more different networks for semantic tags annotation and large-scale music data set unsupervised multi-class annotation algorithm should be considered.

Acknowledgment. Supported by the National Natural Science Foundation of China (Grant No. 61632011); the National Natural Science Foundation of China (Grant No. 61562080); the National Natural Science Foundation of China (Grant No. 61602079)

References

1. BigData-Research. http://www.bigdata-research.cn/content/201606/285. 12 June 2016
2. Casey, M., Veltkamp, R., Goto, M., Leman, M., Rhodes, C., Slaney, M.: Content-based music information retrieval: current directions and future challenges. Proc. IEEE **96**(4), 668–696 (2008)
3. Wang, J., Deng, H., Yan, Q.: A collaborative model of low-level and high-level descriptors for semantics-based music information retrieval. In: International Conference on Web Intelligence and Intelligent Agent Technology, pp. 532–535. IEEE, New York (2008)
4. Buccoli, M., Gallo, A., Zanoni, M., Sarti, A., Tubaro, S.: A dimensional contextual semantic model for music description and retrieval. In: IEEE International Conference on Acoustics, Speech and Signal Processing, pp. 673–677. IEEE, New York (2015)
5. Buccoli, M., Zanoni, M., Sarti, A., Tubaro, S.: A music search engine based on semantic text-based query. In: IEEE International Workshop on Multimedia Signal Processing, pp. 254–259. IEEE, New York (2013)
6. Miotto, R., Lanckriet, G.: A generative context model for semantic music annotation and retrieval. IEEE Trans. Audio Speech Lang. Process. **20**(4), 1096–1108 (2012)
7. Su, J.H., Wang, C.Y., Chiu, T.W., Ying, J.C., Tseng, V.S.: Semantic content-based music retrieval using audio and fuzzy-music-sense features. In: IEEE International Conference on Granular Computing, pp. 259–264. IEEE, New York (2014)
8. Foster, P., Mauch, M., Dixon, S.: Sequential complexity as a descriptor for musical similarity. IEEE Press **22**(12), 1965–1977 (2014)
9. Turnbull, D., Barrington, L., Torres, D., Lanckriet, G.: Towards musical query- by- semantic description using the CAL500 data set. In: International ACM SIGIR Conference on Research & Development in Information Retrieval, pp. 439–446. ACM, New York (2007)

10. Turnbull, D., Barrington, L., Torres, D., Lanckriet, G.: Semantic annotation and retrieval of music and sound effects. IEEE Trans. Audio Speech Lang. Process. **16**(2), 467–476 (2008)
11. Turnbull, D.R., Barrington, L., Lanckriet, G., Yazdani, M.: Combining audio content and social context for semantic music discovery. In: International ACM SIGIR Conference on Research and Development in Information Retrieval, pp. 387–394. ACM, New York (2009)
12. Lee, H., Yan, L., Pham, P., Ng, A.Y.: Unsupervised feature learning for audio classification using convolutional deep belief networks. In: International Conference on Neural Information Processing Systems, pp. 1096–1104. Springer, Heidelberg (2009)
13. Dieleman, S., Brakel, P., Schrauwen, B.: Audio-based music classification with a pretrained convolutional network. In: Proceedings of the ISMIR (2011)
14. Hu, Z., Fu, K., Zhang, C.: Audio classical composer identification by deep neural network. J. Comput. Res. Dev. **51**(9), 1945–1954 (2014)
15. Humphrey, E.J., Cho, T., Bello, J.P.: Learning a robust Tonnetz-space transform for automatic chord recognition. In: IEEE International Conference on Acoustics, Speech and Signal Processing, pp. 453–456. IEEE, New York (2012)
16. Hamel, P., Eck, D.: Learning features from music audio with deep belief networks. In: Proceedings of the ISMIR, pp. 339–344 (2010)
17. Hinton, G., Deng, L., Yu, D., et al.: Deep neural networks for acoustic modeling in speech recognition: the shared views of four research groups. IEEE Sig. Process. Mag. **29**(6), 82–97 (2012)
18. Chawla, N.V., Bowyer, K.W., Hall, L.O., Kegelmeyer, W.P.: SMOTE: synthetic minority over-sampling technique. J. Artif. Intell. Res. **16**(1), 321–357 (2002)
19. Krizhevsky, A., Sutskever, I., Hinton, G.E.: ImageNet classification with deep convolutional neural networks. In: International Conference on Neural Information Processing Systems, pp. 1097–1105. ACM, New York (2012)
20. Lemaitre, G., Nogueira, F., Aridas, C.K.: Imbalanced-learn: a python toolbox to tackle the curse of imbalanced datasets in machine learning. J. Mach. Learn. Res. **18**(17), 1–5 (2017)
21. Chollet, F.: Keras, GitHub repository (2015). https://github.com/fchollet/keras
22. Coviello, E., Chan, A.B., Lanckriet, G.: Time series models for semantic music annotation. IEEE Trans. Audio Speech Lang. Process. **19**(5), 1343–1359 (2011)

A Feature Extraction and Expansion-Based Approach for Question Target Identification and Classification

Wenxiu Xie[1], Dongfa Gao[1(✉)], and Tianyong Hao[1,2(✉)]

[1] School of Information Science and Technology, Guangdong University of Foreign Studies, Guangzhou, China
vasiliky@outlook.com, {gaodf,haoty}@gdufs.edu.cn

[2] Collaborative Innovation Center for 21st-Century Maritime Silk Road Studies, Guangdong University of Foreign Studies, Guangzhou, China

Abstract. Detecting question target words from user questions is a crucial step in question target classification as it can precisely reflect the users' potential need. In this paper we propose a concise approach named as QTF_EE to identify question target words, extract question target features and expand the features for question target classification. Based on two publicly available datasets that are labeled with 50 answer types, we compare the QTF_EE approach with 12 conventional classification methods such as bag-of-words and Random Forest as baseline methods. The results show that the QTF_EE approach outperforms the baselines and is able to improve the question target classification performance to an accuracy of 87.4%, demonstrating its effectiveness in question target identification.

Keywords: Answer type classification · Question target word · Question target feature

1 Introduction

Open-domain Question Answering involves the extraction and rank of correct answers for a given free-text question and has been actively studied all over the world since the Question Answering (QA) track at TREC-8 [1]. As an essential step in QA, answer type classification, also known as Question Target Classification (QTC), is to classify questions into a list of predefined answer type categories. As claimed by previous works [2–4], pinpointing and verifying an accurate answer hinges on identifying the target of intention of a given question to determine the type of sought-after answer and filtering out irrelevant answers from a wide range of answer candidates. For the questions which are short and lack of context, expanding target words especially question semantic feature is necessary. Addressed by Li et al. [5], question classification is a feature-dependent task and semantic information is essentially used to achieve high classification accuracy. They reported that an error of 28.7% can be decreased when semantic features are incorporated into fine-grained classification. Consequently, question target words and their expanded features can benefit semantic constraints on answer type and therefore can reduce the answer candidates for improving answer selection efficiency.

J. Wen et al. (Eds.): CCIR 2017, LNCS 10390, pp. 249–260, 2017.
https://doi.org/10.1007/978-3-319-68699-8_20

Previous research on question answering, e.g. Srihari and Li [6], has shown that it is important to classify questions with respect to their answer types. Particularly, the correct detection of question target words (QTWs) can significantly improve question classification accuracy, since the QTWs are the most informative part of a question [7]. In our previous work [8], Question Target Words are regarded as interrogative words and distinctive words (noun or verb) that directly lead to the answer type or substantially represent users' information need. For example, the target word of a question "*What color is a poison arrow frog?*" is "*color*", which is exactly the answer type as [*Entity/Color*]. However, through the analysis on thousands of question syntactic structures, certain questions may contain useless or even misleading semantic features. For instance, the extracted QTWs from the question "*What is vertigo?*" are "*what*" and "*vertigo*" but the semantic expansion of word "*vertigo*" could lead to the question target category "[*ENTY/Dis.Med*]" since "*vertigo*" is a kind of disease symptom. To deal with this issue, some approaches [4, 5, 9, 10] have been proposed to introduce patterns or predefined templates to assist question classification. Although improvements are shown in their experiments, it still suffers from two common drawbacks: tedious manual work on template creation and limited adaption to general questions.

To that end, this paper presents an automated Question Target Features (QTFs) Extraction and Expansion approach named as QTF_EE to determining the question target deemed essential to answer the question. The QTFs contains 3 kind of features: part-of-speech-based syntactic expansion, hypernym-based semantic expansion, and word cluster-based semantic expansion. To evaluate the performance of the proposed QTF_EE on question classification, 12 widely used classifiers are applied for comparison including SVM, Naïve Bayes Multinomial Classifier, Naive Bayes Classifier, etc. The experiment datasets are two publicly available datasets: UIUC QA dataset containing 5,382 labeled questions and TREC-10 QA dataset containing 500 labeled questions. We further use a number of baseline methods presented in [4, 9–17]. By using the approach on the same two-layer classification taxonomy proposed by Li and Roth [4], our classification accuracy achieves 87.4% on 50 categories on the TREC-10 dataset, increasing about 7% than the baselines.

2 Related Work

For the unique characteristic of open domain questions which are always short and lack of context, a wide variety of features were used by different approaches, e.g., bag-of-words (BOWs) features, WordNet-based features, tree-based features, etc. [18–20]. Though BOWs was a basic document representation in natural language processing, it suffered from a features localization and low coverage problem. To overcome this pitfalls of BOWs, varies syntactic and semantic features expansion technology were investigated. As one representative work, Li and Roth's [4] showed that augmenting the input of classifier with appropriate semantic category information could resulted in significant improvements to classification accuracy. They applied six primitive question feature types including words, POS tags, chunks, named entities, head chunks, and semantically related words. By applying these features in classifier, the accuracy achieved 84.2% for fine-grained classification. The method and the UIUC dataset have inspired many follow-up works on question classification and feature expansion [10].

Later on, Hacioglu and Ward et al. [12] assumed that users' potential information need is an entity and a general category called NA which do not asking for named entities. By applying named entity features and Singular Value Decomposition (SVD)-based transformation to a SVM classifier, their approach obtained an accuracy of 82.0% on fine-grained categories. Mrabet et al. [21] also treated a question topic task as a named entity recognition problem, representing topic recognition as a binary classification task and extending the span of detected topic entities with generic rules. Though their approaches yielded a good performance on classification task by using word itself as feature, yet not all the questions asked for named entities which limited the application scope. Krishnan et al. [13] introduced the notion of the answer type informer span (a short subsequence of question tokens). Using Conditional Random Field (CRF) to identify informer spans and a linear SVM, the method improved classification accuracy to 86.2%. Furthermore, by applying "perfect" informers (hand tagging informer spans), the accuracy of fine-grained classification had further improvement which demonstrated the contribution of question target words. Though these two kind of features were proved to be able to contribute to classification performance, they also had the same tedious manual work problem in feature construction stage.

In addition to semantic features, syntactic structure of the question also demonstrated contribution. In 2007, Nguyen et al. [14] proposed to use sub-trees extracted from the constituency parsing of questions in a boosting model with Maximum Entropy (ME) classifier, formulating question classification task as a tree classification problem without extra semantic features. Their approach achieved an accuracy rate of 91.2% and 83.6% for coarse-grained and fine-grained categories, respectively. However, as their method heavily depended on a parser to acquire the syntactic trees, overall failure was more likely to directly impact classification performance when question syntactic trees were analyzed incorrectly. Since solely applying syntactic trees cannot improve the performance much, many approaches tried to apply both question syntactic and semantic features together. Recently Phuong [10] addressed that typed dependency features are very informative for question classification, helping improve the accuracy of ME classifier on both coarse-grained and fine-grained classification.

Previous work in question classification showed that feature expansion was necessary and crucial to improve classification performance and the key was that the expansion should base on question target. These works motivated us to conduct research on automatic question target identification and feature expansion by referring the existing ideas of dependency relation analysis and WordNet usage from the question classification area even though they were different from task to task.

3 Question Target Feature Extraction and Expansion

A question classification task was defined by Li and Roth [4] as: given a question, maps it to one of predefined categories, which are semantic constraints on the sought-after answer. The formal representation of the research problem can be found in [22].

Regarding Question Target Words (QTWs) as question target feature, how to effectively extract QTWs to prune out irrelevant information that may mislead the classification process is a crucial problem. Therefore, an automatic Question Target Feature Extraction and Expansion (QTF_EE) approach is proposed to detect question target words and expand the semantic features based on the identified QTWs. The QTF_EE approach mainly contains four steps: preprocessing, QTWs detection, target feature expansion, and classification. In order to ensure that the words with different suffices are treated as the same vector, stop word filtering and word stemming are conducted the in preprocess stage. For instance, the words "*connected*," "*connecting*," "*connection*" and "*connections*" in different questions are treated as the same word "*connect*." A principle-based parser is further utilized to obtain QTWs using four strategies. Afterwards, syntactic and semantic features are expanded based on the corresponding QTWs. Finally, the expanded features are sent to a trained classification classifier to obtain question target (answer type) categories from a QTC taxonomy. The taxonomy used in the paper is from Li and Roth [4].

3.1 Question Target Feature Extraction

We define Question Target Feature (QTF) as a type of representation substantially containing semantic context of QTWs and syntactic structure of the question that leading to the expected answer type. This kind of representation is a tuple of interrogative words (IW), question target (QT), and corresponding distinctive words (DW) expanded based on question target. Interrogative words are helpful for identifying coarse-grained answer type. A typical example is that "*who*" questions often indicate QTC as "HUMAN" while "*where*" questions indicate "LOCATION". In this paper, part-of-speech tags are expanded as syntactic feature and hypernyms and word cluster are expanded as semantic feature. Besides, for reinforcing the impact of QT in question classification, a new representation is applied as QT* to differentiate them from BOWs (e.g. QT "*flower*" is represented as "*flower**"). As for the part-of-speech (POS) feature, Industrial-Strength Natural Language Processing tool spaCy[1] is utilized to perform the POS tagging. To obtain the hypernym feature of QTW, WordNet as a popular resource in leveraging semantics is used, where Jeong et al. [23] demonstrated that deep-level WordNet hypernyms are useful for type classification. In WordNet, word senses are organized into hierarchical structures with explicit hypernym relationships, providing a way to obtain hypernym features for QTWs. For instance, the QTW of the question "*What kind of flowers does detective Nero Wolfe raise*?" is "*flower*". The hierarchy for the noun sense of "*flower*" is as: "*flower* → *flowering plant* → *seed plant* → *vascular plant* → *plant* → *organism* → *living thing* → *object* → *physical entity* → *entity*", where "*A* → *B*" represents that *B* is the hypernym of *A*.

To tackle the low coverage problem described in introduction section, a word embedding feature is introduced. Unlike the commonly used semantic feature synonyms and hyponyms, word cluster -based expansion feature can introduce context information to better understand target word. By word cluster-based expansion, words that sharing

[1] https://spacy.io/.

similar surroundings can be clustered into the same or close clusters and therefore a word can be represented by the cluster it belongs to. For instance, QT "*president*" in question "*Who was President of Afghanistan in 1994?*", words "*minister*", "*appointed*", "*cabinet*" and "*presidency*" are in the same cluster as "*president*" which are treated as the same vector in feature space. While in WordNet corpus, the synonyms of "*president*" is only the word itself. Beside, for QT "*animal*" in question "*What animal 's tail is called a brush?*", the same word cluster contains QTWs "rats", "bees", "chicks" and "pups" whereas the hyponyms of QT are "pest", "stunt", "male", "critter" and "darter". Therefore, word cluster-based expansion features can potentially enrich the semantic information of target words. With Wikipedia wikis corpus[2] (11 Gigabit), a word embedding feature is generated by a typical hierarchical clustering tool gensim word2vec[3] which utilizing word representation method proposed by Mikolov et al. [24]. We cluster the word vectors into 1000 classes and represent the QTWs as a number of word clusters.

The formal representation of QTF thus is ***<IW, QT**, DW>***. For instance, the QTF of a question "*What kind of flowers does detective Nero Wolfe raise?*" is <*what, flower**, *flower/NOUN plant 245*>, where *flower/NOUN* denotes POS tag, *plant* denotes the hypernym of *flower,* and *245* denotes the word cluster identifier.

3.2 Question Target Feature Expansion

Question target words, as defined in [8], are the words that precisely represent what the question seeking for. Question target words detection is the most important process since it can directly affect the QTF expansion stage and overall classification performance. With the QTWs, a question can be easily simplified but still maintain similar meaning. For instance, "*What*" and "*flower*" are the QTWs of the question "*What kind of flowers does detective Nero Wolfe raise?*" These two QTWs are appropriate for representing what the question is expecting for. With expanded hypernym "*plant*", the question can be directly linked to the expected answer type "[*ENTY\plant*]". Thus, an essential work is to extract needed QTWs by identifying useful relations. To extract QTW, a principle-based syntactic parser is applied to generate a dependency tree for a given question so as to reveal the potential target. A dependency structure represents dependencies between individual words of a sentence and usually shown as a tree, where nodes represent words and links represent types of syntactic relationship between two linked words.

Due to the manner of question asking and the variety of interrogative words, syntactic structures of different questions may be much different. After analyzing hundreds of both dependency relations and syntactic structures between the manually identified QTWs and other (irrelevant) parts of the questions, four types of cases are identified and analyzed based on our previous work [8], with each of them having a specific processing strategy including: (1) *Strategy 1*: extract QTWs by locating interrogative words; (2) *Strategy 2*: extract QTWs by using interrogative words with preposition relations; (3) *Strategy 3*: extract QTWs by using verb-centered relations; (4) *Strategy 4*: extract QTWs without interrogative word. To better cover more frequent question structures, the forth

[2] https://dumps.wikimedia.org/.
[3] http://radimrehurek.com/gensim/models/word2vec.html.

strategy is designed for the questions are not always led by interrogative words such as imperative questions. For instance, the question "*Name four famous cartoon cats*" does not have interrogative word.

Algorithm 1. Question target feature expansion

Input: question dependency relations *Dependencies,* question center relations *Rels,* Strategy_1 relations *Rels1*, Strategy_2 relations *Rels2*, Strategy_3 relations *Rels3*, Strategy_4 relations *Rels4*, POS expansion, word_cluster, WordNet
for *deps* in *Dependencies* **do**
  If *deps.relation* in *Rels* **then**
    return *deps.relation*
  end if
end for
String *question_root* = center word extracted from *deps.relation*
switch (*question_root*):
  case "interrogative word":
    QTW = **Strategy_1** (*question_root, Rels1*)
  case "interrogative word with preposition relation":
    QTW = **Strategy_2** (*question_root, Rels2*)
  case "verb":
    QTW = **Strategy_3** (*question_root, Rels3*)
  case "not (verb or interrogative word)":
    QTW = **Strategy_4** (*question_root, Rels4*)
end switch
QTFs.append(pos_tagging(*QTW*))
QTFs.append(word_cluster(*QTW*))
QTFs.append(WordNet_hypernym(*QTW*))
Transform *QTW* into *QTW** and assign it to *QTFs*
Output question target features *QTFs*

The QTF expansion process consists of five main steps: dependency tree generation, question root word extraction, information link summarization, QTW extraction, and corresponding distinctive word expansion. The first step is to generate the corresponding dependency tree of the question. The second step is to identify the question root based on the dependency structure to decide which strategy should be used in the next process. For instance, the question "*Name four famous cartoon cats*" does not have interrogative word therefore the Strategy 4 is applied in this process. Based on our previous investigation on hundreds of dependency relations, the question center relation, as denoted as Rels-4, contains two relation patterns "*V:obj1:N*" and "*V:obj2:N*", which are then applied to extract the question root word "*Name*". The third step is to summarize information links with the extracted question root word. According to the Strategy 4 and the above Rels-4, information links leading by root word "*Name*" are generated as: *name* ← *V:obj1:N* → *cartoon* and *name* ← *V:obj2:N* → *cat*. According to the link matching with the question, the QTWs are extracted as "*cartoon*" and "*cat*". The last step is to expand QTFs utilizing spaCy and word2vec tools. The Algorithm 1 shows the detailed expansion procedure.

4 Evaluation and Results

4.1 Datasets

Two publicly available datasets were used to test the effectiveness of QTF_EE approach: **(1) Dataset A**: 5,382 questions with manually annotated question target labels from the University of Illinois at Urbana-Champaign (UIUC)[4]; **(2) Dataset B**: 500 TREC 10 (Text Retrieval Conference)[5] questions with manually annotated question target labels from UIUC. The statistical characteristic of Dataset A is shown in Table 1. Both the datasets were mapped to a taxonomy containing 6 coarse grained categories and 50 fine grained categories. We then split dataset A and B into training and testing datasets for experiments.

Table 1. The statistical characteristics of the Dataset A

Coarse category	# questions	# words	# characters	Maximum # of question chars	Minimum # of question chars	Standard deviation of question chars
ABBR	86	577	2441	94	13	18.104
DESC	1153	8721	37317	193	13	19.177
ENTY	1245	11860	52253	172	21	19.658
HUM	1215	11992	54524	189	14	22.95
LOC	824	7292	33453	151	15	17.537
NUM	859	7935	34404	148	19	17.653

4.2 The Results

Four experiments are conducted to evaluate the effectiveness of the proposed QTF_EE approach for question classification. The evaluation measures are widely-used Accuracy, Precision, Recall and F1-Measure.

The first experiment evaluates how question target classification performance was affected by the size of training datasets to test the scalability of the QTF_EE approach. The used datasets are the same as used in [4]. We randomly divide the Dataset A into 5 training sets containing 1,000, 2,000, 3,000, 4,000 and 5,500 questions. 500 TREC-10 questions are used as testing dataset. Moreover, we compare the approach with existing work including Li and Roth [4], Zhang et al. [11] and Yen et al. [15] as baseline methods on fine-grained categories. The baseline Li and Roth uses a rich expanded feature set including 6 primitive feature types (BOWs, POS tags, chunks, head chunks, named

4 http://cogcomp.cs.illinois.edu/Data/QA/QC/train_5500.label.

5 http://cogcomp.cs.illinois.edu/Data/QA/QC/TREC_10.label.

entities and semantically related words) while the baseline Yen et al. uses 5 feature types (word form, POS, Named-Entity class, term-match degree and token feature). The experiment results with accuracy metric are shown in Table 2. Our method expands only 3 kind of features but achieves the best performance on different sizes of training datset, demonstrating its scalability and effectiveness.

Table 2. The classification performance with different numbers of training questions on the fine-grained categories

# questions	Li and Roth [4]	Zhang and Lee [11]	Yen et al. [15]	QTF_EE
1000	71.0%	65.0%	60.4%	75.0%
2000	77.8%	74.0%	67.6%	81.8%
3000	79.8%	74.8%	82.6%	83.4%
4000	80.0%	77.4%	82.8%	84.0%
5500	84.2%	79.2%	85.6%	87.4%

To evaluate the contribution of QTF on classification task, the second experiment is conducted using 10-fold cross validation. The total 5382 questions from Dataset A are randomly divided into ten even subsets. In each trial, nine of the ten subsets are used for training and the remaining one for testing. Using accuracy, precision, recall, and F1 as four evaluation metrics, we compare QTF_EE with existing above mentioned BOWs feature models. The result presents that the question target feature outperforms the baseline BOWs feature in every round. To better evaluate the effectiveness and robustness of expanded feature set, both QTFs and BOWs are applied to the other 12 classifiers including Naïve Bayes Multinomial Classifier, Naive Bayes Classifier, Bayes Network Classifier, IBk, Attribute Selected Classifier, Filtered Classifier, Randomizable Filtered Classifier, J48, Random Forest, Random Tree, and REP Tree. The experiment results on fine-grained classification are shown in Fig. 1, which shows that question target feature QTFs benefit every classifier than BOWs, demonstrating the contribution of QTFs on question classification tasks.

The third experiment is to evaluate how QTF_EE contribute on all coarse categories. Using the dataset TREC-10, our method is compared with the previous work. Due to the lack of reported evaluation results on coarse categories, we use the work in [15] as baseline. The results are shown in Table 3. Comparing with Yen et al. [15], our QTF_EE obtains lower precision on the category "ABBR" but has higher F1 and Recall on all the other categories. In addition, our method only expands the question target with 3 types of features, which are less than baseline, which uses a bunch of semantic features including POS, word form, token feature, named-entity recognition and term-match (question first noun phrase match, question term exact match, stem match, synonym match, hyponym match, and hypernym match). In addition, based on the result analysis, named entities can cause much noises in classification especially for category "ABBR", "DESC" and "HUMAN". For instance, the named entity of question "*CNN*" in the question "*What does CNN stand for?*" is usually tagged as "ORGANIZATION" in NER, leading to a wrong answer type "[HUM/Group]" (the correct one is "[ABBR/Exp]"). Therefore, how to effectively use extracted named entities to benefit classification tasks is still an unsolved problem.

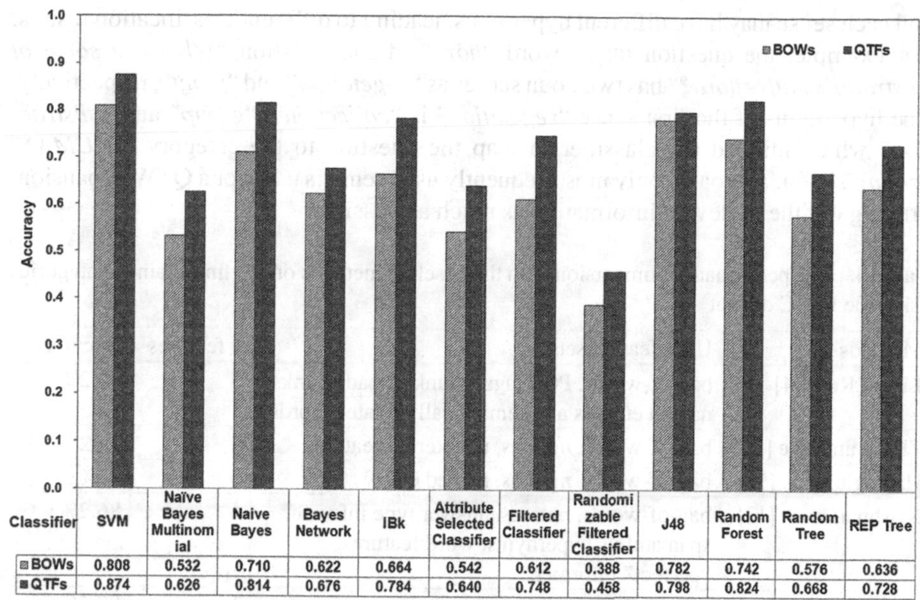

Classifier	SVM	Naïve Bayes Multinomial	Naive Bayes	Bayes Network	IBk	Attribute Selected Classifier	Filtered Classifier	Randomizable Filtered Classifier	J48	Random Forest	Random Tree	REP Tree
BOWs	0.808	0.532	0.710	0.622	0.664	0.542	0.612	0.388	0.782	0.742	0.576	0.636
QTFs	0.874	0.626	0.814	0.676	0.784	0.640	0.748	0.458	0.798	0.824	0.668	0.728

Fig. 1. The performance comparison on fine-grained categories of QTFs and BOWs on 12 different classifiers in accuracy

Table 3. The performance comparsion on the coarse categories using the TREC-10 dataset

Coarse categories	QTF_EE			Yen et al. [15]		
	P(%)	R(%)	F1(%)	P(%)	R(%)	F1(%)
ABBR	77.78	77.78	77.78	100.0	77.78	87.50
DESC	90.91	100.0	95.22	84.15	100.0	91.39
ENTY	85.90	78.72	80.78	98.53	71.28	82.72
HUM	87.32	95.38	91.18	87.50	96.92	91.97
LOC	94.93	91.36	93.01	87.21	92.59	89.82
NUM	99.04	89.38	93.60	97.09	88.50	92.59
Average	91.75	91.20	90.94	92.41	87.84	89.33

The last experiment presents the comparisons of our approach with 10 existing question target classification approaches as baselines collected from 2002 to 2016. Using the same training and testing datasets, we compare the performance on all the fine-grained categories in accuracy. From the result, as presented in Table 4, our approach achieves an accuracy of 87.4% which is the best among all baseline methods. The method with the second highest accuracy was [17], which applies nearly 26 kind of features. As for the third highest accuracy method [13], it uses the same idea of question target words (named as informer span) and applies informer hypernyms from WordNet. The only difference is that it adds all hypernyms of all word senses if the word is noun and within an informer span. However, in certain cases, the target words have multiple noun senses and each sense may have different hypernyms, leading to different classification results. For example, the question target word "*flora*" of the question "*What are some of Australia's native flora?*" has two noun senses as "*vegetation*" and "*plant*", respectively. The hypernyms of the first sense "*vegetation*" is "*collection*", "*group*" and "*abstraction*" which mislead the classifier to map the question to the category "[*HUMAN/group*]". In our approach, only most frequently used sense is used for a QTW expansion, pruning out the irrelevant information as much as possible.

Table 4. The performance comparsion with the baseline methods on the fine-grained categories using the UIUC dataset

Methods	Used feature sets	# features	Accuracy
Li and Roth [4]	bag-of-words, POS tags, chunks, head chunks, named entities and semantically related words	6	84.2%
Zhang and Lee [11]	bag-of-words, *n*grams, tree kernel features	3	79.2%
Hacioglu et al. [12]	bag-of-words, *n*grams, named entity	3	82%
Krishnan et al. [13]	bag-of-words, *n*grams, answer type informer span and its hypernyms, word features, "perfect" informer	6	86.2%
Nguyen et al. [14]	subtree feature	1	83.6%
Phuong et al. [10]	question *wh*-word, unigrams, typed dependencies	3	78.4%
Yen et al. [15]	word form, POS, Named-Entity class, term-match degree and token feature	5	85.6%
Huang et al. [16]	unigram, feature-weighting model	1	85.2%
Hardy et al. [9]	head word, hypernym	2	84.6%
Blunsom et al. [17]	bag-of-words, *n*grams, POS, chunk, named-entity etc.	26	86.6%
QTF_EE	POS, word cluster, hypernyms	3	87.4%

Moreover, for methods [4, 11, 13, 15], their approaches heavily depend on a list of semi-automatically constructed features that are called "RelWords" (related words) and human-made rules. The major limitation of the human-made rules lies on the requirements of human experts' efforts since this is usually tremendous amount of tedious work [14]. The baseline method Huang et al. [16] focuses on the contribution of the question words and obtains relative good results with an accuracy of 85.2% on fine-grained categories. However, it uses two handmade rules which ignores the questions without interrogative words in its constructed unlabeled question collection, e.g., the question "*Name the five positions who are in the line of succession to the presidency.*" and "*Define the Phoenix Club?*" cannot be processed. In contrast, our approach is able to automatically construct the necessary features and is able to deal with the questions without interrogative words. The comparison demonstrates the effectiveness of our approach in question target classification.

5 Conclusions

Aiming at reducing the dimensionality of feature space and tickling low coverage problem, an automatic approach named as QTF_EE for detecting question target and extracting question target feature is proposed. The approach can automatically extract a set of concise but effective features to assist question classification and improve its performance. Based on two publicly available datasets - UIUC dataset and TREC-10 dataset, four experiments are conducted to evaluate the effectiveness of QTF_EE through a comparison with 12 existing methods as baselines. The result demonstrates that the approach can improve the performance of question target classification.

Acknowledgements. This work was supported by National Natural Science Foundation of China (No. 61403088), the programs of Personalized Health Service Public Platform based on Open and Big Data (No. 2014B010118005), Ancient Literature Knowledge base Platform for the Inheritance and Development of Traditional Chinese Medicine (No. 2014A020221039) and Innovative School Project in Higher Education of Guangdong (No. YQ2015062).

References

1. Suzuki, J., Taira, H., Sasaki, Y., Maeda, E.: Question classification using HDAG kernel. In: Proceedings of the ACL 2003 Workshop on Multilingual Summarization and Question Answering, vol. 12, pp. 61–68 (2003)
2. Hovy, E., Gerber, L., Hermjakob, U., Lin, C.Y., Ravichandran, D.: Toward semantics-based answer pinpointing. In: Proceedings of the First International Conference on Human Language Technology Research, pp. 1–7 (2001)
3. Moldovan, D., Paşca, M., Harabagiu, S., Surdeanu, M.: Performance issues and error analysis in an open-domain question answering system. ACM Trans. Inform. Syst. (TOIS) **21**(2), 133–154 (2003)
4. Li, X., Roth, D.: Learning question classifiers. In: Proceedings of the 19th international conference on Computational Linguistics, vol. 1, pp. 1–7. Association for Computational Linguistics (2002)
5. Li, X., Roth, D.: Learning question classifiers: the role of semantic information. Natural Lang. Eng. **12**(3), 229–249 (2006)
6. Srihari, R., Niu, C., Li, W.: A hybrid approach for named entity and sub-type tagging. In: Proceedings of the Sixth Conference on Applied Natural Language Processing, pp. 247–254 (2000)
7. Loni, B.: A survey of state-of-the-art methods on question classification. Literature Survey, p. 54. Published on TU Delft Repository (2011)
8. Hao, T., Xie, W., Xu, F.: A WordNet expansion-based approach for question targets identification and classification. In: Sun, M., Liu, Z., Zhang, M., Liu, Y. (eds.) CCL 2015. LNCS, vol. 9427, pp. 333–344. Springer, Cham (2015). doi:10.1007/978-3-319-25816-4_27
9. Hardy, H., Cheah, Y.N.: Question classification using extreme learning machine on semantic features. J. ICT Res. Appl. **7**(1), 36–58 (2013)
10. Le-Hong, P., Phan, X.-H., Nguyen, T.-D.: Using dependency analysis to improve question classification. In: Nguyen, V.-H., Le, A.-C., Huynh, V.-N. (eds.) Knowledge and Systems Engineering. AISC, vol. 326, pp. 653–665. Springer, Cham (2015). doi:10.1007/978-3-319-11680-8_52

11. Zhang, D., Lee, W.S.: Question classification using support vector machines. In: Proceedings of International ACM SIGIR Conference on Research and Development in Information Retrieval, pp. 26–32. ACM (2003)
12. Hacioglu, K., Ward, W.: Question classification with support vector machines and error correcting codes. In: Proceedings of HLT-NAACL 2003, vol. 2, pp. 28–30 (2003)
13. Krishnan, V., Das, S., Chakrabarti, S.: Enhanced answer type inference from questions using sequential models. In: Proceedings of the Conference on Human Language Technology and Empirical Methods in Natural Language Processing, pp. 315–322 (2005)
14. Le Nguyen, M., Tri, N.T., Shimazu, A.: Subtree mining for question classification problem. In: Proceedings of IJCAI, pp. 1695–1700 (2007)
15. Yen, S.J., Wu, Y.C., Yang, J.C., Lee, Y.S., Lee, C.J., Liu, J.J.: A support vector machine-based context-ranking model for question answering. Inf. Sci. **224**, 77–87 (2013)
16. Huang, P., Bu, J., Chen, C., Qiu, G.: An effective feature-weighting model for question classification. In: Proceedings of IEEE International Conference on Computational Intelligence and Security, pp. 32–36 (2007)
17. Blunsom, P., Kocik, K., Curran, J. R. Question classification with log-linear models. In: Proceedings of International ACM SIGIR Conference on Research and Development in Information Retrieval, pp. 615–616. ACM (2006)
18. Moldovan, D., Badulescu, A., Tatu, M., Antohe, D., Girju, R.: Models for the semantic classification of noun phrases. In: Proceedings of the HLT-NAACL Workshop on Computational Lexical Semantics, pp. 60–67 (2004)
19. Hao, T., Agichtein, E.: Finding similar questions in collaborative question answering archives: toward bootstrapping-based equivalent pattern learning. Inf. Retrieval **15**(3–4), 332–353 (2012)
20. Zhang, H., Chow, T., Wu, J.: Organizing books and authors using multi-layer SOM. IEEE Trans. Neural Networks Learn. Syst. **27**(12), 2537–2550 (2016)
21. Mrabet, Y., Kilicoglu, H., Roberts, K., Demner-Fushman, D.: Combining open-domain and biomedical knowledge for topic recognition in consumer health questions. In: Proceedings of AMIA Annual Symposium, p. 914 (2016)
22. Hao, T., Xie, W., Chen, C., Shen, Y.: Systematic comparison of question target classification taxonomies towards question answering. Commun. Comput. Inform. Sci. **568**, 131–143 (2015)
23. Jeong, Y., Myaeng, S.-H.: Using WordNet hypernyms and dependency features for phrasal-level event recognition and type classification. In: Serdyukov, P., Braslavski, P., Kuznetsov, S.O., Kamps, J., Rüger, S., Agichtein, E., Segalovich, I., Yilmaz, E. (eds.) ECIR 2013. LNCS, vol. 7814, pp. 267–278. Springer, Heidelberg (2013). doi:10.1007/978-3-642-36973-5_23
24. Mikolov, T., Sutskever, I., Chen, K., Corrado, G.S., Dean, J.: Distributed representations of words and phrases and their compositionality. In: Advances in Neural Information Processing Systems, pp. 3111–3119 (2013)

Combine Non-text Features with Deep Learning Structures Based on Attention-LSTM for Answer Selection

Chang'e Jia, Chengjie Sun, Bingquan Liu, and Lei Lin(✉)

Harbin Institute of Technology, Harbin 150001, China
{cejia,cjsun,liubq,linl}@insun.hit.edu.cn

Abstract. Because of the lexical gap between questions and answer candidates, methods with only word features cannot solve Answer Selection (AS) problem well. In this paper, we apply a LSTMs with Attention model to extract the latent semantic information of sentences and propose a method to learning non-text features. Besides, we propose an index to evaluate the sorting ability of models with the same accuracy value. Our model achieved the best accuracy and F1 performance than other known models, and the ranking index results, including MAP, AvgRec and MRR index's result, are after only KeLP system and Beihang MSRA system in SemEval-2017 Task 3 Subtask A.

Keywords: LSTMs · Attention · Answer selection · Non-text features

1 Introduction

AS is the crucial intermediate link of Question Answering (QA) between Information Retrieval (IR) and Answer Fusion, and it is directly related to the quality of Answer Fusion. In the previous study, AS faced two major problems: (i) There is rarely intersect in lexical level between the question and answers, which is called "lexical chasm" [1,2]; (ii) Some sentences in correct answers are irrelevance with questions.

Traditional methods are most based on lexical features and non-text features, while some lexical features are rely on basic natural language processing tools. [3] adopt a pairwise neural network architecture with machine translation features, sentence-pairs similarity features and task special features. [4] applied tree kernels to questions and answers' syntactic tree, which is obtained by parsing tools. [5] calculated the similarity of tokens on question-answer pairs with the same part of speech, which using the tagger tools. Basic natural language processing tools that are not particularly accurate, would lead to error transfer in feature extraction. In recent years, Deep Learning has made considerable strides in sentence-pair relevance. [6] tried a stacked Bidirectional Long Short-Term Memory (BiLSTM) model to get a state-of-the-art result in AS. [7] employed a Convolutional Neural Network (CNN) model with precise designs and a relevance

J. Wen et al. (Eds.): CCIR 2017, LNCS 10390, pp. 261–271, 2017.
https://doi.org/10.1007/978-3-319-68699-8_21

method for vectors to rank short text pairs. [8,9] used a listwise approach, which treated all answers in a question's candidate answer list as a whole; They combined question-answer pairs' joint representations, encoded by a CNN model, to a sequence, and put it as the input of an LSTM model. Besides, some researchers have made some improvements to the original model. [10] proposed three inner-attention-RNN models aimed at the attention bias problems in the traditional attention model; [11] added a latent distribution as the representations of questions, which allows to deal with the ambiguity inherent and learn pair-specific representations, in the attention model.

Deep learning models can be a good understanding of the potential semantics of the sentence, while non-semantic features play a noticeable role in ranking tasks sometimes. [6,7] both provided methods to combine deep learning model with artificial features. [6] treated the result of stack BiLSTM model as a feature like artificial features, while [7] add the artificial features to the penultimate hidden layer of deep learning models directly. In our work, we applied an LSTM model with attention mechanism to get a semantic relevance value, a non-text features extraction model to obtain a relevance score between question

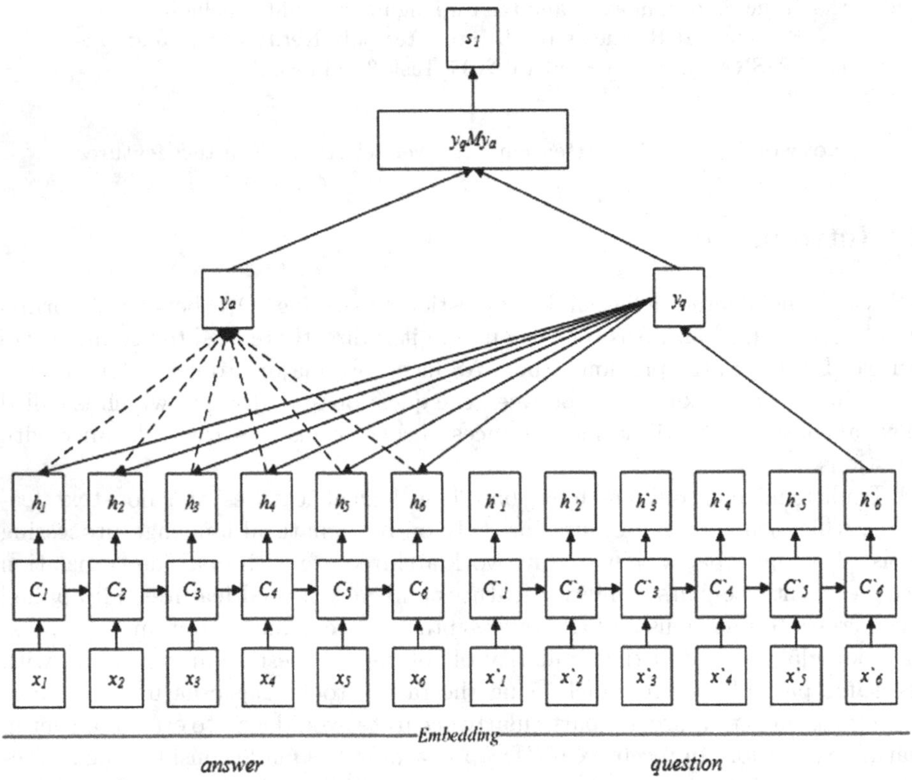

Fig. 1. Answer selection model structure chart based on Attention-LSTM.

and answer, and combine them finally. In addition, we notice that models have the same accuracy value but different ranking results. In the light of this problem, we proposed an indicator based on two lever ranking tasks to evaluate the ranking ability of classification models.

To sum up, we make two contributions in this work: (1) We provided a method to extract non-text features by models, which can be combined well with the deep learning models. (2) We put out an indicator to quantify the sorting ability of classification models combine with the accuracy of models.

2 Model

In this work, we employed an LSTM model [12] with attention mechanism, which is used in [13], to encode sentences. As Fig. 1 shows, the encode part contains three parts: word embedding, LSTM and Attention.

2.1 Word Embedding

The first step that via Machine learning to solve QA problems is to convert words to vectors. The most simplest ways is one-hot representation, but it has a serious problem that we cannot obtain any information between two words. Distributed representation, also called embedding, is a good solution to this problem. The approaches used to initialize embedding layer are random initialization and initializing with a well trained word vector, such as Glove [14].

2.2 LSTMs

Recurrent Neural Network (RNN) is skillful in solving sequence problems, such as neural language. However, RNN is suffer from exploding gradients and vanishing gradients. LSTMs [12] and GRUs [15] (Gated Recurrent Units) were explicitly designed to deal with vanishing gradients and efficiently learn long-range dependencies through a gating mechanism.

LSTMs are defined as follows:

$$i_t = sigmoid(W_i x_t + U_i h_{t-1} + b_i) \tag{1}$$

$$f_t = sigmoid(W_f x_t + U_f h_{t-1} + b_f) \tag{2}$$

$$o_t = sigmoid(W_o x_t + U_o h_{t-1} + b_o) \tag{3}$$

$$\tilde{C}_t = tanh(W_c x_t + U_c h_{t-1} + b_c) \tag{4}$$

$$C_t = i_t \odot \tilde{C}_t + f_t \odot C_{t-1} \tag{5}$$

$$h_t = o_t \odot tanh(C_t) \tag{6}$$

With i, f and o are input gate, forget gate and output gate, $\tilde{C}$ and C are candidate state value and current state in memory cell, h_t is the output we needed at time t, and σ is the sigmoid function that squashes a value between 0 and 1. The input gate and the forget gate control respectively how much of the newly state and previous state to the internal state for current. The output gate defined the flow of the internal state exposing to the external network.

2.3 Attention

The idea of LSTM that treating a sentence as a sequence and processing a token at a time is similar with human reading. However, people will be attracted to some words, which provide bigger contribution than others, and attention mechanism imitate the characteristics of human reading.

Let $Y \in \mathbb{R}^{k \times l}$ be a matrix making up of output vectors of answers that produced by LSTMs mentioned above, $h_q \in \mathbb{R}^k$ is the questions' representations which is the output from the last time step of LSTMs. Attention mechanism is defined as follows.

$$M_t = tanh(WY + (Uh_q + b_m) \otimes e_l) \tag{7}$$

$$\alpha = softmax(wM_t + b) \tag{8}$$

$$r = Y\alpha^T \tag{9}$$

Where W, $U \in \mathbb{R}^{k \times l}$ are trained projection matrices, and $w \in \mathbb{R}^k$ is a trained parameter vector. In order to reduce parameters, we take r as the representation of answers.

2.4 Relevance Calculation

After getting the representation of question and answers, we need to calculate the relevance between question and answer's vector. Most researchers applied similarity method. For example, [16] used nine methods to obtain similarity of two vectors, and [13] via cosine similarity to all structure they mentioned. However, it is inappropriate that evaluating relevance of question-answer pairs' vector because of less similarity between questions' vector and answers' vector.

In our work, we follow the approach of [7], which is also applied by [17]. It is defined as Eq. 10.

$$sim(x_1, x_2) = sigmoid(x_1^T M x_2) \tag{10}$$

Where x_1 and x_2 are sentence vectors, and M is a metric that maps two vectors to a scalar.

2.5 Non-text Features

Non-textual information plays an important role in Answer Selection Task [3]. However, it is difficult in extracting non-text feature based on deep learning models at present. Besides, the extraction of non-text features are mostly based on experience, which may ignore some useful information. For example, users in Qatar Living have these properties: accuracy of answering questions, areas of expertise, probability of advertising and so on. We usually turn a blind eye to those features, while focus on more direct features such as whether the person answering the question is same as the one who asked it. In this work, we put four kinds of non-text features upon the ranking model:

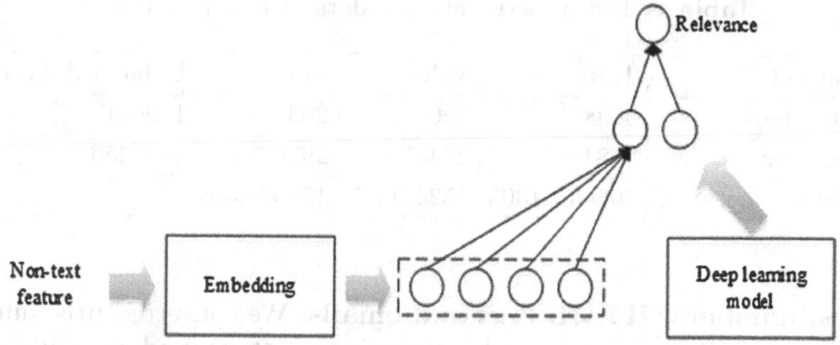

Fig. 2. Connection method of non-text features and deep learning models

- **User features.** Including the question user features and answer user features.
- **Answer position features.** The position of the answer from the candidate answer list.
- **Sentence length features.** The number of word in sentence.
- **Problem category.** Each question has a category, and question in different category has different answer accuracy.

Inspired by [18], which mapped the word position to a vector initialized randomly, we obtained the user features, answer position features, problem category features and sentence length features by looking up their embeddings, which keeping updated during all the train times and is coupled together with the output layer of deep learning models. The connect method is as Fig. 2 shown.

3 Experiments

We made experiments on the data provided by SemEval-2015 Task 3[1], SemEval-2016 Task 3[2] and SemEval-2017 Task 3[3]. The train data contains all data from SemEval-2015 Task 3 [19] and TRAIN-PART1/2 from SemEval-2016 Task 3 [20]; the validation data is the development data in SemEval-2016 Task 3, and the test data in SemEval-2017 Task 3 [21] is as the test data in our work. Table 1 shows the statistics of three dataset.

3.1 Data Pretreatment

The principle of data preprocessing is storing more amounts of original information. The details of data processing are as follows:

- **Removing Catchwords.** Users may add a catchwords, like "**If winter comes, can spring be far behind**" which is useless to understand the original means of a sentence, to the end of questions or answers.

[1] http://alt.qcri.org/semeval2015/task3/.
[2] http://alt.qcri.org/semeval2016/task3/.
[3] http://alt.qcri.org/semeval2017/task3/.

Table 1. The statistics of three dataset in experience

Dataset	Train	Valid	Test	Unlabeled data
Questions	5898	500	293	189941
Answers	37848	2440	2930	1894456
Positive/negative	16543/21305	1523/1407	1329/1941	–/–

- **URLs, numbers, HTML tags and emails.** We converted urls, numbers, HTML tags and emails into <url>, <number>, <tag> and <email> respectively.
- **Repeated punctuations and letters.** All repeated punctuations and letters are turned into one.
- **Special symbol.** Special symbols are removed except for letters and punctuations.

It should be noted that we case sensitive and do not remove stop words.

3.2 Parameter Setting

We use 200-dimension word vectors trained by Word2Vec in gensim [22] on unlabeled dataset to initialize the word embedding layer. The parameter *window* and min_count of Word2Vec are set 5 and 20. In order to avoid over-fitting, we applied dropout [23] to all layers in our model. We use RMSprop [24] for optimization with a learning rate of 0.0001, and employed the cross entropy as the cast function The detail of all parameter settings is shown in Table 2.

Table 2. Parameter setting in model

Layer	Parameter
Sentence length	100
User number	47801
Sentence length	101
Position number	11
Question category number	150
Embedding dim	5
Vocabulary number	123353
Embedding dropout	0.5
LSTM dim	200
LSTM dropout	0.3
Attention dim	200
Attention dropout	0.5

In order to avoid the influence of non-text features to sentence encoder models' parameters updating, We train the Attention-LSTM model firstly and the whole answer selection model lastly. During the training time, the model, which has the best MAP performance on valid dataset, would be saved.

4 Results

Table 3 shows our three different models' performance on SemEval-2016 Task 3 development dataset.

Table 4 shows different systems' results on SemEval-2017 Task 3 test dataset, which can be found in SemEval-2017 Task 3 website[4]. The baseline 1 is the result of an IR system provided by SemEval-2017 Task 3 organizers, and the baseline 2 is the random systems' result.

Table 3. Results on SemEval-2016 Task 3 development dataset

Model	MAP	AvgRec	MRR	Acc	P	R	F1
LSTM	0.6581	0.8523	0.7071	0.7410	0.6105	0.6284	0.6193
LSTM + attention	0.6563	0.8525	0.7159	0.7402	0.6045	0.6504	0.6266
LSTM + attention + non-text	0.6646	0.8646	0.7301	0.7439	0.6156	0.6284	0.6219

5 Results Analysis

As shown in Tables 3 and 4, our model has the best Accuracy and F1 performance than other models, and the ranking index's results, including MAP, AvgRec and MRR index's result, are after KeLP system and Beihang MSRA system.

Table 4. Results of different systems on SemEval-2017 Task 3 test dataset

Model	MAP	AvgRec	MRR	Acc	P	R	F1
Baseline 1 (IR)	0.7261	0.7932	0.8237	–	–	–	–
Baseline 2 (random)	0.6230	0.7056	0.6874	0.5270	0.5315	0.7597	0.6254
1_{st} (KeLP)	**0.8843**	0.9379	**0.9282**	0.7389	0.8730	0.5824	0.6987
2_{nd} (Beihang-MSRA)	0.8824	**0.9387**	0.9234	–	–	–	–
3_{rd} (IIT-UHH) [25]	0.8688	0.9204	0.9120	0.7270	0.7337	0.7452	0.7394
4_{th} (*lanman*)	0.8672	0.9262	0.9145	0.7843	0.8409	0.7216	0.7767
LSTM	0.8482	0.9157	0.8886	0.7717	0.8112	0.7308	0.7689
LSTM + attention	0.8552	0.9209	0.8931	0.7730	0.7975	0.7551	0.7757
LSTM + attention + non-text	0.8752	0.9336	0.9196	**0.7870**	0.8362	0.7341	**0.7818**

[4] http://alt.qcri.org/semeval2017/task3/index.php?id=results.

By comparing the results of the Attention-LSTM model and Attention-LSTM with non-text features model's on both validation dataset and test dataset, we can see that the latter's ranking index results outperforms the former: about 1 point on validation dataset and 2 points on test dataset. All these prove non-text features play an important role in the ranking task and the approach to learn non-text features is effective.

As shown in Table 3, although the ranking index results of LSTM and LSTM with attention have little difference on validation dataset, the ranking index results, showed in Table 4, of the latter is much better than the former on the test dataset. We guess that may be caused by different positive and negative proportions of different dataset, and this also shows that the generalization ability of Attention-LSTM model is stronger than LSTM model.

We also find that models have same Accuracy value but ranking values. To explain this phenomenon, we propose a index named Intersection probability (IP). We think the IP index, to a certain extent, can reflect the sorting ability of a model.

It is generally known that the mean of the random variable is the expected value (or location), and the standard deviation of the random variable is the volatility (or spread). Here let $\overline{x}$ be the mean of x and $\sigma(x)$ be the standard deviation of x, the range of x can be defined in Eq. 11.

$$r\_x = \overline{x} \pm \sigma(x) \tag{11}$$

Assume that there is a dataset of a two-level ranking task, and the predicted value ranged from 0 to 1 of the dataset is as shown in Fig. 3(a). If we took it as a classification task and 0.5 as the dividing line between positive and negative cases, the blue dot is the correctly classified samples and the red point is the misclassified classified samples. We take the samples, which is divided into positive, out alone to analysis their range. As shown in Fig. 3(c), the red/green line is the mean of predicted values of the correctly/misclassified classified samples, and the red/green area is the range of the correctly/misclassified classified samples' predicted value. So the intersection region between the red area and the green area reflected the probability that the predicted value of a misclassified classified sample is bigger than a correctly classified samples. As we can see, the bigger the IP is, the smaller the ranking results. It is defined in Eq. 12. In the course of the experiment, we fold the axes in the dividing line and calculate the IP value then.

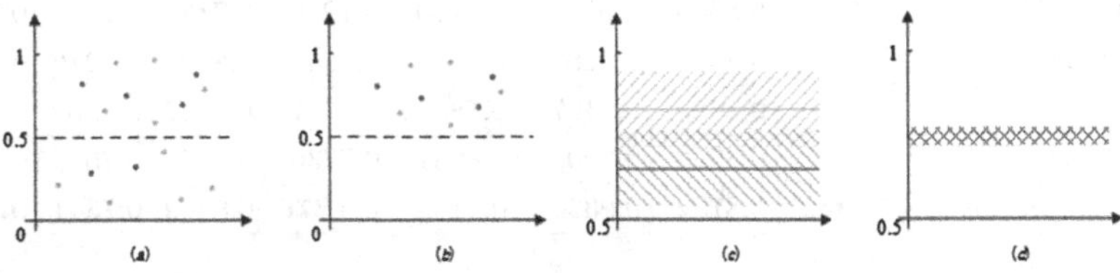

Fig. 3. IP index (Color figure online)

Table 5. IP values of different models on SemEval-2016 Task 3 development dataset

Model	$mean_{err}$	$mean_{corr}$	std_{err}	std_{corr}	IP	Acc
LSTM	0.20968	0.32395	0.11385	0.12873	0.12832	0.7410
LSTM + attention	0.20460	0.32542	0.11273	0.13077	0.12269	0.7402
LSTM + attention + non-text	0.12518	0.27762	0.07405	0.09884	0.02045	0.7439

Table 6. IP values of different models on SemEval-2017 Task 3 test dataset

Model	$mean_{err}$	$mean_{corr}$	std_{err}	std_{corr}	IP	Acc
LSTM	0.21041	0.31650	0.10955	0.14120	0.14466	0.7717
LSTM + attention	0.17433	0.33564	0.12612	0.12654	0.09134	0.7730
LSTM + attention + non-text	0.17310	0.26096	0.08530	0.10233	0.09977	0.7870

$$IP = (\overline{x_{err}} + \sigma(x_{err})) - (\overline{x_{corr}} - \sigma(x_{corr})) \tag{12}$$

But when getting down to the details, we can see that we cannot compare two models had different accuracy according the IP index, because different model may have the same IP value but different accuracy (For instance, there are two different models that one has two misclassified examples performance on a dataset and another has two correctly classified samples performance on the same dataset, they may have the same IP value but different ranking results).

The Tables 3 and 4 show the IP value and the middle results of LSTM model, Attention-LSTM model and Attention-LSTM with non-text features model on valid data and test data. As shown in the table, the third model has the smallest IP value on valid data, Attention-LSTM's IP value is smaller than LSTM's. The differences in accuracy between them is very small, and the same phenomenon, which confirms our previous statement, also appeared on the test set (Tables 5 and 6).

6 Conclusion

In this paper, we used an attentive neural network based on LSTM models to encode question-answer pairs. Due to many parameters in deep learning models and small labeled data we have, the model we employed is often over-fitting if we didn't take steps. We adopt dropout to Embedding, LSTM and Attention layers, and set large values to dropout. Experiment proves that dropout can effectively prevent over fitting.

In order to make full use of non-text features, we did not use the traditional method of artificial feature extraction, but the idea of embedding, which maps a feature to a high dimension and updates it during train time.

At last, we demonstrated the relationship between the classification results and ranking values, and proposed an approach to evaluate the sorting ability of classification models in the two-lever ranking tasks.

Acknowledgment. This work is sponsored by the National High Technology Research and Development Program of China (2015AA015405) and National Natural Science Foundation of China (61572151 and 61602131).

References

1. Berger, A.L., Caruana, R., Cohn, D., Freitag, D., Mittal, V.O.: Bridging the lexical chasm: statistical approaches to answer-finding, pp. 192–199 (2000)
2. Riezler, S., Vasserman, A., Tsochantaridis, I., Mittal, V., Liu, Y.: Statistical machine translation for query expansion in answer retrieval (2007)
3. Guzman, F., Marquez, L., Nakov, P.: Machine translation evaluation meets community question answering, pp. 460–466 (2016)
4. Filice, S., Croce, D., Moschitti, A., Basili, R.: KeLP at SemEval-2016 task 3: learning semantic relations between questions and answers. Proc. SemEval **16**, 1116–1123 (2016)
5. Mihaylov, T., Nakov, P.: SemanticZ at SemEval-2016 task 3: ranking relevant answers in community question answering using semantic similarity based on fine-tuned word embeddings. In: Proceedings of SemEval, pp. 879–886 (2016)
6. Wang, D., Nyberg, E.: A long short-term memory model for answer sentence selection in question answering. In: ACL, vol. 2, pp. 707–712 (2015)
7. Severyn, A., Moschitti, A.: Learning to rank short text pairs with convolutional deep neural networks, pp. 373–382 (2015)
8. Zhou, X., Hu, B., Chen, Q., Tang, B., Wang, X.: Answer sequence learning with neural networks for answer selection in community question answering, pp. 713–718 (2015)
9. Lin, X., Wang, Y.X.X.: ICRC-HIT: a deep learning based comment sequence labeling system for answer selection challenge. In: SemEval-2015, p. 210 (2015)
10. Wang, B., Liu, K., Zhao, J.: Inner attention based recurrent neural networks for answer selection. In: The Annual Meeting of the Association for Computational Linguistics (2016)
11. Miao, Y., Yu, L., Blunsom, P.: Neural variational inference for text processing. In: Proceedings of ICML (2016)
12. Hochreiter, S., Schmidhuber, J.: Long short-term memory. Neural Comput. **9**(8), 1735–1780 (1997)
13. Tan, M., Santos, C.d., Xiang, B., Zhou, B.: LSTM-based deep learning models for non-factoid answer selection. arXiv preprint arXiv:1511.04108 (2015)
14. Pennington, J., Socher, R., Manning, C.D.: Glove: global vectors for word representation. EMNLP **14**, 1532–1543 (2014)
15. Chung, J., Gulcehre, C., Cho, K., Bengio, Y.: Empirical evaluation of gated recurrent neural networks on sequence modeling. arXiv preprint arXiv:1412.3555 (2014)
16. Feng, M., Xiang, B., Glass, M.R., Wang, L., Zhou, B.: Applying deep learning to answer selection: a study and an open task. In: 2015 IEEE Workshop on Automatic Speech Recognition and Understanding (ASRU), pp. 813–820. IEEE (2015)
17. Bordes, A., Weston, J., Usunier, N.: Open question answering with weakly supervised embedding models, pp. 165–180 (2014)
18. Zeng, D., Liu, K., Lai, S., Zhou, G., Zhao, J.: Relation classification via convolutional deep neural network (2014)

19. Nakov, P., Marquez, L., Magdy, W., Moschitti, A., Glass, J., Randeree, B.: SemEval-2015 task 3: answer selection in community question answering. In: Proceedings of the 9th International Workshop on Semantic Evaluation (SemEval), vol. 15, pp. 269–281 (2015)
20. Nakov, P., Màrquez, L., Moschitti, A., Magdy, W., Mubarak, H., Freihat, A.A., Glass, J., Randeree, B.: SemEval-2016 task 3: community question answering. In: Proceedings of SemEval, vol. 16 (2016)
21. Nakov, P., Hoogeveen, D., Marquez, L., Moschitti, A, Mubarak, H., Baldwin, T., Verspoor, K: SemEval-2017 task 3: community question answering. In: Proceedings of the 11th International Workshop on Semantic Evaluation. Association for Computational Linguistics (SemEval), Vancouver, vol. 17 (2017)
22. Rehurek, R., Sojka, P.: Software framework for topic modelling with large corpora. In: Proceedings of the LREC 2010 Workshop on New Challenges for NLP Frameworks. Citeseer (2010)
23. Srivastava, N., Hinton, G.E., Krizhevsky, A., Sutskever, I., Salakhutdinov, R.: Dropout: a simple way to prevent neural networks from overfitting. J. Mach. Learn. Res. **15**(1), 1929–1958 (2014)
24. Tieleman, T., Hinton, G.: Rmsprop: divide the gradient by a running average of its recent magnitude. COURSERA: neural networks for machine learning. Technical report, p. 31 (2012)
25. Nandi, T., Biemann, C., Yimam, S.M., Gupta, D., Kohail, S., Ekbal, A., Bhattacharyya, P.: IIT-UHH at SemEval-2017 task 3: exploring multiple features for community question answering and implicit dialogue identification. In: Proceedings of the 11th International Workshop on Semantic Evaluation (SemEval), Vancouver, vol. 17, pp. 91–98 (2017)

Author Index

Printed in the United States
By Bookmasters